The Behaviour of Short Fatigue Cracks

Conference Committees

International Committee

A. F. Blom (*Sweden*)
M. W. Brown (*UK*)
E. R. de los Rios (*Secretary*)
J. D. A. Dominguez (*Spain*)
J. Foth (*Germany*)
R. Galatolos (*Italy*)
K. J. Miller (*(Chairman*)
A. Plumtree (*Canada*)

L. Remy (*France*)
J. Schijve (*The Netherlands*)
I. F. C. Smith (*Switzerland*)
S. Suresh (*USA*)
D. Taylor (*Ireland*)
I. Verpoest (*Belgium*)
B. Weiss (*Austria*)
K. Yamada (*Japan*)

Local Organising Committee

M. W. Brown
K. J. Miller (*Chairman*)

E. R. de los Rios
N. Parkes (*Secretary*)

Editorial Panel

J. P. Bailon
W. J. Baxter
C. J. Beevers
A. F. Blom
G. A. D. Briggs
R. Brook
J. Byrne
E. R. de los Rios
L. Edwards
F. Guiu
E. Hay
C. Howland
M. N. James
J. M. Kendall
M. Kermani
J. E. King

J. F. Knott
J. Mendez
K. J. Miller
J. Petit
A. Plumtree
L. Remy
I. F. C. Smith
R. A. Smith
R. Stickler
R. W. Suhr
D. Taylor
T. H. Topper
R. J. H. Wanhill
B. Weiss
J. Yates

Editors of Proceedings

K. J. Miller

E. R. de los Rios

The Behaviour of Short Fatigue Cracks

Edited by
K. J. Miller
and
E. R. de los Rios

EGF Publication 1

Mechanical Engineering Publications Limited
London

First published 1986

© 1986 Fatigue of Engineering Materials Limited

ISBN 0 85298 614 7

Photoset by Paston Press, Norwich
Printed in Great Britain at the University Press, Cambridge

Contents

An Introduction

Thirty-four papers have been gathered together in this first attempt to collectively examine several important aspects of short fatigue crack behaviour, but before discussing the order of presentation of these papers the editors would like to define what is meant by a 'short' fatigue crack and why it needs to be considered as different from a fatigue crack which is not short.

Linear elastic fracture mechanics (LEFM) analysis of crack tip stress fields permits a rationalization of fatigue crack growth behaviour based on the *elastic* stress intensity factor range $\Delta K \{=Y\Delta\sigma\sqrt{(\pi a)}\}$* because very long cracks in large engineering structures can be duplicated in their behaviour by testing smaller samples in a laboratory. This correspondence occurs if the similitude of the term $\Delta\sigma\sqrt{a}$ is maintained in both cases, since LEFM predicts that both cracks will have identical (or near identical) stress–strain crack tip fields. The reason for similar behaviour is that only small scale yielding conditions apply at the tip of the cracks in both cases and the plastic zone being so small does not seriously perturb the crack tip elastic field. Therefore, by using small laboratory samples to determine the rate of fatigue crack growth, the fatigue lifetime of cyclically loaded components and structures can be evaluated.

However, similitude conditions are frequently difficult to achieve in the laboratory. The well known compact tension specimen geometry has a 7 per cent error in its elastic field analysis – an acceptable error to engineers – and so it is not surprising that different behaviour patterns soon came to be recognized from tests on widely different specimen geometries. Anomalous crack growth rates were frequently recorded as being faster than LEFM analyses would predict and a LEFM characterization of a material was seen to be only a lower bound solution in terms of the ΔK parameter.

Why did some cracks grow faster? The answer lies in the loss of similitude which occurs when stress levels are too high and small scale yielding conditions are exceeded *and/or* when the cracks are so small as to be seriously affected in their behaviour by microstructural features.

Kitagawa and Takahashi, at the Second International Conference on the Mechanical Behaviour of Materials in Boston in 1976, presented a figure similar to Fig. 1. The line given by ΔK_{th} represents the threshold condition below which a crack should not grow if LEFM assumptions are valid. Obviously they are invalid when small scale yielding conditions are exceeded and this occurs to a greater or less extent when the term $\Delta\sigma$ exceeds about two thirds of

---

* For a description of symbols see Fig. 1.

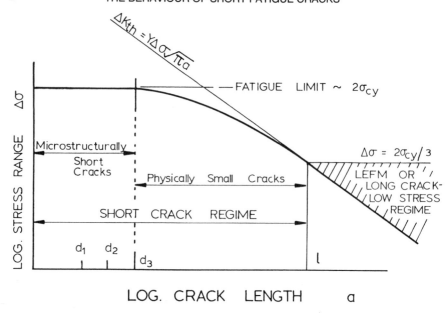

Fig 1 Schematic of the Kitagawa–Takahashi curve showing the thresholds between propagating and non-propagating cracks

the cyclic yield stress, σ_{cy}, in a reversed stress test. A second line on the Kitagawa–Takahashi plot is the fatigue limit itself, which can be approximated to the cyclic yield stress range. Obviously LEFM is not applicable at these levels of stress. An examination of the figure reveals why cracks can grow at levels less than ΔK_{th}, but it should also be appreciated that cracks have been reported as growing on surfaces of plain specimens at stress levels below the fatigue limit. However, such cracks eventually stop propagating.

Several other lines can be drawn on the plot, i.e., d_1, d_2, d_3, etc., to represent the size of microstructural units such as grain sizes, inclusion spacing, precipitation spacing, surface finish, etc., and these too will be expected to affect crack growth behaviour. But at crack lengths beyond microstructural effects, i.e., $a > d_3$, it is to be expected that a continuum mechanics approach will represent crack growth behaviour. However, LEFM analyses of crack tip fields may not be of sufficient accuracy at these stress levels to describe fatigue crack growth behaviour permitting correspondence between large structures and small laboratory specimens.

In the papers that follow researchers and designers examine all of these factors to determine why cracks smaller than the crack length *l* behave as they do, and they attempt to rationalize that behaviour in order to help assess the implications of short crack growth.

Finally it will be noted, from the above remarks, that fatigue cracks fall into three categories:

(1) microstructurally short cracks: $a < d_3$;
(2) long cracks sometimes termed as LEFM-type cracks: $a > l$;
(3) transition length cracks: $d_3 < a < l$.

The first category of short cracks require high stresses for continuous propagation, i.e., $\Delta\sigma > 2\sigma_{cy}$. The second category is essentially for cracks in low elastic stress fields. They too can behave as short cracks with anomalously high crack growth rates should the stress level be raised above $2\sigma_{cy}/3$. The third category is also a short crack. From these remarks it will be appreciated that most cracks can be classed as short cracks, but when the applied stress range levels are less than $2\sigma_{cy}/3$ a special case occurs which is frequently referred to in the papers that follow as 'long-crack' growth behaviour.

Perhaps it would be wise if we did not assume short cracks to be the ones with anomalous behaviour, but rather the long cracks, which would be more accurately described as low stress propagating cracks.

K. J. Miller
E. R. de los Rios

Sheffield, April 1986

Summary

The thirty-four papers in this volume follow a natural sequence of presentation which begins with three papers related to design considerations, and terminates with two papers discussing recent developments in acoustic microscopy techniques that will further advance our understanding of short crack behaviour.

Design considerations

A note of caution is introduced by *Smith* who reminds researchers that designers give relatively little attention to problems associated with cyclic loading. This regretted fact reinforces his message for more industrially relevant projects which can be of more direct assistance to designers. In this context he lists the various models currently available to describe short crack growth and presents several topics deserving the attention of researchers. *Wanhill* continues by discussing the problem of short cracks in aerospace structures and links their importance to developments in non-destructive inspection (NDI) techniques. In particular he points out that the change in design philosophy from the fail-safe approach to a damage-tolerance evaluation requires a better understanding of the behaviour of small cracks, notably in aeroengine discs and blades. He concludes by equating the long-term durability of metallic airframe structures to a better appreciation of short crack growth in details such as fastener holes. The last paper in this section presents an overview, useful to designers, of how several parameters can affect the behaviour of short crack growth. *Blom and his co-workers* discuss two high-strength aluminium alloys with differing microstructures and properties. Experiments are reported on different geometries and slight notches, different test frequencies and R (stress ratio) values. Of interest is the report that the fatigue limit of 7475 Al alloy is higher than 2024 Al alloy, although the reverse situation applies to their respective long crack growth threshold values. They conclude that microstructurally short cracks can only partially be related to crack closure phenomena and that differences in initiation sites and early crack growth of the two alloys are associated with different features of the micro-structures.

Material and specimen preparation effects

The following section of five papers concerns surface finish, grain size, and powder-metallurgical materials. The paper by *Suhr* concentrates on an old problem, namely the effect of surface preparation on fatigue life. This investigation separates out surface residual stress effects by prior heat treatment and shows that critical surface flaw sizes can involve an interaction between surface roughness, inclusion size, and inclusion orientation. The study relates to a low alloy steel (tempered bainite) taken from a forged rotor and so is of extra value to designers, especially since it provides data on mean stress effects, two types of loading, and six different types of surface preparation.

Grain size effects are well chronicled by *Radhakrishnan and Mutoh* who compare the behaviour of two very different steels, a type 304 stainless steel and a low carbon steel, each material having three different grain sizes. The relationships expressed between grain size and fatigue limits, fatigue limits and yield stress, yield stress and grain size, and, finally, between grain size and threshold ΔK values, present a balanced appraisal of the interdependence of these parameters. However, a suggestion previously proposed by Taylor, that the demarcation between short and long crack behaviour occurs when the crack length $a = 10d$ (d = grain size) will promote much discussion.

Bolingbroke and King examine a high strength aluminium alloy in two conditions, one being underaged, the other being overaged. Although both treatments produce the same yield strength, they create different precipitate morphologies which have different effects on short crack growth. The authors report that periodical interruptions in crack growth progressively diminish in intensity as crack length increases, indicating that microstructural influences decrease. What is controversial is that the ΔK parameter describes the behaviour of microstructural short cracks including those subjected to stress levels as high as 90 per cent of the yield stress.

Of great interest is the paper by *Fathulla, Weiss, and Stickler* who compare Mo and MoTiZr powder metallurgical materials; both metals being in various deformed and heat treated states. Short crack growth is characterized by a marked reduction in growth rate when cracks encounter microstructural features such as grain boundaries. Crack growth subsequently enters a transition period before the final phase described by LEFM. The final paper in this section is by *Soniak and Remy*, who give their attention to the advanced turbine disc material Astroloy, which is manufactured by the powder metallurgy route. They rationalize their results by invoking crack closure arguments and plotting both long and small cracks against an effective stress intensity factor range which appears to be adequate for crack lengths greater than 500 microns in a material that is composed of both large and small grains of 50 and 20 microns diameter, respectively.

Environmental aspects

The section on environmental aspects begins with a paper by *Mendez, Violan, and Gasc*, who make a valuable contribution by studying the effect of both air and vacuum environments on the development of surface short cracks in fine grained polycrystalline copper. Initiation sites and propagation modes are tabulated as well as microcrack lengths and density in both environments at three different levels of applied stress. They note the number of cycles to form a one grain boundary long crack (which occurs relatively quickly) and the period taken to allow the crack to propagate beyond the first obstruction to growth (a more substantial period). It appears that environment plays a major role in this second phase and on the subsequent propagation phase.

The second paper in this group, presented by *Petit and Zeghloul*, provides information on a 7075-T651 high strength aluminium alloy tested in air and in vacuum. They describe an accurate technique to assess both closure and opening levels at the crack tip and, hence, are able to report that in air, at low R values, an intermediate growth behaviour occurs due to closure inhibiting water vapour effects. The effective stress intensity factor range is used to rationalize the results.

Finally *Miranda and Pascual* examine exceedingly fine, high density surface cracks in 304 stainless steel induced by the presence of hydrogen. The hydrogen had been subsequently reduced in content by a degassing operation. As a consequence of this treatment, short fatigue cracks grew sub-surface at the interface between the ductile core and the hard but microcracked exterior.

Microstructural considerations

The group representing microstructural effects consists of five papers. Obviously all papers throughout this volume stress the importance of microstructure in studies concerned with the behaviour of very small cracks, but the following papers have been chosen for this group because microstructural aspects are dominant in comparison with other considerations.

Baxter examines the growth of persistent slip bands (psbs) in 6061-T6 aluminium which is fatigued and photographed in a photoemission-microscope. The psbs elongate across a surface grain by the sequential addition of further extrusions with each new extrusion being associated with a particular cell in the psb microstructure. The question that this paper raises is whether or not the linking of surface extrusions is a reflection of the linking of closed microcracks too difficult to see. Of interest is the fact that the strain in the psb decreases as the psb elongates. *Mulvihill and Beevers* also produce an interesting paper from a study of short crack growth in an aluminium alloy fatigued inside an electron microscope. Grain boundary initiation takes place in the precipitate-free zone and retardation occurs at inclusions, triple points, and at transitions from intergranular to trans-granular growth. A cast aluminium alloy structure was studied by *Plumtree and Schafer*, who found that triple points, particle decohesion, dendritic structure, eutectic matrix, and particle cleavage had roles to play in the short crack growth phase. The critical distance limiting the short crack growth phase appeared to be the spacing of triple points in the interdendritic spaces. Initiation occurred at silicon particles, either by debonding, primarily at low stresses, or cracking at high stresses. Their evidence on short crack growth behaviour illustrates that crack growth rate is independent of the stress intensity factor at a high strain range (1 per cent) while at low strain range (0.36 per cent) it does not account for the decreasing crack growth rate period. *Howland* studied single crystals of a precipitation strengthened superalloy; in particular the growth of cracks on specific slip planes radiating from a corner edge notch. Stress ratio effects are presented and K is not a satisfactory parameter to characterize the mixed mode crack growth behaviour. The last paper in this section is a combination of the work done by *Kendall, James, and Knott* who first review the types of defects of importance to engineers. They then present experimental data for two types of steel. The first study on a weldable alloy steel having a tempered martensitic structure, is over a stress ratio range of 0.2 to 0.7, the second study, at $R = 0.5$, is on a material having a ferrite–pearlite microstructure. They suggest the use of a closure-free ΔK_c value to describe short crack behaviour.

Notches

The study of short cracks is always associated with relatively high stress levels, i.e., stress levels that invalidate the assumptions invoked for LEFM analyses. In engineering practice, these high stress level conditions frequently occur at stress concentration features, and so many researchers, for both academic and practical reasons, study short crack behaviour in the roots of notches. The first three summaries below discuss the effects of surface micronotches.

The paper by *Yamada, Kim, and Kunio* is a welcome contribution from the Japanese school. They report on the differences in fatigue limits between plain and micronotched samples, in both annealed and non-annealed states, of three different carbon content steels. Briefly the 0.84% steel has a critical surface defect micronotch of 230 μm diameter, whilst an un-notched sample can produce a surface crack length of 340 μm. Both these dimensions can be decreased in their respective cases to 85 μm and 80 μm if the samples are annealed. For a 0.36%C steel, the comparative figures for annealed samples are 90 μm (micronotched) and 172 μm (un-notched). From such data it follows that tolerant (i.e., non-propagating) defect sizes are geometry, heat treatment, and microstructure sensitive, the latter point being emphasized in the 0.36%C steel samples, where the critical defect size is a function of the distance of the micro defect from the nearest pearlite colony.

The second paper in this series is by *Murakami and Endo* who provide much reference material. Their thesis is that surface hardness is one of two important parameters required to assess fatigue resistance. Hardness values range from 70 HV (70/30 brass) to 720 HV (maraging steel) for the fourteen different materials reviewed. The second parameter of importance is the square root of the area of the defect which, in the present study, covers a range of 16–949 microns. They find that, from all the materials examined, two simple equations can predict the fatigue limit and the threshold value of the long crack stress intensity factor range.

Oni and Bathias also study micronotches. They employ a carefully calibrated pulsed d.c. electric-potential crack monitoring technique to study two high strength, low alloy, martensitic steels. They note that several cracks form in the shallow notch root either from Al_2O_3 inclusions or, in the case of the higher strength material, from notch profile irregularities.

The paper by *Hay and Brown* starts a series of papers on larger notches. They report on notches under torsion, a persistent problem in the design of rotating shafts. They present data from different authors who provide details for different materials, temperatures, and specimen sizes. Because an elastic mode III stress intensity factor cannot unify the data, Hay and Brown present an elasto-plastic strain intensity factor which provides a conservative upper bound solution for crack growth rates for all the conditions studied. They also indicate three zones of behaviour from factory-roof type fracture surfaces commonly seen at low strain intensity levels, to flat surfaces generated at high strain intensity levels. The former zone is due to the formation of several mode I type cracks, the latter due to mode III crack growth behaviour.

Central circular notches have been studied in an aluminium alloy and a steel by *DuQuesnay, Topper and Yu*, who note that the fictitious crack length, l_0, which they introduced in 1979 to modify the stress intensity factor, is not a material constant as first thought, but is dependent on stress range and minimum stress. Nevertheless, they use the term l_0 to calculate fatigue notch factors for sharp (small circular) notches below a critical notch size, itself a function of l_0, which gives an approximate theoretical prediction (± 15 per cent error) when compared to experimental results for the two materials studied. The paper by *Hussey, Byrne, and Duggan* also presents data on circular notches. Experimental results on Waspaloy and Nimonic 105 are compared with theoretical predictions based on finite element solutions which are themselves compared with more simple analytical solutions derived by other workers. *Foth and his co-workers* have studied surface crack growth in the roots of notches in a 2024-T3 aluminium alloy. While the paper notes that microstructural cracks cannot be evaluated by LEFM analyses, they state that physically small, semi-elliptical cracks can be evaluated in terms of K within an arbitrary stress field. *Lahor, Sehitoglu, and McClung* use a plane-stress, elastic–plastic finite element method to assess opening and closing behaviour of a crack at one R ratio and one stress level in one material. They record good agreement between experiment and finite element results.

Verreman, Bailon, and Masounave study the practical problem of short crack growth at the toe of fillet welds having a notch angle of 135 degrees. They successfully use a strain gauge system to monitor crack advance and to determine the instant of crack opening. This system can detect 10–20 μm cracks in the heat affected zone at the weld toe. Finite element analyses in the toe area determines the size of the plastic zones to which short crack growth is related via an effective stress intensity factor range. Finally they state that, for the six different stress levels reported, the crack initiation phase can be neglected.

Fracture mechanics

The seven papers of the proceedings devoted to fracture mechanics cover perhaps the most important section in the book for two basic reasons. First, a mathematical treatment permits an evaluation of the most important parameters associated with short crack growth and their inter-relationships with both microstructural details on the one hand and micro-mechanics on the other. Secondly, if an accurate description of short crack growth can be formulated based on the physical processes operating during the short crack growth phase, then the duration of this phase can be ascertained. Both of these aspects of fracture mechanics will help designers and researchers alike, the latter to produce materials with increased fatigue resistance, the former to have a far more accurate and hence safer estimation of fatigue life.

The section opens with a paper by *Guiu and Stevens* who advocate an alternative approach to understanding nucleation, short, and long crack growth from thermodynamic considerations. They formally present the paradoxical situations associated with the growth of both short and long cracks, but show that, by extending the classical Griffith approach to include a local Helmholtz free energy term, an explanation is found to describe the real behaviour of cracks, i.e., cracks that apparently nucleate with great ease on the surface of a material, but which can then stop, having reached a situation in which the driving force is insufficient for further growth. Both of these phenomena have been frequently observed in experimental studies and the authors conclude that further studies on the strain energy release rate approach are required if short crack behaviour is to be properly understood.

The second paper is by *Brown* who examines the boundary conditions between short, long, and non-propagating cracks from a mechanistic viewpoint. By providing contours of constant crack growth rate, including a near zero (threshold) base, he constructs a fatigue fracture-mode map that encompasses six zones which are stress level and crack length dependent, namely: (i) LEFM mode I type cracks, (ii) EPFM mode I type cracks, (iii) EPFM mode III type cracks, (iv) microstructural mode II stage I type cracks, (v) mode II crystallographic type cracks, and (vi) non-propagating modes I/II type cracks. His fracture map is similar to the Kitagawa–Takahashi map, but with more information, especially regarding fatigue limit conditions and the growth of cracks below microstructural barriers. The second paper by *Hobson, Brown, and de los Rios* considers a medium carbon steel. From theoretical considerations, using dimensional analysis arguments, they derive two crack growth equations, one for each of two types of short crack, a microstructurally short crack and a physically short crack. The first equation includes a microstructural parameter, while the second equation is developed for continuum mechanics analyses and is expressed in terms of elastic–plastic parameters. In a transition period between the two short crack growth phases the crack growth rate is the sum of both equations which are better represented on linear scales rather than the more usual log–log plots. The initial period of crack growth is shown to be negligible. Both types of short cracks grow at rates higher than those predicted by LEFM.

The work by *Nisitani and Goto* could have been reported in the notch effects section of this volume since the authors used 0.1 mm surface micro-notches in their study of small crack growth. However, the formulation of their results firmly placed this study into the fracture mechanics section. They develop two laws, one for short and one for long cracks, the former being applicable at stress levels greater than one-half of the yield stress of the plain carbon (0.45%) steel tested. Of interest is the fact that the exponent in their small crack growth law, $dl/dN = \sigma_a^n l$, is approximately 8. This equation is not dissimilar to the equations quoted in the previous two papers where the stress term is replaced by the plastic strain range.

Taylor, in discussing the limitations of fracture mechanics, states that LEFM analyses are not applicable below a critical crack length which is a function of both yield stress and grain size (readers should refer to the work of Radhak-rishnan) and by citing and carefully plotting data from twelve references, Taylor shows that this length may be $10d$ (d = grain diameter) as originally proposed or possibly $10r_p$ (r_p = plastic zone size).

The final two papers in this section are concerned with cumulative damage and the role of short crack growth in the evolution of that damage. *Miller, Mohamed, and de los Rios* consider the accumulation of damage below the fatigue limit and how it adversely affects subsequent damage accumulation above the fatigue limit as the stress range increases throughout a test. Using separate equations for microstructurally short and physically small cracks derived from constant stress range tests, they show that the classical approaches to cumulative damage are inaccurate and that it is essential to take account of crack growth below conventionally determined fatigue limits. It is also shown to be important to determine the most effective barrier to crack growth, this barrier being dependent on the applied stress level.

Bouksim and Bathias study the superposition of major and minor cycles of loading and consider the propagation of a crack to a length of 200 μm (which they call initiation) in an aluminium alloy with a pancake shaped grain measuring $500 \times 200 \times 20$ μm. This initiation phase includes the growth of a crack to a length of 10 μm that is predominantly controlled by the large stress cycles ($R = 0$ type) of each block. Should large stress cycles be interspaced with small stress fluctuations, then these minor cycles have a considerable influence on the continued propagation of the crack. An interesting way of assessing the rate of accumulation of damage is to split the block stress pattern into positive and negative components and thence derive a crack growth rate equation from the sum of the components.

Acoustic microscopy

The final section looks to the future and the development of acoustic microscopy. *Briggs, de los Rios, and Miller* briefly discuss the propagation and scattering of Rayleigh waves and how they are affected by surface and near surface cracks. They discuss the advantages of the technique over other methods, which include a resolution better than a micron, the ability of observing closed cracks not detectable by other means, the determination of crack shape, and, finally, there being no necessity for replication and etching. Two examples of short crack detection, using a single acoustic lens system, are presented, one in 316 stainless steel, and one in an Al-Si bearing material.

The final paper is written by the Stanford team of *London, Shyne, and Nelson* who, with a different acoustic system involving a transmitting lens and a receiving lens, give additional reasons for the need to develop this technique. They examine 4140 steel in four tempers and report on the accurate and continuous determination of closure stress levels and the application of acoustic microscopy to other materials, e.g., ceramics. Two of their short crack results will stimulate much discussion because these cracks grew at a lower rate than the long crack growth data provided by conventional LEFM tests on the quenched and low temperature-tempered 4140 steel. Higher temperatures removed this anomaly.

K. J. Miller
E. R. de los Rios

Sheffield, May 1986

DESIGN CONSIDERATIONS

I. F. C. Smith*

Applying Fatigue Research to Engineering Design

REFERENCE Smith, I. F. C., **Applying Fatigue Research to Engineering Design,** *The Behaviour of Short Fatigue Cracks*, EGF Pub. 1 (Edited by K. J. Miller and E. R. de los Rios) 1986, Mechanical Engineering Publications, London, pp. 15–26.

ABSTRACT This paper reviews some crack initiation and short crack propagation models, and examines their impact on engineering design methods. Many industries are not able to use some proposals due to a lack of relevant information and inspection difficulties. Further work is needed in many areas. When fatigue research is planned and carried out, recognition of practical applications is important for the development of rational design guidelines.

Introduction

Considerable differences exist between a scientific investigation and an engineering design. Frequently, a scientific investigation is motivated by a need to explain observed behaviour. Usually, the investigation concentrates on a comparison of theoretical models with average experimental data. After an iterative process of adjustments to the model(s) and further testing, the investigation concludes whether or not the theory agrees with the experimental data. In this way, a greater understanding of fatigue behaviour is achieved.

An engineering design is motivated by a demand for a structure or a component, and design models are based upon upper or lower bounds of test data. The designer begins with simple models using approximate methods. If the component is fatigue sensitive, a more sophisticated assessment is performed and testing may be carried out. The goal of most engineering designs is to achieve a high probability of survival at low cost. Figure 1 is a schematic showing the base-line differences between a scientific investigation and an engineering design.

Considering such differences, it is understandable that a model which performs well during a scientific investigation may not necessarily improve an engineering design. For example, a fatigue assessment of a surface scratch of average depth on a given component may require a sophisticated short crack analysis. However, a deeper scratch which can be assessed satisfactorily using a simpler model may determine the fatigue life.

The significance of recent fatigue research has been discussed previously for specific design philosophies (1)(2). Generally, it is assumed that specialists are available to perform the fatigue assessment. However, many fatigue

* ICOM – Steel structures, Swiss Federal Institute of Technology (EPFL), 1015 Lausanne, Switzerland.

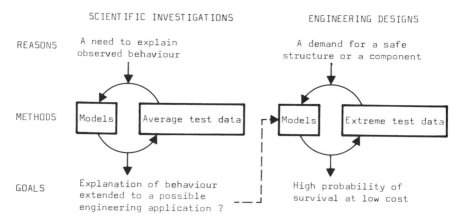

Fig 1   Typical base-line characteristics of scientific investigations and engineering designs

sensitive components and structures are not assessed by design specialists. For example, cranes, ships, engine components, farm machinery, pressure vessels, bridges, marine structures, and automotive parts may be assessed by an engineer who has only a rudimentary knowledge of fatigue, Fig. 2. Such designers invariably rely on design guidelines in order to carry out the fatigue assessment. Many of these documents recommend the use of empirical relationships which bear little resemblance to the theoretical models currently examined in scientific investigations.

Over the past thirty years, much fundamental work has increased our understanding of the mechanisms which determine the fatigue life of engineering materials and structures. Advances have been reported within three research areas: fatigue crack nucleation, short crack growth, and long crack propagation. The next thirty years should be a period where these advances are consolidated into guidelines for general use by design engineers. However, a useful synthesis of these three fields, followed by the drafting of design guidelines for a range of design problems, will be difficult to achieve.

This paper reviews some models proposed for crack initiation and short crack behaviour and investigates their applicability to engineering problems. The conditions necessary for new design guidelines are discussed and opportunities for further work are identified.

Terminology

Fatigue terminology varies widely among scientists and engineers. Certain definitions are assumed in this paper; in order to avoid confusion, a short list of terms is given below.

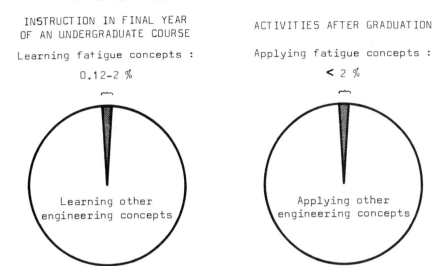

INSTRUCTION IN FINAL YEAR
OF AN UNDERGRADUATE COURSE

Learning fatigue concepts :

0.12-2 %

Learning other
engineering concepts

ACTIVITIES AFTER GRADUATION

Applying fatigue concepts :

< 2 %

Applying other
engineering concepts

Fig 2   A designer's exposure to fatigue concepts

Crack: a sharp discontinuity which grows in size upon application of fatigue loading. This loading must exceed a critical level determined by stresses, environment, material properties, and geometrical parameters.

Crack initiation: the fatigue life resulting from a strain-controlled fatigue test of a smooth unnotched specimen.

Crack nucleation: the fatigue life expended in the formation of a crack.

Designer: an engineer whose exposure to fatigue concepts may be described by Fig. 2. Such engineers use design guidelines for fatigue assessments.

Design guidelines: standards, recommendations, directives, handbooks, notes, specifications, requirements, commentaries, and codes published by international, national, industrial, private, and professional organizations. Drafting committees are made up of fabricators, consultants, contractors, suppliers, inspection experts, and design specialists.

Design specialist: an engineer or scientist with research experience. A specialist should be competent to evaluate and use models proposed in the scientific literature and to provide technical input during the drafting of design guidelines.

Long crack propagation (LCP):   fatigue crack growth which is similar to the behaviour described by linear elastic fracture mechanics (LEFM). Near-threshold crack growth, measured using standard LCP specimens, and crack growth in the elastic region of the material affected by a notch is included in this definition.

Short crack propagation (SCP):   fatigue crack growth which cannot be described by LEFM.

Fatigue research

Early fatigue research concentrated on empirical correlations based on component testing. Over 100 years ago, the use of laboratory specimen testing began to replace full scale tests, thus introducing the similarity (similitude) problem and the subsequent need for representative models. Early models such as the stress range–fatigue life relationship, the Goodman diagram, the fatigue reduction factor and Miner's rule were developed. These relationships still enable satisfactory fatigue assessments in many situations. Nevertheless, fatigue failures still occur and economic pressures for efficient designs remain strong. In addition, new challenges such as energy conservation and increased litigation risks have further encouraged modern fatigue research.

For many years, crack nucleation and LCP remained separate and distinct disciplines with specific applications for particular structures, components, and materials. During the past decade, SCP studies have provided opportunities to unify the two disciplines. To date, the relative roles of crack nucleation, SCP, and LCP are not clear for a range of components. However, many studies have provided proposals and some models are reviewed below.

Crack initiation

Many models have been proposed for crack nucleation and early crack growth. Each approach has its own set of definitions and corresponding applications. The strain-controlled fatigue life of a smooth unnotched specimen, crack initiation, is the basis of many engineering models. The model described herein was chosen due to the large amount of design-oriented research which supports its application.

The principal parameter, strain range, was originally studied by Bauschinger around 1880. In the 1950s, it was revived independently by Manson and Coffin for use in low-cycle fatigue. This parameter is divided into plastic and elastic strain components in a formula which includes fatigue life. These two components contribute to 'damage' which accumulates until failure occurs. When plasticity is negligible, this formula simplifies to the relationship proposed by Basquin in 1910. The Manson–Coffin formula, sometimes known as the low-cycle fatigue equation, has formed the foundation for many subsequent studies and became the subject of an ASTM standard in 1977.

Researchers in this area have studied cyclic stress–strain behaviour, notch analysis, temperature effects, mean stress, variable amplitude modelling, and computer simulation using a wide range of materials. Important contributions include the determination of the cyclic stress–strain curve, cumulative damage rules, approximations of the cyclic strain at a stress concentration, and the development of sophisticated testing systems. References (3) and (4) are detailed reviews. The development of a formula which determines the fatigue strength reduction factor for rough surfaces is a recent advance (5).

Limitations of crack initiation models include difficulties associated with the definition of 'damage', high-cycle fatigue, surface roughness, corrosion, and multiaxial effects (3). Also, the percentage of crack initiation spent in SCP is not included in most test data. Proposals for the crack length of a so-called initiated crack vary from zero to several millimetres. Such lengths depend upon the crack growth model which is used to calculate the remaining fatigue life.

Short crack propagation

SCP models were developed to give a more detailed description of the 'damage' modelled by the crack initiation approach. Short cracks may be micro-structurally short due to material characteristics, or mechanically short due to crack tip conditions which are mechanically different from those characteristic of long cracks, e.g., plasticity, crack closure, etc. Often, these two types of SCP were not studied independently and much confusion exists in the literature. Nevertheless, microstructurally short cracks usually dominate when cracks are less than a few grain diameters long and when they are subject to stress ranges near the fatigue limit. On the other hand, mechanically short cracks can be several millimetres in length and they are most obvious when the cyclic stresses are much higher than the fatigue limit.

Frost (6) recognised the importance of microstructurally short cracks before 1960 in his study of non-propagating fatigue cracks. He defined a minimum crack length 'of the order of the grain size' where the propagating stress no longer varied with crack length, but converged on the fatigue limit. Kitagawa and Takahashi (7) attempted to obtain data missing from Frost's original study, thereby examining SCP in the absence of bulk plasticity. These studies aroused considerable interest in the behaviour of short cracks at stress ranges near the plain fatigue limit.

Amplified growth rates due to bulk plasticity were modelled in 1965 by Boettner et al. (8) who proposed a strain intensity factor which accounted for cracks growing in cyclic plastic strain fields. An approximate analysis showed that if the initial crack length is the same for all conditions, the integration of the growth law results in Coffin's law for crack initiation. This study is believed to be the first attempt to model short crack growth in terms of the product of crack length and an elastic–plastic loading parameter. Later, Pearson (9) explored the limitations of LEFM and demonstrated that linear elastic model-ling underestimated the growth rates of short cracks.

Since these early studies, theoretical models have been proposed by many researchers. Table 1 contains a list of investigations grouped into three categories according to the type of model used and the principal parameters considered. The majority of the models in Categories 1 and 3 describe microstructurally short cracks, whereas Category 2 is concerned mainly with mechanically short cracks.

Table 1   **Models for short crack propagation**

Category	Parameters	References
① Modified elastic fracture mechanics	$\Delta K_{\text{th}}(a)$ $\sqrt{(D\rho)}$ l_{o} ΔK_{eff} $f(\Delta\sigma) + f(\Delta K)$	(7)(10) (11)–(13) (14)–(17) (15)(18) (19)
② Elastic–plastic growth models	CI^{*} $\Delta K(\Delta\varepsilon)$ $\Delta K_{\text{eff}}(\Delta\varepsilon)$ ΔJ $\gamma_{\text{p}}^{2}a$	(20)–(24) (8)(22)(25)(26) (24) (26)–(29) (30)
③ Microstructural fracture mechanics	$(d - a)^{1-\alpha}a_{\alpha}$ $\Delta K^{n}\left\{1 - k(\Phi)\left(\dfrac{D - 2X}{D}\right)^{m}\right\}$ $\dfrac{\tau(L - a)}{L}$ $\tau\sqrt{(aR)}$ $R\,\Delta\varepsilon_{\text{p}}^{1/c}$	(31) (32) (33) (34) (35)

* CI: Crack initiation model is used to represent the growth of short cracks to some crack length.

Most investigations are placed in either Categories 1 or 2. In Category 2 the first parameter, CI, refers to studies which employed the crack initiation model to represent short crack growth. Some proposals account for both micro-structurally and mechanically short cracks using unified models. For example, Cameron and Smith (22) suggest that the crack initiation model, CI, can represent crack lengths smaller than the constant, l_{o} (14), and a strain based approach (25) is proposed for longer cracks. Others studies concentrated on only one type of SCP. A comprehensive study of the merits of these models is hindered by a lack of compatible data, although attempts have been made for some proposals (13)(24)(26)(35)–(37).

The characteristics of the parameters employed differ from category to category. In Category 1, they are far field or global values and consequently, they can be determined simply. In order to give greater accuracy in non-elastic cases, Category 2 models require the knowledge of an elastic–plastic parameter. This demands a more detailed analysis than prescribed by the models in Category 1. Category 3 introduces parameters such as grain size, d, and grain orientation, Φ, in order to model cracks less than a few grain diameters in length. In a recent review, Miller (38) has proposed that crack nucleation occurs immediately, and thus the fatigue limit is defined by a crack growth threshold which is determined by microstructural parameters.

The impact of SCP models on fatigue design assessments

The fatigue strength calculation completes only part of a fatigue design. Figure 3 gives an example of the amount of design time spent evaluating the fatigue strength in relation to other factors. Uncertainty in loading data and material properties as well as the effect of the consequences of failure, the costs, quality assurance, and in-service monitoring must be taken into account. Note that in considering these factors, a designer employs less than 2 per cent of his total design time, see Fig. 2.

One design criterion is the verification of a certain probability that the effect of fatigue loading is less than the fatigue strength – at minimum cost. The consequences of failure usually determine the required probability of survival. Consideration of both this criterion and the characteristics of the factors in Fig. 3 may preclude a rational use of SCP models. For example, the loads may be so poorly defined that increased sophistication in the fatigue strength calculation changes neither the probability of survival nor the final cost of the component.

The occurrence of SCP

Another difficulty may arise upon identifying those cases when short crack behaviour should be expected. Many welded structures contain large discontinuities and suffer low service stresses, and thus it is easily determined that no short crack analysis is needed. Other cases are not as clear.

For example, the fatigue critical discontinuity is usually the deepest discontinuity in a notch and its depth may be more than five times the centre-line

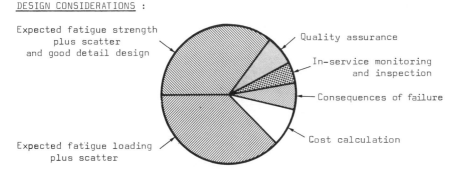

DESIGN CONSIDERATIONS :

Expected fatigue strength
plus scatter
and good detail design

Quality assurance

In-service monitoring
and inspection

Consequences of failure

Cost calculation

Expected fatigue loading
plus scatter

DESIGN CRITERIA :

A certain probability that the effects of :

FATIGUE LOADING < FATIGUE STRENGTH (at minimum cost).

Fig 3 An example of the relative importance of various considerations in a fatigue design assessment

average value measured at the same location. Furthermore, this discontinuity may have a root radius of less than one micron (**39**). It is not known if there is a relationship between surface roughness and the deepest discontinuity for a range of materials and surface finishes.

Another case is related to the service conditions of the component. Machined components become worn, scratched, and corroded during service due to fretting, poor lubrication, misuse, scouring, corrosive attack, etc. A SCP model may not be applicable to the fatigue strength calculation if the component is exposed, accidentally or otherwise, to service conditions which introduce surface discontinuities.

Nevertheless, improvements in fabrication techniques, load analyses, stress measurement, quality assurance, and new design methods are increasing the number of situations where SCP models are useful design tools.

Design methods and SCP models

A damage tolerant design method (fail-safe, fitness for purpose, etc.) provides the greatest justification for the application of propagation models to life and remaining life calculations. Designs are improved by emphasising quality assurance and in-service inspection, Fig. 3. These methods use data taken from measurements on the actual component and, as a result, a more precise estimate of its behaviour under fatigue loading is possible. In addition to improving fatigue strength calculations, damage tolerant design methods enable great flexibility during quality evaluation and structural repair.

Unfortunately, advances in related fields, such as crack detection, do not coincide with the development of SCP models. Accounting for human error, present non-destructive inspection (NDI) technology is rarely able to detect discontinuities which behave as short fatigue cracks. Previous papers (**1**)(**2**)(**40**) have discussed the significance of NDI with respect to damage tolerance design.

SCP models may be useful when a design method *other* than damage tolerant design is employed. Accurate models are essential when the similitude of existing data is in doubt and prototype testing under service loading is not practicable. Also, SCP models may provide important design tools even if they are not directly used in the fatigue strength calculation. For example, SCP models can be useful for selecting materials, changing existing designs and approximating inspection intervals. In these cases, the initial crack size need not necessarily be detectable by NDI.

Guidelines for fatigue design assessments

The options available to a designer are limited by the sophistication of design guidelines. Some of these guidelines are based on the stress range–cycle life approach developed by Wöhler over 100 years ago, e.g., (**41**).

Simple LEFM calculations often result in very conservative life predictions;

they can be more than an order of magnitude less than test results. This is especially true for high-cycle fatigue. Although LCP guidelines have existed for some years, e.g., (42), only a small amount of design information exists for near-threshold crack growth at stress concentrations.

One recent attempt to draft LCP guidelines resulted in a list of warnings against misuse (43). This document recognises that in many large steel structures, information concerning load levels, lack of fit, thermal stresses, differential settlement, and fabrication stresses is unavailable. Thus, it is recommended that all applied compressive loading is included in the analysis and that the intrinsic threshold stress intensity is used in calculations – unless the mean stress can be determined. Additional prudence is recommended for other factors such as plasticity, stress analysis, variable amplitude loading, and crack size.

Some studies have examined the importance of SCP by identifying those cases where SCP can be ignored, even though it occurs. For example, Taylor and Knott (44)(45) have suggested an inherent discontinuity equal to the deepest influence of microstructurally short crack effects. Cameron and Smith (22) have identified cases where LEFM calculations do not change the fatigue life by more than 50 per cent.

In general, greater sophistication can be hindered by cautious drafting committees which rely more upon the practical experience of their members than upon the conclusions of scientific investigations. Design specialists are often in a minority on drafting committees.

Before changes are made in any guideline, most committees wish to establish that:

- present engineering practice is unsafe, costly, or inappropriate;
- an easier way does not exist;
- the principles behind the change can be explained in simple terms;
- the change is acceptable to designers.

Short crack investigations which establish these points are rare. The absence of such information contributes to a lack of understanding between the scientist and the designer.

Further work

Despite concentrated research activity, little information is filtering down to the non-specialist. Further work is needed in many fields before design guidelines can be written. Without guidelines, many industries cannot profit from recent progress. A short list of possible research topics is given below.

LCP topics requiring further work include:
- crack growth laws and material constants in the near-threshold crack growth regime;

- a compendium of stress intensity factors for surface cracks in engineering notches, including various welded joints and bolted assemblies;
- experimental methods for the determination of the threshold stress intensity factor;
- crack retardation and acceleration due to variable amplitude loading;
- the intrinsic threshold stress intensity factor;
- the effect of residual stresses at notches;
- crack growth from notches for numerous stress ratios;
- crack growth at high and low temperatures;
- crack growth laws for corrosion enhanced fatigue crack growth, etc.

Considering the amount of research which is needed in the domain of LCP, it is understandable that much work on SCP remains to be completed. Some of the more practical areas include:

- surface finish and initial discontinuities;
- comparative studies of SCP models and the crack lengths at which they are relevant;
- identification of situations when LEFM cannot be used;
- elastic–plastic stress analysis for cracked geometries;
- short crack growth and variable amplitude loading;
- experimental data for a range of materials and environments;
- the influence of fabrication processes;
- identification of relevant microstructural units for a range of materials, etc.;
- statistical analysis.

In the future, the importance of crack propagation models will increase with technical developments and greater expertise in related fields such as:

- non-destructive inspection technology;
- fatigue loading, impact, and vibration;
- stress analysis;
- in-service monitoring;
- safety and reliability analysis, etc.

Conclusions

Although many new models are proposed for calculating fatigue strength, designers are rarely able to make use of them. Scatter in design parameters and service conditions may not justify increased sophistication. Damage tolerent design methods have improved the significance of SCP and LCP models but frequently, crack lengths cannot be detected reliably. Nevertheless, the signifi-cance of new models will increase with improved fabrication methods and advances in related fields. Research which considers the information necessary for drafting design guidelines will help many industries exploit the possibilities offered by modern fatigue research.

Acknowledgements

The paper was funded by the Swiss National Science Foundation. The author is grateful to Professor M. A. Hirt for helpful discussions and to the staff at ICOM for their help in the preparation of the document.

References

(1)　WANHILL, R. J. H. (1981) Some case studies and the significance of fatigue thresholds, *First International Conference on Fatigue Thresholds*, Stockholm (EMAS, Warley, UK), pp. 953–965.

(2)　SCHIJVE, J. (1984) The practical and theortical significance of short cracks. An evaluation, *Fatigue 84* (University of Birmingham, UK), pp. 751–771.

(3)　MORROW, J. and SOCIE, D. F. (1981) The evolution of fatigue crack initiation life prediction methods. Materials, experimentation and design in fatigue, *Proceedings of Fatigue 81* (Westbury House, Guildford, UK), pp. 3–21.

(4)　COFFIN, L. F. (1984) Low cycle fatigue – A thirty year perspective, *Fatigue 84* (University of Birmingham, UK), pp. 1213–1234.

(5)　LAWRENCE, F. V., HO, N. J., and MAZUMDAR, P. K. (1980) Predicting the fatigue resistance of welds. Fracture Control Program, Report No 36, University of Illinois, USA.

(6)　FROST, N. E. (1959) A relation between the critical alternating propagation stress and crack length for mild steel, *Proc. Instn mech. Engrs*, **173**, 811–827.

(7)　KITAGAWA, H. and TAKAHASHI, S. (1976) Applicability of fracture mechanics to very small cracks or the cracks in the early stage, *Proceedings of the Second International Conference on the Mechanical Behaviour of Materials*, Boston, USA, pp. 627–631.

(8)　BOETTNER, R. C., LAIRD, C., and McEVILY, A. J. (1965) Crack nucleation and growth in high strain-low cycle fatigue, *Trans. Met. Soc. AIME*, **233**, 379–385.

(9)　PEARSON, S. (1975) Initiation and fatigue cracks in commercial aluminium alloys and the subsequent propagation of very short cracks, *Engng Fracture Mech.*, **7**, 235–247.

(10)　ROMANIV, O. N., SIMINKOVICH, V. N., and TKACH, A. N. (1981) Near-threshold short fatigue crack growth, *First International Conference on Fatigue Thresholds*, Stockholm (EMAS, Warley, UK), pp. 799–807.

(11)　SMITH, R. A. and MILLER, K. J. (1977) Fatigue cracks at notches. *Int. J. Mech. Sci.*, **19**, 11–22.

(12)　CAMERON, A. D. and SMITH, R. A. (1981) Upper and lower bounds for the lengths of non-propagating cracks, *Int. J. Fatigue*, **3**, 9–15.

(13)　HUSSEY, I. W., BYRNE, J., and DUGGAN, T. V. (1984) The influence of notch stress field on the fatigue crack growth threshold condition, *Fatigue 84* (University of Birmingham, UK), pp. 807–816.

(14)　EL HADDAD, M. H., SMITH, K. N., and TOPPER, T. H. (1979) Fatigue crack propagation of short cracks, *J. Engng mater. Tech.*, **101**, 42–46.

(15)　TANAKA, K. and NAKAI, Y. (1983) Propagation and non-propagation of short fatigue cracks at a sharp notch, *Fatigue Engng mater. Struct.*, **6**, 315–327.

(16)　HARKEGARD, G. (1981) An effective stress intensity and determination of the notch fatigue limit, *First International Conference on Fatigue Thresholds*, Stockholm (EMAS, Warley, UK), pp. 867–879.

(17)　DUGGAN, T. V. (1981) Influence of notch-geometry on fatigue threshold, *First International Conference on Fatigue Thresholds*, Stockholm (EMAS, Warley, UK), pp. 809–826.

(18)　LIAW, P. K. and LOGSDON, W. A. (1985) Crack closure: an explanation for small fatigue crack growth behaviour, *Engng Fracture Mech.*, **22**, 115–121.

(19)　SAXENA, A., WILSON, W. K., ROTH, L. D., and LIAW, P. K. (1985) The behaviour of small fatigue cracks at notches in corrosive environments, *Int. J. Fracture*, **28**, 69–82.

(20)　SOCIE, D. F., MORROW, J., and CHEN, W. C. (1979) A procedure for estimating the total fatigue life of notched and cracked members, *Engng. Fracture Mech.*, **11**, 851–860.

(21)　MAZUMDAR, P. K., CHEN, W. C., and LAWRENCE, F. V. (1981) An analytical study of the fatigue notch effect, *First International Conference on Fatigue Thresholds*, Stockholm (EMAS, Warley, UK), pp. 845–865.

(22) CAMERON, A. D. and SMITH, R. A. (1982) Fatigue life prediction for notched members, *Int. J. Pres. Ves. Piping*, **10**, 205–207.

(23) DOWLING, N.E. (1979) Fatigue at notches and the local strain and fracture mechanics approaches, *ASTM STP 677*, pp. 247–273.

(24) LEIS, B. N. (1985) Displacement controlled fatigue crack growth in inelastic notch fields: implications for short cracks, *Engng Fracture Mech.*, **22**, 279–293.

(25) EL HADDAD, M. H., SMITH, K. N., and TOPPER, T. H. (1979) A strain based intensity factor solution for short fatigue cracks initiating from notches, *ASTM STP 677*, pp. 274–289.

(26) STARKEY, M. S. and SKELTON, R. P. (1982) A comparison of the strain intensity and cyclic J approaches to crack growth, *Fatigue Engng mater. Struct.*, **5**, 329–341.

(27) DOWLING, N. E. (1977) Crack growth during low-cycle fatigue of smooth axial specimens, *ASTM STP 637*, pp. 97–121.

(28) EL HADDAD, M. H., DOWLING, N. E., TOPPER, T. H., and SMITH, K. N. (1980) J integral applications for short fatigue cracks at notches, *Int. J. Fracture*, **16**, 15–30.

(29) OBRTLIK, K. and POLAK, J. (1985) Fatigue growth of surface cracks in the elastic-plastic region, *Fatigue Fracture Engng Mater. Structures*, **8**, 23–31.

(30) IBRAHIM, M. F. E. and MILLER, K. J. (1980) Determination of fatigue crack initiation life, *Fatigue Engng Mater. Structures*, **2**, 351–360.

(31) HOBSON, P. D. (1982) The formulation of a crack growth equation for short cracks, *Fatigue Engng Mater. Structures*, **5**, 323–327.

(32) CHAN, K. S. and LANKFORD, J. (1983) A crack-tip strain model for the growth of small fatigue cracks, *Scripta Met.*, **17**, 529–532.

(33) DE LOS RIOS, E. R., TANG, Z., and MILLER, K. J. (1984) Short crack behaviour in a medium carbon steel, *Fatigue Engng Mater. Structures*, **7**, 97–108.

(34) DE LOS RIOS, E. R., MOHAMED, H. J., and MILLER, K. J. (1985) A micro-mechanics analysis for short fatigue crack growth, *Fatigue Fracture Engng Mater. Structures*, **8**, 49–63.

(35) REGER, M., SONIAK, F., and RÉMY, L. (1984) Propagation of short cracks in low cycle fatigue, *Fatigue 84* (University of Birmingham, UK), pp. 797–806.

(36) SURESH, S. and RITCHIE, R. O. (1984) Propagation of short fatigue cracks, *Int. Met. Rev*, **29**, 445–476.

(37) LANKFORD, J. (1985) The influence of microstructure on the growth of small fatigue cracks, *Fatigue Fracture Engng Mater. Structure*, **8**, 161–175.

(38) MILLER, K. J. (1984) Initiation and growth rates of short fatigue cracks, *Fundamentals of deformation and fracture* (Cambridge University Press, UK), pp. 477–500.

(39) SMITH, I. F. C. (1982) *Fatigue crack growth in a fillet welded joint*, Ph.D. thesis, Cambridge University.

(40) WANHILL, R. J. H. (1984) Engineering significance of fatigue thresholds and short fatigue cracks for structural design, *Fatigue 84* (University of Birmingham, UK), pp. 1671–1681.

(41) BS 5400: Part 10 (1980) *Steel, concrete and composite bridges, Code of practice for fatigue* (British Standards Institution, London).

(42) BSI PD 6493 (1980) *Guidance on some methods for the derivation of acceptance levels for defects in fusion welded joints* (British Standards Institution, London).

(43) Recommendations for the Fatigue Design of Steel Structures, European Convention for Constructional Steelwork (1985). Avenue Louise 326, Bte 52, B-1050 Brussels, Belgium.

(44) TAYLOR, D. and KNOTT, J. F. (1981) Fatigue crack propagation behaviour of short cracks; the effect of microstructure, *Fatigue Engng Mater. Structures*, **4**, 147–155.

(45) TAYLOR, D (1982) Euromech Colloquium on short fatigue cracks, *Fatigue Engng Mater. Structures*, **5**, 305–309.

R. J. H. Wanhill *

Short Cracks in Aerospace Structures

REFERENCE Wanhill, R. J. H., **Short Cracks in Aerospace Structures,** *The Behaviour of Short Fatigue Cracks*, EGF Pub. 1 (Edited by K. J. Miller and E. R. de los Rios), 1986, Mechanical Engineering Publications, London, pp. 27–36.

ABSTRACT The practical engineering significance of short fatigue cracks in aerospace structures is examined with respect to the design and operating requirements of safety and durability. It is shown that this significance is presently limited to the safety of some engine parts, notably discs and blades, and the durability of metallic airframe structures.

Introduction

Short crack growth is the subject of much recent research into fatigue of metallic materials and composites. The study of short cracks is undoubtedly important for improved understanding of the fatigue process and development of materials with better resistance to fatigue. However, the practical engineering significance of short crack behaviour appears to be limited (**1**). This will be illustrated and explained with respect to aerospace structures in the present paper. To do this, it is first necessary to consider the design and operating requirements of safety and durability of fatigue-critical aerospace structures.

Safety and durability of fatigue-critical aerospace structures

Structural fatigue design philosophies

Initially the only philosophy for designing against fatigue of aerospace structures was the safe-life approach, which means designing for a finite service life during which significant fatigue damage will not occur.

In the 1950s the fail-safe philosophy evolved and was first applied to civil transport aircraft. The fail-safe approach requires designing for an adequate service life without significant damage, but also enabling operation beyond the actual life at which such damage occurs. However, it must be shown that the damage (cracks or flaws) will be detected by routine inspection before it propagates to the extent that residual strength falls below a safe level.

Since 1970 the United States Air Force (USAF) has developed the damage tolerance approach (**2**). This philosophy differs from the original fail-safe approach in two major respects:

(1) the possibility of cracks or flaws already in a new structure must be accounted for;

---

* National Aerospace Laboratory NLR, Amsterdam, The Netherlands.

(2)   structures may be inspectable or non-inspectable in service, i.e. there is an option for designing structures that are not intended to be inspected during the service life.

Inspectable structures can be qualified either as fail-safe or as slow flaw growth structures, for which initial damage must grow slowly and not reach a size large enough to cause failure between inspections. Non-inspectable structures, according to the USAF, NASA, and ESA (2)–(4), may still be classified as damage tolerant provided they can be qualified for slow flaw growth, which in this case means that initial damage must not propagate to a size causing failure during the design service life. However, this classification is debatable: civil aviation authorities place non-inspectable slow flaw growth structures in the (undesirable) safe-life category (5).

All these design philosophies are in use, not only for different types of aerospace vehicle, but also for different areas in the same structure. An attempt at a general classification is given in Table 1. Although this situation is rather confusing, the important point is that there is a trend to try and increase damage tolerance design in all structural areas.

Definitions of safety and durability

So far, safety has been discussed without qualification. However, actual design balances performance requirements against economic factors such that the probability of failure during the design life is less than some acceptable value. Since all structures deteriorate in service, i.e., the probability of failure increases with time, there is a safety limit. This is the time beyond which the risk of failure is unacceptable unless preventive actions are taken.

For safe-life and non-inspectable slow flaw growth structures the necessary preventive action is – in theory – retirement from service. In fact, the situation is more complicated. It may be possible to extend the service life by structural audit. This involves inspection or reassessment of known 'hot spots' and repair or replacement of damaged or suspect areas. This approach is made possible usually by advances in non-destructive inspection (NDI) techniques and improved understanding of the accumulation of fatigue damage in the structure. The result is that an originally safe-life design becomes, to some extent, amenable to assessment using damage tolerance principles.

For inspectable fail-safe and slow flaw growth structures the preventive action is, in the first instance, repeated inspection, followed by repair or replacement if required and feasible. Only when repair or replacement are not feasible, or when the frequency of inspection becomes uneconomic, need the structure be withdrawn from service.

Consideration of preventive action for ensuring safety leads to the requirement of durability. The necessity for a structure to be durable means, primarily, that the economic life (including any inspections, repairs, or replacements) should equal or exceed the design life. Current practice generally bases the

Table 1 Current application of structural fatigue design philosophies

Design approach		Defects assumed for new structure	Structural fatigue design category	Structural items/Types of vehicle				
				Airframes	Engines	Landing gear	Pressure vessels	Bolts
Safe-Life	Non-inspectable or no planned inspections	No	Original Safe-Life	• most general aviation and military aircraft • helicopters	all	all		all except STS* payloads
	Planned inspections	No	Original Fail-Safe	• pre-1980s civil transports • some helicopter components				
Damage Tolerance	Planned inspections	Yes	Fail-Safe or Slow Flaw Growth	• modern civil transports and military aircraft • space shuttle orbiter	some gas turbine discs			$\phi \leq 8$ mm for STS* payloads
	Non-inspectable or no planned inspections	Yes	Slow (Safe) Flaw Growth Life	• some areas on F-16, B-1 and space shuttle orbiter • most STS* payloads	some turbine items (blades, discs)	space shuttle orbiter	space shuttle orbiter and STS payloads	$\phi > 8$ mm for STS* payloads

* STS = Space Transportation System.

economic life on full-scale and component test results, in particular the frequent occurrence of cracking. However, analyses are being developed, notably by the USAF (6)(7), to enable quantifying the economic life at the design stage. The most advanced analyses are concerned with the widespread initiation and growth of small cracks at fastener holes in metallic airframe structures, since such cracks are one of the most common maintenance problems. The upper limit of crack size that determines durability is defined on the basis of economic repair, e.g., the largest radial crack that can be cleaned up by reaming a fastener hole to the next fastener size. This is followed by installation of an appropriate oversize fastener.

Significance of short fatigue cracks

It must be stated right away that short fatigue cracks have no practical engineering significance for composite structures now or in the foreseeable future. The presence of short 'cracks' and their growth and coalescence during fatigue to form macroscopic defects in composites are of fundamental importance to development of more fatigue resistant composite materials, but there is no way in which such early damage accumulation may be quantified for engineering use.

On the other hand, short fatigue cracks are, or may be, practically significant for metallic structures. As is well known, there is considerable evidence that short fatigue cracks in metals grow at faster rates and lower nominal ΔK values than those predicted from macrocrack growth data (8). These apparent and unfavourable anomalies are found typically for cracks with governing dimensions less than 0.5 mm. Notable exceptions are cracks in some large grain size engine materials, e.g., (9).

The practical engineering significance of short fatigue cracks in metallic aerospace structures is the subject of the remainder of this paper.

Short fatigue cracks and the safety of metallic aerospace structures

Figure 1 gives an overview of the potential relevance and importance of short fatigue crack growth for the safety of metallic structures. The flow chart logic takes into account the structural fatigue design categories, non-destructive inspection (NDI) capabilities, and types of service load histories.

Broadly speaking, short fatigue cracks are potentially significant for safety only if a sufficiently high level of NDI is feasible, i.e., possible and economically justifiable. In more detail this means:

(1) For new damage tolerance structures pre-service NDI must be capable of detecting short cracks or flaws with high reliability. This is a necessary minimum requirement because in-service NDI usually has lesser capabilities.

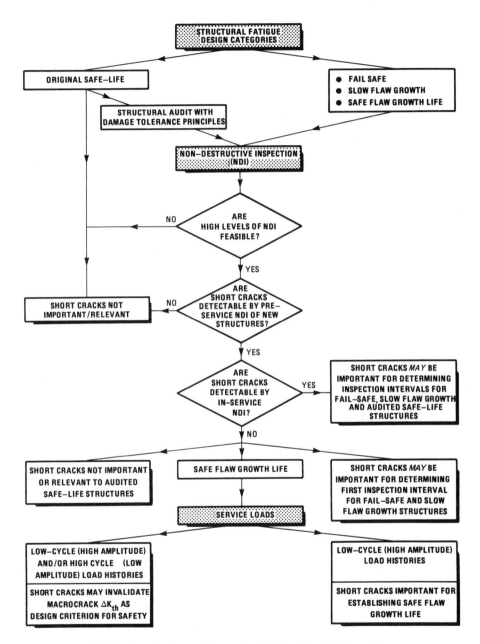

Fig 1 Short fatigue crack growth and safety of metallic structures

(2) For safe-life structures audited using damage tolerance principles in-service NDI must be capable of detecting short cracks or flaws with high reliability.

In fact, the current situation is that, except for a few special cases in engines – particularly military gas turbines – the feasible levels of NDI for aerospace structures correspond to relatively large flaw sizes that must be assumed immediately after inspection (2)–(4)(10). This is illustrated for airframes, the space shuttle orbiter, and Space Tranportation System (STS) payloads in Table 2. All initial damage sizes are beyond 0.5 mm, which is about the limit of the short crack regime for most materials.

Table 2 does not include aircraft landing gear and engines, which are traditionally designed and operated according to the original safe-life approach. For aircraft landing gear this situation is likely to continue for the foreseeable future. But for engines there is considerable effort to introduce the damage tolerance approach, e.g., (11).

Blades and discs are generally the most fatigue-critical items in engines. Short cracks are important for safe damage tolerance design and operation, but for different reasons. Fatigue loads on blades are mainly large numbers of low amplitudes. Thus ΔK_{th} (the threshold stress intensity factor range for fatigue crack growth) may be used as a design criterion for preventing high cycle fatigue failure due to, for example, thermal fatigue cracks in coatings, cooling holes, and other stress concentrations. However, the value of ΔK_{th} obtained from macrocrack growth data could be unconservative if the transition to high-cycle fatigue occurs in the short crack regime.

On the other hand, fatigue in engine discs is mainly a low-cycle, high amplitude problem. Critical crack lengths may be close to, or even in, the short crack regime. In such cases the behaviour of short cracks is of primary importance for application of damage tolerance principles to disc lifing.

Short fatigue cracks and the durability of metallic aerospace structures

When short cracks are important for safety they are also important for the durability of a structure. But even when short cracks are not important for safe damage tolerance design and operation they may be important for durability analyses.

As mentioned earlier, the most developed durability analyses concern the widespread initiation and growth of small cracks at fastener holes in metallic airframe structures. The way in which such analyses are done is illustrated schematically in Fig. 2. Crack propagation curves are obtained from visual and fractographic measurements on test components and specimens and are extrapolated analytically (using macrocrack-based crack growth models) to 'initial crack lengths'. These fictitious crack lengths are called Equivalent Initial Flaw Sizes (EIFS). The values and statistical distributions of EIFS define the initial fatigue quality and scatter in fatigue life, and these parameters are used

Table 2 Current well-defined safety requirements for assumed initial damage in aerospace structures (2)–(4)

Types of flaw		Aspect ratio (a/c)	Flaw size a(mm) to be assumed immediately after inspection					In-service inspection of USAF airframes with special NDI
			New structures with pre-service inspection capabilities					
			USAF airframes with high standard NDI		Space shuttle orbiter except engines		STS payloads with high standard NDI	
Description	Geometry		Fail-Safe	Slow Flaw Growth	High Standard NDI	Special NDI		
Surface flaw		1.0 / 0.2	1.27	3.18	1.9	0.635	1.9 / 0.65	6.35
Corner flaw		1.0 / 0.2					1.9 / 0.65	
Through crack			2.54	6.35			1.9	12.7
Embedded flaw					2.54	1.19		
Through edge crack						1.9		
Corner flaw at a hole		1.0 / 0.2	0.51	1.27	1.27	1.19	2.5	6.35 mm beyond fastener head or nut
Surface flaw in bore of hole		1.0 / 0.2					2.5 / 1.25	
Through crack at a hole			0.51	1.27	1.27		2.5	6.35 mm beyond fastener head or nut

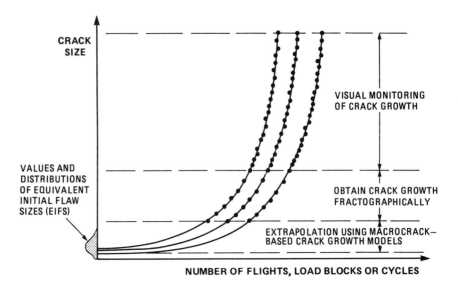

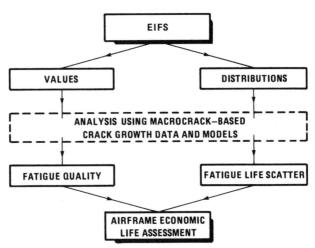

Fig 2   Schematic of the procedure for the EIFS concept of durability analysis

in assessing the economic life of the airframe. Details of the analysis procedure, which is quite complex, are given, for example, in (6)(7).

Apart from the complexity of analysis the EIFS approach appears straightforward. However, extrapolation to initial crack lengths relies on macrocrack-based crack growth data and models, whereas the EIFS values are usually well within the short crack regime. The actual behaviour of short cracks is greatly influenced by a number of factors, including crack size, shape, and location

(12), local stress–strain fields at notches (13) and fastener holes, fretting (14), load history (15), fastener fit (16) and hole preparation (e.g., cold working), and material microstructure (17). Present knowledge is inadequate to account for these factors quantitatively. This means that EIFS values and distributions apply only to the particular set of conditions for which they are derived.

A better understanding of the apparently anomalous behaviour of short cracks would enable modification of analytical modelling and extrapolation and provide a more certain basis for the EIFS approach. Some progress has been made but much remains to be done (7). Thus it may be concluded that short crack growth is of primary importance for durability analyses of metallic airframe structures.

Conclusions

The practical engineering significance of short fatigue cracks in aerospace structures is limited. At the present time there are two areas in which short fatigue crack behaviour is of interest or importance:

(1)   safe damage tolerance design and operation of some engine components, notably discs and blades;
(2)   durability analysis of widespread cracking at fastener holes in metallic airframes.

Even in these areas the current importance of short cracks is mainly restricted to military aircraft, whose performance requirements place greater demands on structural integrity. This situation may change, but only gradually. If short cracks are to become generally significant for safety it will be necessary to achieve major advances in feasible NDI capabilities. With respect to durability, short cracks are potentially important for analysis of widespread cracking in both civil and military metallic airframe structures.

References

(1)   WANHILL, R. J. H. (1984) Engineering significance of fatigue thresholds and short fatigue cracks for structural design, *Fatigue 84* (Engineering Materials Advisory Services, Warley, West Midlands, UK), pp. 1671–1681.
(2)   *Military Specification Airplane Damage Tolerance Requirements.* MIL-A-83444 (USAF), July 1974.
(3)   FORMAN, R. G. and HU, T. (1984) Application of fracture mechanics on the space shuttle, *Damage Tolerance of Metallic Structures: Analysis Methods and Applications, ASTM STP 842* (American Society for Testing and Materials, Philadelphia), pp. 108–133.
(4)   *European Fracture Control Guidelines for STS Payloads: Draft.* ESA PSS-01-401, European Space Agency, September 1984.
(5)   SWIFT, T. (1983) Verification of methods for damage tolerance evaluation of aircraft structures to FAA requirements, *Proceedings of the Twelfth ICAF Symposium* (Centre d'Essais Aéronautique de Toulouse, France), pp. 1.1/1–1.1/87.
(6)   RUDD, J. L., YANG, J. N., MANNING, S. D., and GARVER, W. R. (1982) Durability design requirements and analysis for metallic airframes, *Design of Fatigue and Fracture Resistant Structures, ASTM STP 761* (American Society for Testing and Materials, Philadelphia), pp. 133–151.

(7) RUDD, J. L., YANG, J. N., MANNING, S. D., and YEE, B. G. W. (1983) Probabilistic fracture mechanics analysis methods for structural durability, *Behaviour of Short Cracks in Airframe Components,* AGARD-CP-328 (Advisory Group for Aerospace Research and Development, Neuilly-sur-Seine, France), pp. 10-1–10-23.

(8) SURESH, S. and RITCHIE, R. O. (1984) The propagation of short fatigue cracks, *Int. Met. Rev.,* **29**, 445–476.

(9) BROWN, C. W. and HICKS, M. A. (1983) A study of short fatigue crack growth behaviour in titanium alloy IMI 685, *Fatigue Engng Mater. Struct.,* **6**, 67–76.

(10) DILL, H. D. and SAFF, C. R. (1978) Environment–load interaction effects on crack growth. Report AFFDL-TR-78-137, Air Force Flight Dynamics Laboratory, Wright-Patterson Air Force Base, Dayton, Ohio, USA.

(11) NETHAWAY, D. H. and KING, T. T. (1980) F100(3) engine structural durability and damage tolerance assessment final report. Report FR-10481-9, Pratt and Whitney Aircraft Group, West Palm Beach, USA.

(12) CIRCLE, R. L. and CONLEY, F. M. (1980) A quantitative assessment of the variables involved in crack propagation analysis for in-service aircraft, AIAA Paper 80-0752 (American Institute of Aeronautics and Astronautics).

(13) DUGGAN, T. V. (1981) Fatigue integrity assessment, *Int. J. Fatigue,* **3**, 61–70.

(14) ALIC, J. A. and KANTIMATHI, A. (1979) Fretting fatigue, with reference to aircraft structures, SAE Paper 790612 (Society of Automotive Engineers).

(15) COOK, R., EDWARDS, P. R., and ANSTEE, R. F. W. (1981) Crack propagation at short crack lengths under variable amplitude loading, *Proceedings of the Eleventh ICAF Symposium* (National Aerospace Laboratory NLR, Amsterdam, The Netherlands), pp. 2.8/1–2.8/29.

(16) STONE, M. and SWIFT, T. (1979) Future damage tolerance approach to airworthiness certification, *Proceedings of the Tenth ICAF Symposium* (Belgian Aeronautics Administration, Brussels, Belgium), pp. 2.9/1–2.9/25.

(17) SCHIJVE, J. (1981) Differences between the growth of small and large fatigue cracks. The relation to threshold K-values. Report LR-327, Delft University of Technology Department of Aerospace Engineering, Delft, The Netherlands.

A. F. Blom,* A. Hedlund,† W. Zhao,‡ A. Fathulla,‖ B. Weiss,‖ and R. Stickler‖

Short Fatigue Crack Growth Behaviour in Al 2024 and Al 7475

REFERENCE Blom, A. F., Hedlund, A., Zhao, W., Fathulla, A., Weiss, B. and Stickler, R., **Short Fatigue Crack Growth Behaviour in Al 2024 and Al 7475**, *The Behaviour of Short Fatigue Cracks*, EGF Pub. 1 (Edited by K. J. Miller and E. R. de los Rios) 1986, Mechanical Engineering Publications, London, pp. 37–66.

ABSTRACT The present investigation consisted of a study of fatigue crack initiation and growth in two high-strength technical Al alloys (Al 2024 and Al 7475) differing in mechanical properties and microstructure. Emphasis was on experiments at low load amplitudes slightly below and above the fatigue limit. Attempts were made to investigate the effect of stress ratios ($-1 \leq R \leq 0.7$) and to provide quantitative information on both the short crack and long crack growth behaviour for identical specimen material, orientation, and test conditions. To determine closure stresses a dynamic compliance technique was developed and the crack opening phenomena followed by microscopy methods. Initiation and growth of fatigue cracks were found to be strongly dependent on microstructural features (second phase particles, grain boundaries). A comparison of the growth behaviour of both microstructurally-short and physically-short cracks with that of long cracks shows that the peculiarities of short crack growth can be partly related to crack closure phenomena.

Introduction

Since the early observations by deLange (**1**) on the apparently anomalous growth of fatigue microcracks and the subsequent findings of Pearson (**2**) that fatigue cracks show the characteristic long crack growth behaviour only beyond a certain crack length, the phenomenon of short crack behaviour has attracted considerable interest. The magnitude of research efforts and the increasing number of pertinent publications reveal not only academic interest but also technical relevance and engineering consequences of the short crack problem (**3**)(**4**).

The state of knowledge on the behaviour of short fatigue cracks has been documented and reviewed in several publications (e.g., (**5**)–(**11**)) and in contributions to the Euromech colloquium 151 on the creation and behaviour of such cracks, e.g., (**12**).

In spite of the world-wide research effort, some aspects of short crack behaviour still appear ill understood (**13**). Information on initiation sites in pure metals and technical alloys is controversial (e.g., (**2**)(**5**)(**14**)–(**19**)). In contrast to long fatigue cracks (LC) there exists no adequate definition of the growth mechanisms, the crack tip stress/strain fields, or the interactions with

* The Aeronautical Research Institute of Sweden, Bromma, Sweden.
† The Royal Institute of Technology, Stockholm, Sweden.
‡ The Chinese Aeronautical Establishment, Beijing, People's Republic of China.
‖ University of Vienna, Austria.

microstructural features, for short fatigue cracks (SC). The specificities in SC growth have been associated with closure effects which vary with increasing crack length (11)(15)(20), and mechanisms which give rise to closure with increasing crack length have been proposed (21)–(24). Considerable attention was given to the interaction of SC with crystallographic features, the effects of microstructure (12)(25)(26)(43), and the effects of crack deflection at grain boundaries (27)–(29). Only a little information can be found about the effects of stress ratios on SC behaviour (11)(25). Investigations on SC growth in notches (18)(19)(30) indicate no difference to plain specimens as long as the tests are carried out at low stress amplitudes (negligible notch plasticity effects).

Many investigators found it convenient to present SC growth data in combinaton with LC data as a function of the stress intensity range, in spite of the fact that the applicability of the K concept to cracks smaller than a critical size was already questioned by Kitagawa (38). The breakdown of the similitude requirements of LEFM in the case of SCs was pointed out in several publications (e.g., (3)(6)(8)).

Various hypotheses have been proposed to quantify the transition from SC to LC behaviour (9)(31)–(37). Based on a diagrammatic presentation of log stress-range versus log crack-length first suggested by Kitagawa (38), three regions of growth behaviour can be differentiated, i.e.: (i) the SC region in which the presence of a SC does not affect the plain specimen fatigue limit; (ii) a transition region in which the threshold stress for fatigue crack growth is gradually lowered with increasing crack length (EPFM considerations); and (iii) the region in which the growth of a fatigue crack follows LEFM relations, i.e., the LC region. Several authors have employed such a type of presentation for a general description of SC behaviour, but, little quantitative information is available with respect to the extent of these three regions and, in particular, how variations in stress ratio influence the respective border lines (39). As indicated in Table 1, LC growth behaviour in Al alloys was reported by various authors to occur beyond a crack length of 100–300 μm. Recent experimental results (16)(17) indicate that the border between SC and LC behaviour can be defined uniquely by applying the value of the effective threshold stress intensity, assumed to be a material parameter which can be determined experimentally and should remain unaffected by the stress ratio.

A considerable amount of information regarding SC behaviour has been published for engineering high-strength Al alloys, as summarized in Table 1. It can be seen that most of these investigations were carried out at relatively high stress amplitudes approaching the yield strength of the alloys, and in general for only a relatively small number of loading cycles. Thus, experimental values listed for fatigue limits and threshold stress intensities may be higher than the true limiting values. A direct comparison of SC and LC behaviour is frequently lacking for the same material.

The present investigation consisted of a detailed study of fatigue crack initiation and growth in two high-strength technical Al alloys differing in

Table 1 Published information on short crack growth behaviour in Al alloys

Reference	Alloy	Testing conditions			da/dN m/c	Method of SC observation	Remarks
		$\sigma_{al}/\sigma_{0,2}$	R	Frequency (Hz)			
de Lange (1)	26ST	—	—	—	10^{-7}–10^{-5}	replica SEM	non-propagating cracks 5–10 grain diameters
Pearson (2)	DTD 5050 (~7075-T651) L65 (~2014-T3)	0.4–0.6	0.1 to 0.78	25	$>10^{-9}$	LM	LC > 127 μm
Morris (15)	2048-T851	0.6–0.9	-1	5	—	in situ SEM	observation of residual closure
Morris et al. (21)(40)(41)	2219-T851	0.7–0.9	-1	5	—	LM, LM-stereo imaging	microcrack closure crack tip plasticity
Kung, Fine (19)	2024-T4, 2124-T4	0.65–1.16	-1	—	2×10^{-8}	in situ LM	LC > 300 μm
James et al. (24)(26)(42)	2219-T851 7075-T6	0.6–0.9	0, -1	5	$>10^{-9}$	—	LC > 5–15 grain diameters
Nisitani, Takao (5)	Al-alloy annealed Al-alloy hardened	2.3 0.6	-1 -1	50 50	— —	— —	initiation at grain boundaries
Hirose, Fine (30)	PM-X 7091 (MA87)	0.3–0.8	-1	25	—	in situ, replica-SEM	Strain distribution, COD
Lankford et al. (29)(44)–(46)	7075-T651, -OA, -T6, 6061-T6	0.8	0.1	1–5	$>2 \times 10^{-10}$	SEM-stereo imaging	influence of notches topography of short cracks transition length to LC between 220–250 μm significance of $\Delta K_{th,eff}$
Foth et al. (18)	2024-T3	(high)	0	—	$>2 \times 10^{-6}$	LM	
Fathulla et al. (16)(17)	2024-T3	0.29–0.52	-1	20000	5×10^{-13} 5×10^{-11}	in situ LM replica-SEM	

mechanical properties and microstructure. Emphasis was put on performing the cyclic experiments at low load amplitudes slightly below and above the high-cycle fatigue limit or the threshold stress intensity. Attempts were made to investigate the effect of mean stress, and to provide quantitative information on SC and LC growth behaviour for identical specimen material, orientation, and test conditions. For the determination of closure stresses a dynamic compliance technique was developed and the crack opening phenomena followed by microscopy methods. Experimental data were compared to FEM calculations on the extent and shape of plastic zones at the tips or in the wake of short and long fatigue cracks.

Specimen materials and experimental techniques:

In a continuation of previously published studies on the long crack (LC) and short crack (SC) growth behaviour (**16**)(**17**) further measurements were carried out on the high-strength alloy Al 2024-T3. Specimens from the same lot were available for detailed investigations of the closure effect and the transition from SC to LC growth. In addition, specimens from a batch of the alloy Al 7475-T761 were selected for a comparative study. This alloy is known to exhibit higher strength, fracture toughness, and, in particular, higher stress corrosion cracking resistance than alloy Al-2024.

Composition, heat treatment, grain dimensions, and tensile properties of the specimen materials supplied in plate form (2 and 6 mm thickness) are listed in Table 2.

The thermomechanical pretreatment of the plate material resulted in a typical pancake-type grain structure. Both alloys were found to contain several phases, in agreement with published information (**47**) coarse particles were identified by EDAX analysis to correspond to the following intermetallic compounds.

Al 2024: insoluble coarse particles of $Al_{12}(Fe,Mn)_3Si$ and Al_7Cu_2Fe;
          partially soluble Mg_2Si, Al_2Cu_2Mg;
          and, in recrystallized material, $Al_{20}Cu_2Mg$.
Al 7475: insoluble coarse particles of Al_7Cu_2Fe;
          partially soluble Mg_2Si, dispersoid of $Al_{12}Mg_2Cr$.

No attempts were made to investigate the nature and morphology of the hardening precipitates (i.e., Al_2Cu_2Mg in Al 2024 and $Mg(Al,Cu,Zn)_2$ in Al 7475 (**47**)).

Pertinent information on test procedures and specimen geometries for both alloys are summarized in Table 3 and are briefly discussed below.

(*i*) For the conventional S–N fatigue tests (using a 'LF' Schenck servo-hydraulic test system at $-1 \leq R \leq 0.75$, and 40–80 Hz, specimens with cylindrical gauge sections were machined from the plate material. However at low stress amplitudes (near the fatigue limit) experiments were performed on a

Table 2   Specimen materials and properties

Chemical composition (weight %) Element	Al 2024 (nominal)	Al 7475 (actual)
Cu	4.5	1.8
Mg	1.5	2.3
Mn	0.6	0.5
Si	0	0.04
Zn	0	5.8
Cr	0	0.22
Fe	—	0.10
Ti	—	0.003
Al balance	balance	balance

Heat treatment	Al 2024	Al 7475
Commercial	T3	T761

Grain size (mean intercept method) (μm)			
Alloy	Al 2024	Al 7475	
Plate thickness (mm)	6	2	6
Orientation:			
longitudinal, L	120	150	350
Transverse, T	65	80	150
Short transverse, ST	25	20	20

Tensile properties Alloy	Al 2024	Al 7475			
Plate thickness (mm)	6	2		6	
Orientation	L	L	T	L	T
$\sigma_{0.2}$ (MPa)	345	463	449	482	475
σ_{UTS} (MPa)	450	510	514	526	543
A (%)		14	15	15	12
HV			161		170
Dynamic Young's Modulus (GPa)	72.5			70.5	71.4

'HF' resonance system ($R = -1$, test frequency 20 kHz) (**48**). HF-SN data were obtained with specimens of either a cylindrical gauge section or with specimens of a rectangular cross section containing a mild notch. In the latter case the stress amplitudes listed in Table 3 correspond to the surface stress in the root of the notch, determined experimentally with miniature strain gauges. These values were found to be in good agreement with values calculated by a stress concentration factor computed for this notch geometry by finite element methods; see Fig. 1.

(*ii*)   The long-crack (LC) growth behaviour was studied at LF frequencies with standard compact-tension (CT) and centre-notched (CN) specimens. Fatigue crack growth data were determined in accordance with the proposed

Table 3   Test methods and specimen geometries

Test method	Stress ratio	Specimen geometry (plate thickness)	Specimen orientation	
			Specimen axis	Crack growth direction
S–N data (endurance limit)[*]				
LF, $N_{max} = 1 \times 10^7$	$-1 < R < 0.75$	cylindrical (6 mm)	L	T
HF, $N_{max} = 2 \times 10^8$	$R = -1$	cylindrical (6 mm)	L	T
		(gauge section 4 mm diam.)	T	L
LC growth data (da/dN–ΔK, ΔK_{th}, $\Delta K_{th,eff}$)[†]				
LF	$-1 < R < 0.75$	CT, CN (6 mm)	L	T
			T	L
HF	$R = -1$	CN (6 mm)	L	T
			T	L
SC growth data (c–N, dc/dN–c)				
LF	$R = 0.05$	side notch (2 mm)	L	ST
HF	$R = -1$	side notch (6 mm)	L	ST

L   rolling direction
T   transverse directon
ST   short transverse direction
[*] Al 7475 tested only at $R = 0.05$.
[†] Al 2024 tested only with stress axis in the L direction.

ASTM standard procedure (49)(50). For the HF tests CN specimens were used, containing either a through-thickness starting notch or a lancet-shaped surface notch giving rise to semi-elliptical crack growth (ASTM Standard Practice E740-80 (51). Threshold values corresponding to fatigue crack growth rates of less than 1×10^{-13} m/cycle were obtained by a strain shedding technique (52). The computational procedure developed for the HF tests have been reported for through-thickness cracks (50) and for semi-elliptical surface cracks (52). Following ASTM recommendations, only the tensile part of the stress cycle at negative R values was taken for the calculation of ΔK_{th}.

(*iii*) The crack closure stress (defined here as the first contact of crack surfaces at diminishing cyclic amplitudes) and the effective threshold stress intensity values were determined by the test procedure described in reference (50).

(*iv*) The SC growth behaviour was studied with full-thickness plate specimens of rectangular cross-section. Cylindrical side notches were machined into one of the narrow sides of the specimens in order to reduce the surface area to be scanned during cyclic loading. To calculate the surface stresses in the notch root

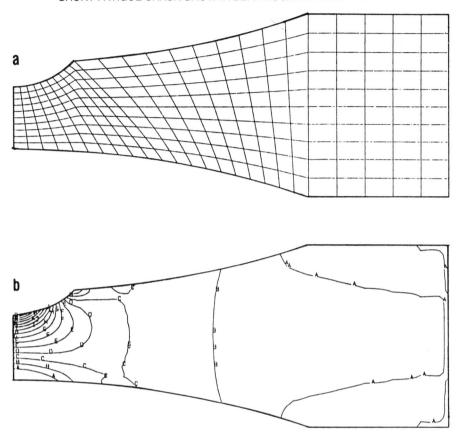

Fig 1   Stress determination of a side-notched specimen with $K_t = 2.468$
    (a)   Finite element discretization
    (b)   An isostress plot

we used stress concentration factors computed by FEM. The selected notch geometries resulted in nominal K_t values between 1.1 and 2.97. For the low stress amplitudes applied in these tests (maximum surface stress always smaller than the yield strength) it was assumed that effects of notch-root plasticity can be neglected.

To observe the initiation and to follow the growth behaviour of short fatigue cracks under LF test conditions a SEM-replica method was applied. Plastic replicas were prepared of the specimen surface after appropriate numbers of loading cycles and micrographs recorded of corresponding surface areas. During the HF tests the specimen surface could be directly examined at high resolution under a light microscope. This is possible because of the fact that, for specimens excited to resonance vibrations at the lowest 'eigen' frequency, the maximum in strain amplitude in the mid-section of the specimen coincides with the location of zero displacement. The microscopic observations were continuously recorded on video-tape (synchronously with pertinent experimental data) to permit a post-test quantitative evaluation. The SC growth measurements were extended to more than 10^7 loading cycles for each applied stress level. In this way information on retardation or halting of the crack growth, and on the interaction of the advancing crack with microstructural features could be collected with high sensitivity.

Experimental results

S–N data

Fatigue life curves determined by HF tests for both alloys at $R = -1$ are shown in Fig. 2. The data points obtained by LF testing fall within the respective

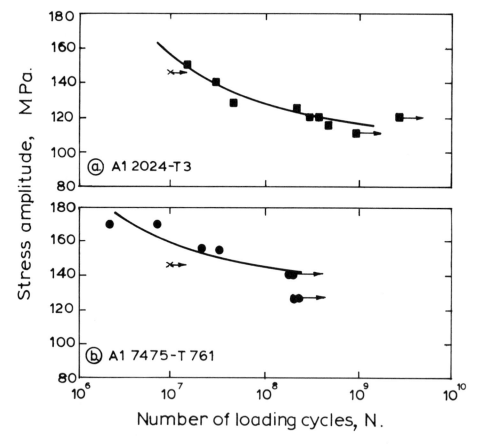

Fig 2   S–N data at $R = -1$ and room temperature, test frequency 20 kHz. Crosses indicate unbroken specimen tested at 50 Hz. Stress axis parallel to L direction

Table 4   Effect of stress ratio on fatigue strength of Al 2024 and Al 7475

Alloy	Al 2024		Al 7475	
Test method	LF	HF	LF	HF
N_{max}	1×10^7	2×10^8	1×10^7	2×10^8
Endurance limit (MPa) at:				
$R = -1$	147	128	142*	148
$R = 0.05$	117		115	
$R = 0.5$	76		80*	
$R = 0.75$	45		45*	

* Taken from literature.

scatterbands, indicating the absence of a pronounced frequency effect on fatigue life. Microscopy examination of the specimen surface in the gauge section of unfailed specimens revealed the presence of non-propagating micro-cracks. The S–N curves reveal that a true fatigue limit may not exist or may not be reached prior to 10^8 loading cycles.

An extensive investigation of the effects of test frequency in the applied frequency range (53)(54) has shown that, for ductile fcc materials, a small frequency effect on the fatigue limit and the threshold stress intensity of long cracks can be attributed mainly to thermally activated processes. For high-strength materials the frequency effect in the near-threshold regime can be considered as practically negligible (50).

The effect of the stress ratio R on fatigue life of Al 2024 and Al 7475 can be deduced from the values listed in Table 4.

Long crack growth and threshold behaviour

Fatigue crack growth curves for Al 2024 tested at various R values have been published previously (50). The FCG behaviour of Al 2024 and Al 7475 is shown in Fig. 3.

The effect of stress ratio, specimen orientation, and test frequency on the LC threshold stress intensity of both alloys can be deduced from Fig. 4. It can be seen that for the same orientation the threshold values of the stronger Al 7475 fall below that of Al 2024 over the whole range of R.

This difference is also apparent in the values of the effective threshold stress intensity plotted for both alloys in the same diagrams. It should be noted, however, that the values of ΔK_{th} and $\Delta K_{th,eff}$ could only be determined with an accuracy not better than 10 per cent.

Metallographic observations do not show any resolvable plastic zones associated with the crack advancing near threshold in both alloys. In Al 7475 a tendency to localized decohesion along glide planes (resulting from the preced-ing cold work) can be recognized. An evaluation of the fracture topography,

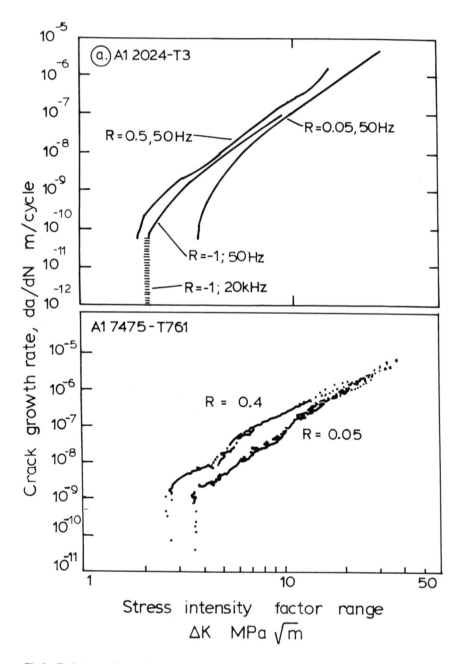

Fig 3 Fatigue crack growth data for various stress ratios and test frequencies determined at room temperature. The crack plane in Al 7475 is parallel to the L direction

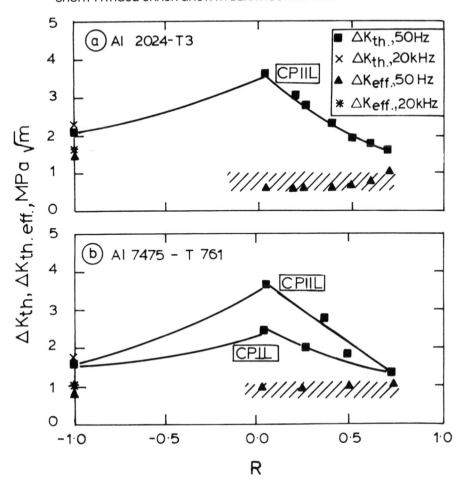

Fig 4 Effect of stress ratio, specimen orientation, and test frequency on threshold stress intensity and effective threshold stress intensity

however, reveals a difference in the crack path irregularity between the two alloys. This irregularity can be expressed by a roughness parameter (**55**), i.e., the ratio of the actual length of the crack at the specimen surface and the length of the crack projected onto a straight line normal to the stress direction, l_{act}/l_{proj}. This roughness parameter was determined for Al 2024 as 1.15 and for Al 7475 as 1.05. Thus, the differences between threshold and effective threshold values in Al 2024 and Al 7475 appear mainly to be due to various degrees of geometrically induced closure, with oxide induced closure a possible contributory factor.

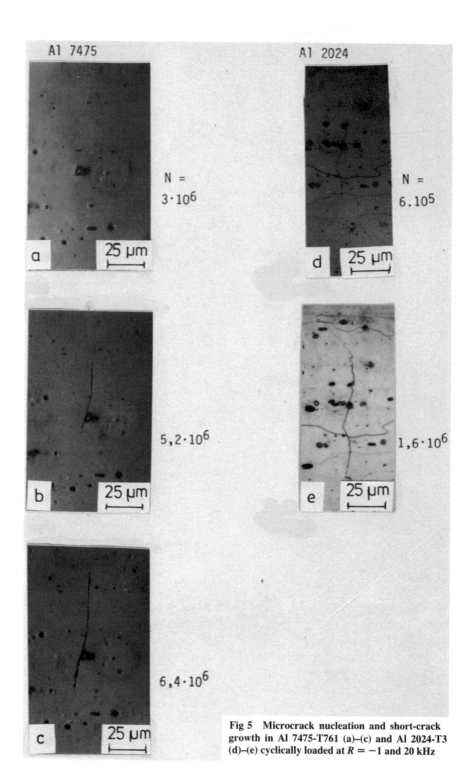

Al 7475

Al 2024

$N = 3 \cdot 10^6$

$N = 6 \cdot 10^5$

25 µm

25 µm

a

d

$5,2 \cdot 10^6$

$1,6 \cdot 10^6$

25 µm

25 µm

b

e

$6,4 \cdot 10^6$

25 µm

c

Fig 5 Microcrack nucleation and short-crack growth in Al 7475-T761 (a)–(c) and Al 2024-T3 (d)–(e) cyclically loaded at $R = -1$ and 20 kHz

Short crack initiation and growth

Microscopy observations during $R = -1$ fatigue loading (see Fig. 5), revealed that crack nucleation occurs in Al 2024 at fractured bulky (Fe,Si containing) intermetallic particles, the fracturing taking place during fatigue. In Al 7475 crack nucleation proceeded occasionally along a particle/matrix interface, but also inside a grain and apparently related to a slip band. Cracking of inter-metallic particles during fatigue loading was not observed in Al 7475.

The growth of such microcracks in Al 2024 is represented in the c–N and dc/dN–c curves of Fig. 6. Here c represents the half-length of the surface crack which was assumed to be semi-elliptical. The penetration, a, of these surface cracks, was determined by fractographic evaluation of specimens ruptured after completion of the short-crack measurements. A shape factor a/c decreasing from 0.9 to 0.6 for fatigue cracks increasing from $c = 20$ to $c = 300\ \mu$m was found for Al 2024. The np line of Fig. 6(a) represents an initially non-propagat-

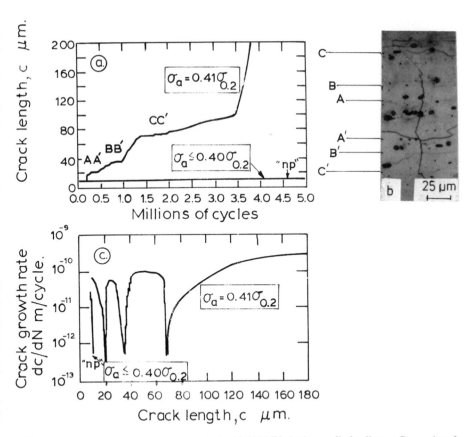

Fig 6 Growth behaviour of short cracks in Al 2024-T3 during cyclic loading at $R = -1$ and 20 kHz. Crack plane parallel to ST

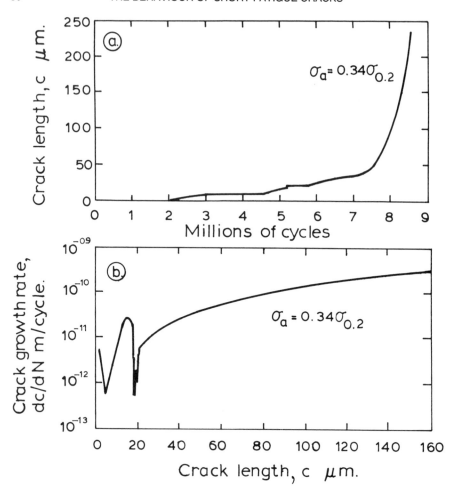

Fig 7   Growth behaviour of short cracks in Al 7475-T761 during cyclic loading at $R = -1$ and
20 kHz. Crack plane parallel to ST

ing microcrack which continued growth after a slight increase in stress amplitude (interacting with microstructural features such as grain boundaries) up to a length of 110 μm as shown in the micrograph, Fig. 6(b).

Similar curves for Al 7475 at $R = -1$ are shown in Fig. 7 and for $R = 0.05$ in Fig. 8 for two different stress levels. These stress levels correspond to crack initiation after 10^4 and 10^5 loading cycles, respectively, and thus fall above the HF S–N curve. The equivalent stress levels for the tests at $R = -1$ are only slightly larger than the fatigue limit at $N = 5 \times 10^8$ of the corresponding S–N curve. An evaluation of the effect of stress levels on the short crack growth behaviour must take into account that the S–N curve for $R = -1$ is above that

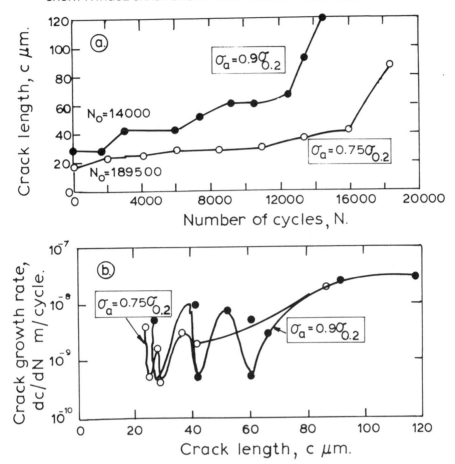

Fig 8 Growth behaviour of short cracks in Al 7475-T761 during cyclic loading at $R = 0.05$ and 50 Hz for two stress amplitudes. Crack plane parallel to ST

for $R = 0.05$. A non-uniform crack tip advance can be recognized up to a characteristic crack length which is considerably shorter for Al 7475 as compared to Al 2024. Beyond this length the cracks appear to advance uninhibited. The microscopic observations indicate that the advance of both tips of a crack at the specimen surface does not always occur at equal rates.

Crack growth rates deduced from the above $c–N$ curves are also plotted as function of crack length for Al 7475 in Figs 7 and 8. The higher cyclic stress levels in the $R = 0.05$ tests result in faster crack growth rates than in the case of $R = -1$. Retardation of the crack advance occurred when a crack tip approached a grain boundary. The degree of this retardation can be deduced from the numbers of loading cycles required to initiate a resumption of crack growth in the next grain. In the case of the low-amplitude HF tests at $R = -1$

retardation was observed to extend up to more than 10^6 cycles, while in the LF tests at $R = 0.05$ at the higher cyclic amplitudes the grain boundary retardation amounted typically only to 10^3 cycles. Consequently, the minima in the respective LF growth-rate curves are less pronounced, with intermittent maxima somewhat higher due to the higher stress amplitude.

A cessation of the interaction of the advancing crack with microstructural features (predominantly grain boundaries) can be recognized in the dc/dN curves at characteristic values of crack length, depending on stress amplitude and stress ratio.

Topographical features of short fatigue cracks

Optical and electron microscopy observations revealed typical features of the crack path which appear to characterize short crack behaviour and the transition from short crack to long crack growth. A clear transition in the crack advance from Stage I (transcrystalline crystallographic) to Stage II (trans crystalline non-crystallographic) with an intermediate transition region (localized crystallographic Stage II) for Al 2024 could also be deduced from the micrographs for Al 7475; Fig. 9. For Al 2024 this transition was found to

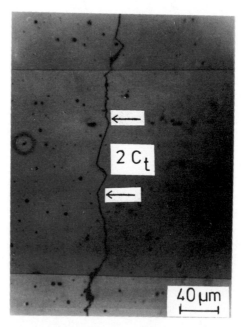

Fig 9   Micrograph of a short crack in Al 7475–T761 after cycling at $R = -1$ and 20 kHz at a stress level of $\sigma_a = 0.34\sigma_{0.2}$. The transition length $2c_T$ is indicated by arrows and marks the change from transcrystalline Stage I to crystallographic Stage II crack propagation

Table 5 **Characteristic crack lengths for transition from short-crack to long-crack growth behaviour**

Alloy	R	Stress amplitude (MPa)	N_{max}	ΔK_{th} $(MPa\,m^{1/2})$	$\Delta K_{th,eff}$ $(MPa\,m^{1/2})$	c_{eff} (μm)	c_T (μm)	c_2 (μm)	c_1 (μm)
Al 2024	−1	128	5×10^8	2.15	1.3	90	110	140	400
Al 7475	−1	148	2×10^8	1.62	0.9	30	45	—	—
	0.05	356	1.8×10^5	—	—	—	40	—	—

correspond to a characteristic crack length, apparently not related to the grain dimensions (**17**). The measured transition length c_T for alloys Al 2024 and 7475 are listed in Table 5. It is interesting to note that this characteristic transition length compares reasonably well with the critical crack length indicated in the dc/dN–c curves from constant amplitude fatigue tests at which the marked retarding effect of grain boundaries ceases.

Discussion

Transition from short-crack to long-crack growth

Pertinent test results for Al 2024 are summarized in the Kitagawa-type diagram of Fig. 10. It can be seen that the sloping line computed for ΔK_{th} of a semi-elliptical crack intersects the horizontal line corresponding to the fatigue limit ($N = 5 \times 10^8$) at a crack length, termed c_0 in the literature, of approximately 240 μm. As will be shown in the following this mathematical value appears to have little physical meaning. Comparable results were also found for through-thickness cracks.

If we assume that the horizontal line resembles the true fatigue limit we may surmise the absence of propagating SCs below this line. Indeed, metallographic observations failed to reveal even non-propagating microcracks below this stress level. Above this stress level microcracks were observed which exhibited characteristic SC behaviour (crystallographic transcrystalline through entire grains) in the stress range up to approximately 1.5 times the fatigue limit.

As described earlier (**16**)(**17**) the threshold behaviour of such SCs was investigated by the following test sequence. A specimen was loaded at consecutively increasing amplitudes until crack nucleation and slow short crack growth to a crack length by 80 μm had occurred. Then the amplitude was lowered to approximately half of the fatigue limit, at which no further crack growth could be detected. A step-wise increase in loading amplitudes after each 10^7 cycles was repeated until a resumption of crack growth became noticeable; the corresponding stress amplitude was recorded as the threshold for growth of this particular SC. The test sequence was repeated for the same crack at incremental length values. The results are indicated in Fig. 10(a) by data points. It is

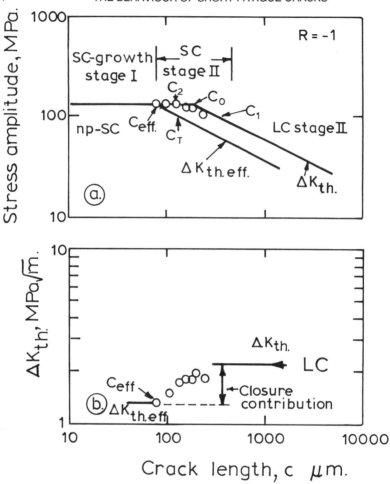

Fig 10 Summary of observations on crack growth behaviour in Al 2024-T3
(a) A Kitagawa stress-amplitude versus crack-length diagram. Data points indicate stress amplitudes for resumption of growth of the same 'short crack' at successively increasing lengths
(b) Apparent threshold stress intensities for the 'short crack' data points shown in Fig. 10(a)

interesting to note that resumption of crack growth occurred initially independently of crack length just above the fatigue limit. Beyond a crack length of approximately 140 μm (c_2 in Fig. 10(a)), a reduction in threshold stress became apparent. An extrapolation of the curve interconnecting these data points showed that the ΔK_{th} line is approached at a crack length of approximately 400 μm, that is, c_1 in Fig. 10(a). For cracks longer than this value the growth behaviour should obey conventional LEFM criteria.

The results show that SC growth below the fatigue limit takes place only in the narrow field between c_2 and c_1. Otherwise, SC growth behaviour (Stage I) prevailed in a small range above the fatigue limit, but only slightly to the right of the ΔK_{th} line, at which crack propagation changed to crystallographic Stage II. Longer cracks proceeded by the non-crystallographic Stage II mode.

Short crack growth and effective threshold stress intensity

Presuming the absence of closure effects for SCs it appears plausible that the SC growth behaviour ceases at a crack length corresponding to the intersection of the sloping line calculated for the experimental value of $\Delta K_{th,eff}$ and the fatigue limit line, as shown by c_{eff} in Fig. 10(a). We propose that the LEFM crack growth laws can be applied to cracks longer than c_{eff}, provided $\Delta K_{th,eff}$ is used instead of ΔK_{th}

Replotting the threshold stress intensity values as a function of crack length one finds a fall-off from the LC–ΔK_{th} at a crack length corresponding to c_1 along a line intersecting the $\Delta K_{th,eff}$ level at the crack length of $c_{eff} = 90$ μm (Fig. 10(b)). It should be pointed out, however, that the ΔK_{th} calculations have been carried out using an experimentally determined shape factor for cracks decreasing from c_1 to c_{eff}.

A further indication for a transition from SC to LC growth is apparent from the metallographic observations of crack path topography. As described above, crystallographic crack growth across entire grain diameters (at approximately 45 degrees to the stress axis, Stage I) changes with increasing length first to crystallographic Stage II (along segments of crystallographic planes zigzagging along a plane normal to the stress direction) and finally to conventional Stage II (non-crystallographic transcrystalline, essentially normal to the stress direction). The transition from Stage I to Stage II was found to be associated with a characteristic crack length, $c_T = 110$ μm. This transition length should again be related to the minimum length beyond which LEFM relationships may be applied to characterize crack growth.

A comparison of the crack growth behaviour, dc/dN, with the calculated ΔK_{th} values as a function of crack length, Fig. 11, shows clearly the relationship between cessation of anomalous short-crack growth (retarding effect of grain boundaries) at a crack length slightly larger than c_{eff}. Up to a crack length c_{eff} characteristic short crack growth occurs, indicating reduced effect of closure. Between c_{eff} and c_1 transition to long crack behaviour takes place. A summary of data on critical crack lengths is included in Table 5. The differences between c_{eff}, c_T, and c_2 may be due to the differing degrees of sensitivity of the respective test methods and actually may correspond to one and the same characteristic crack dimension. Moreover, part of the inconsistencies may be due to the fact that all the test results cannot be obtained from the same specimen, although all tests were carried out with specimens of the same orientation.

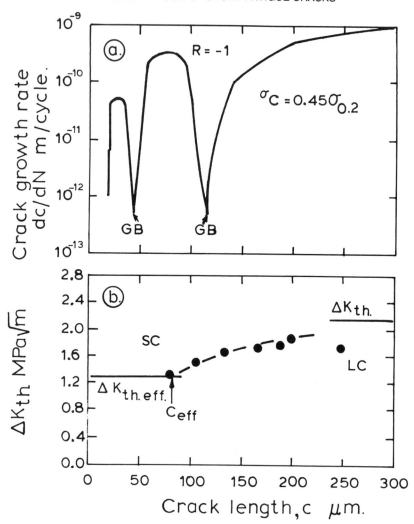

Fig 11   Relation between growth behaviour and threshold stress intensity values of a 'short crack' with successively increasing length in Al 2024-T3

The test result for both alloys are compared in Fig. 12, which shows that the value of c_{eff} for Al 7475 is considerably smaller than for Al 2024, providing an explanation for the experimental difficulties encountered in monitoring SC growth behaviour in Al 7475.

To reveal the influence of the stress ratio on SC behaviour an attempt was made to plot the experimental results for the range of $-1 \leq R \leq 0.75$ in a Kitagawa-type diagram. Following a procedure employed by Usami and Shida

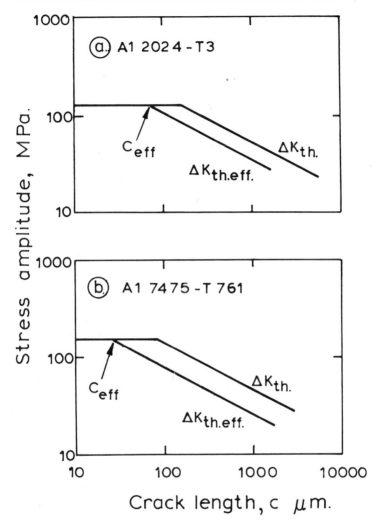

Fig 12 Comparison of crack growth behaviour in the two alloys. Note the significantly smaller value for c_{eff} in Al 7475

(**39**) the LF data for Al 2024 are presented schematically in Fig. 13 in a modified maximum-stress versus crack size diagram. The sloping lines were calculated according to the relation

$$\Delta K_{th}/(1 - R) = K_{th,max} = \sigma_{max} \cdot F(c^{1/2})$$

(where F is the geometry function) while the horizontal lines represent the fatigue limit as a function of the maximum stress (alternating stress plus mean

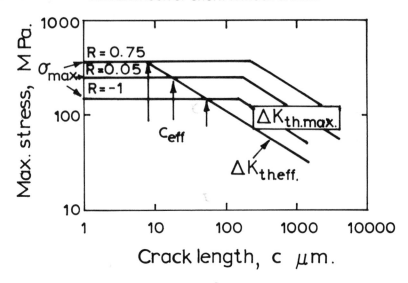

Fig 13   Semi-schematic presentation of the effect of stress ratio in Al 2024-T3

stress). It is interesting to note that in this presentation the intersection of both sets of lines occurs almost independent of the R value at a comparable crack length, c_0.

Under the assumption that the effective threshold stress intensity, $\Delta K_{\text{th,eff}}$, corresponds to the range between zero (at crack opening) and the maximum stress intensity actually encountered at the crack tip, $K_{\text{th,eff,max}}$, the sloping 'effective' line can be plotted according to the relation

$$K_{\text{th,eff,max}} = \sigma_{\text{eff,max}} \cdot F(c^{1/2})$$

The intersection of this 'effective' line (which corresponds to the whole range of R values) with the respective fatigue limits gives then the values of the critical crack length, c_{eff}, which resemble the border between SC and LC growth behaviour.

Non-closure of short cracks

Although at the present crack closure stresses cannot be measured with sufficient accuracy for cracks shorter than 500 μm, an indication of the closure effect can be obtained from the onset of a residual crack opening under zero load. As shown in the micrographs of surface replicas prepared at successive growth intervals of a particular short crack in Al 7475 (Fig. 14), lobes on the replicas associated with portions of the microcrack can be taken as an indication of non-closure of the mating fracture surfaces. The length of the cracks at which such lobes can be recognized first in the replicas is just slightly shorter than the measured transition length, $2c_{\text{T}}$.

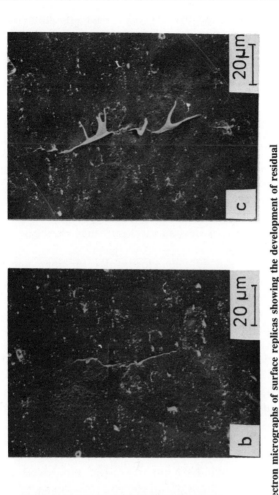

Fig 14   Scanning electron micrographs of surface replicas showing the development of residual crack opening with increasing length of a short crack in Al 7475-T761 cyclically loaded at $R = 0.05$ and 50 Hz and at a stress amplitude of $\sigma_a = 0.9\sigma_{0.2}$

(a) $N = 20\,000$ cycles
(b) $N = 23\,000$ cycles
(c) $N = 26\,500$ cycles

For a comparison of our test results on crack opening phenomena with data published by Lankford (**45**) one should take into account that Lankford's experiments were carried out at much higher stress levels, thus giving a considerably increased notch tip plasticity than in the case of our low-stress loading.

Plastic zones associated with short cracks

As discussed above there is experimental evidence for a characteristic transition length, c_{eff}, from microstructurally short cracks to elasto-plastic short cracks, and another characteristic transition length, c_1, to long cracks obeying a conventional LEFM growth relationship. The development of closure between c_{eff} and c_1 is expected to include not only plasticity-induced closure but also roughness-induced closure. In some cases, depending on material and environmental combinations, oxide induced closure may also be a contributing factor.

To verify, at least qualitatively, such closure development, and to obtain another estimate of the length c_{eff}, we shall consider in the following the pure plasticity effects of short, as compared to long, cracks. The results to be discussed are valid for Al 2024-T3 and are described in detail in (**56**). The calculations were carried out for a three-point bend specimen under a state of plane strain (which is supposed to be prevalent at the growth rates applied in the present investigation) and subjected to a constant cyclic load with a stress ratio of $R = 0$.

The numerical procedure described earlier (**32**) consists of elastic–plastic finite element calculations taking into account the propagation of a crack by releasing the crack tip node, changing the boundary conditions, and solving the contact problem occurring at the crack surfaces. In the computer program an initial isotropic strain hardening was assumed; the finite elements were two-dimensional triangles with cubic base functions. The initial crack length was taken as 10 μm and the size of the crack tip element was chosen to be 1.25 μm. Since the mesh in the vicinity of the crack tip was fine, it was expected that the correct crack tip singularity would be obtained.

In Fig. 15 some of the computed results of plastic zone sizes and shapes are shown. In Fig. 15(a) the plastic zone is shown for a very short crack of 10 μm length subjected to a nominal stress intensity range of $\Delta K = 2.31$ MPa$\sqrt{m}$. These results should be compared to results for a long crack with a length equal to one half of the specimen width and subjected to approximately the same driving force as for the short crack, $\Delta K = 2.56$ MPa$\sqrt{m}$, as shown in Fig. 15(b). We find that the local plastic flow is much easier in the case of the short crack than for the long crack, probably attributable to a reduced plastic constraint. It is also interesting to note the entirely different shapes of the plastic zones of short and long cracks. To produce approximately the same extent of the plastic zone as for the short crack at $\Delta K = 2.31$ MPa$\sqrt{m}$, a nominal driving force of $\Delta K = 6.4$ MPa$\sqrt{m}$ has to be applied to the long crack, Fig. 15(c).

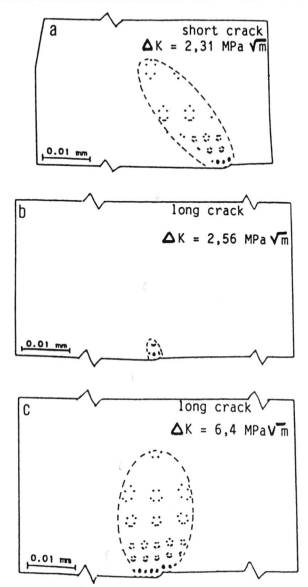

Fig 15   Shape and extent of plastic zones associated with the tips of short and long fatigue cracks from FEM computations

All the nominal stress intensity ranges mentioned above were calculated from the elastic displacement field by LEFM techniques.

As shown in Fig. 15(a), the plastic zone is formed in the direction of the maximum octahedral shear stress ahead of the crack tip, whereas no plasticity is contained in the wake of the short crack. This may indicate that the crack

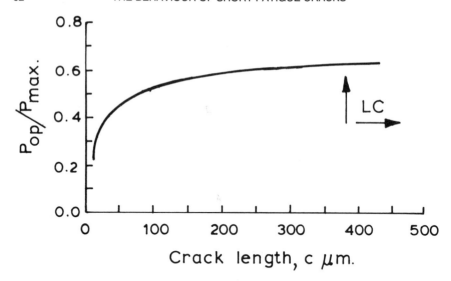

Fig 16   **Development of crack closure (P_{op}/P_{max}) with increasing crack length. Results of FEM computations**

favours a continued growth in slip planes oriented in this direction, a fact which is corroborated by experimental observations. If, instead, we consider the case of the long crack, we may predict that the crack would favour a conventional Stage II growth and that extensive plasticity would be left in the wake of the propagating crack. This in turn would cause plasticity-induced closure.

By performing a series of calculations with growing cracks (**56**) it could be shown that the development of crack closure can be calculated and plotted in terms of the load ratio P_{op}/P_{max}, as shown in Fig. 16. We find that the development of crack closure occurs rapidly for short cracks with increasing crack length, and that the saturation level for steady state behaviour occurs at a crack length of 400 μm. This value is in excellent agreement with our experimental result of c_1 for Al 2024-T3, as discussed above. The development of crack closure shown in Fig. 16 is obviously only an estimate of the physical situation which, in reality, involves local anisotropy, crystallographic features, roughness, etc. The trend, however, matches very well with the observed experimental behaviour and the estimate of a saturation or transition length, c_1, to conventional LEFM behaviour is expected. It is also shown to be correct since this is the length which defines the boundary beyond which an overall continuum treatment involving self-similarity is permissible.

Summary and conclusions

This investigation was carried out to provide a comparison of the high-cycle fatigue properties, the initiation, and growth of microstructurally short to long fatigue cracks, and the threshold behaviour at various stress ratios for two technical high strength Al alloys, Al 2024-T3 and Al 7475-T6. The test results and the conclusions from this study can be summarized as follows.

(1)   The $S-N$ data indicate that for symmetrical loading the fatigue limit is not reached prior to $N = 10^8$ in both alloys. The effect of specimen orientation (L or T) on the fatigue limit is minor; the stronger alloy 7475 exhibits a higher fatigue limit than alloy 2024.

(2)   Long-crack threshold stress intensity values depend on the applied stress ratio (in the range $-1 \leq R \leq 0.75$). In contrast, the effective threshold stress intensities appear to fall within a single scatter band. The effect of specimen orientation on threshold is particularly noticable for R values near zero. For the alloy 7475 both long-crack threshold values fall below those of alloy 2024.

(3)   In tests carried out near the fatigue limit the initiation of microcracks was observed to occur in alloy 2024 at bulky intermetallic particles cracked during cyclic loading, while in alloy 7475 crack initiation was observed either at the particle/matrix interface or along slip lines present in the as-received microstructure. The initially transcrystalline microcracks were invariably oriented at an angle of approximately 45 degrees to the stress axis.

(4)   Threshold stresses for microcracks growing within a single grain were observed, but only at very small loading amplitudes.

(5)   Microcracks up to a characteristic critical length were observed to grow only at stress amplitudes at, or slightly above, the fatigue limit. Retardation or halts of the advancing crack tips occurred predominantly by interaction with grain boundaries; at low stress amplitudes cracks were found to remain halted for 10^7 cycles. Slight increases in stress amplitude caused a resumption of the crack advance, accelerating after traversing the blocking grain boundary and decelerating again at the approach of the next grain boundary. The extent of retardation and the maximum growth rate between these interactions depended sensitively on cyclic amplitude and stress ratio. The number of repetitions of this interaction sequence was related to the grain dimensions in the crack growth direction.

(6)   The anomalies typical for short fatigue cracks extended up to a characteristic transition length which was found to depend on the alloy but not on grain dimensions or specimen orientation. Up to this transition length short-crack growth occurred only at, or slightly above, the fatigue limit; the growth behaviour of these short cracks cannot be described by LEFM terms. Cracks beyond this length were found to resume growth at stress levels progressively lowered with increasing crack length, until crack growth took place at stress levels corresponding to the threshold stress intensity of long cracks. Thus, the transition length approximates the limit for true short-crack growth behaviour.

(7)   The transition length can be calculated from the effective threshold stress intensity value and the stress level corresponding to the fatigue limit. This fact clearly indicates the governing role of closure in the short crack behaviour. It is interesting to note that the transition length is much shorter for alloy 7475 than for alloy 2024.

(8)   The transition length can also be deduced from corroborating observations of changes in the fracture path topography with increasing crack length, i.e., from crystallographic stage I (single crack plane through entire grains) to crystallographic stage II (transcrystalline crystallographic zigzagging along a line normal to the stress axis). The final transition to non-crystallographic stage II indicates the crack length for which conventional LEFM-considerations are applicable.

(9)   A Kitagawa-type diagram reveals the influence of differences in fatigue strength and long-crack threshold values. The limiting length for short-crack behaviour is uniquely defined by the line corresponding to the effective stress intensity of long cracks. To demonstrate the effects of varying stress ratios, plotting of the test results in a modified diagram with the maximum stress rather than the stress range as abscissa is suggested. In this way the anticipated reduction of transition length with increasing stress ratio is revealed by the intersections with the line corresponding to the effective threshold stress intensity.

(10)   The experimental and computational results of this investigation imply that, for design considerations, the LEFM relations can be applied to small fatigue cracks exceeding the transition length (in the order of 150 μm in high-strength Al alloys) if the effective threshold values for long cracks are taken into account. In his way uncertainties caused by varying stress ratios are eliminated. Further investigations, however, are required to understand the true short-crack growth mechanism, which appears to be of rather academic interest because it concerns crack lengths below engineering detectability.

Acknowledgements

The investigations were partly supported by the Swedish Board for Technical Development, by the Fonds zur Förderung der wissenschaftlichen Forschung, and by the Hochschuljubiläumsstiftung der Stadt Wien, Austria. The authors thank Dr A. Hadrboletz for calibration measurements.

References

(1)   de LANGE, R. G. (1964) Plastic replica methods applied to a study of fatigue crack propagation, *Trans. AIME*, **230**, 644–648.
(2)   PEARSON, S. (1975) Investigation of fatigue cracks in commercial Al alloys and subsequent propagation of very short fatigue cracks, *Engng Fracture Mech.*, **7**, 235–247.
(3)   SCHIJVE, J. (1984) The practical and theoretical significance of small cracks. An evaluation, *Fatigue 84* (EMAS, Warley), p. 751.

(4) JEAL, R. H. (1985) The specification of gas turbine disc forgings, *Metals Mater.*, **1**, 528–533.
(5) NISITANI, H. and TAKAO, K. I. (1981) Significance of initiation, propagation, and closure of microcracks in HCF of ductile materials, *Engng Fracture Mech.*, **15**, 445–456.
(6) MILLER, K. J. (1982) The short crack problem, *Fatigue Engng Mater. Structures*, **5**, 223–232.
(7) RITCHIE, R. O. and SURESH, S. (1982) Mechanics and physics of the growth of small cracks, Proc. AGARD 55th SMP meeting, Toronto.
(8) RITCHIE, R. O. (1983) Fracture mechanics approach to fatigue crack propagation, *Encyclopedia of Materials Science and Engineering* (Pergamon Press, Oxford).
(9) MILLER, K. J. (1984) Initiation and growth rates of short fatigue cracks. *Eshelby Memorial Conference (IUTAM)* (Cambridge University Press), pp. 477–500.
(10) SURESH, S. and RITCHIE, R. O. (1984) Propagation of short fatigue cracks, *Int. Met. Rev.*, **29**, 445–476.
(11) PINEAU, A. (1984) Short fatigue cracks and crack closure, *Proc. ECF-5*.
(12) TAYLOR, D. (1982) EUROMECH 151 colloquium on short fatigue cracks, *Fatigue Engng Mater. Structures*, **5**, 305–309.
(13) JAMES, M. R., MORRIS, W. L., and ZUREK, A. K. (1983) On the transition from near-threshold to intermediate growth rates in fatigue, *Fatigue Engng Mater. Structures*, **6**, 293–305.
(14) FATHULLA, A., WEISS, B., STICKLER, R., and FEMBÖCK, J. (1985) The initiation and growth of short cracks in pm-Mo and Mo-alloys, proc. 11th Int. Plansee Seminar, paper RM 11, p. 45.
(15) MORRIS, W. L. (1977) The early stages of fatigue crack propagation in Al 2048-T851, *Met. Trans, 8A*, 589–596.
(16) FATHULLA, A., WEISS, B., and STICKLER, R. (1984) Initiation and propagation of short cracks under cyclic loading near threshold in technical alloys, Proc. Spring Meeting French Metals Society, Paris, p. 182.
(17) FATHULLA, A., WEISS, B., and STICKLER, R. (1984) Initiation and propagation of short cracks under cyclic loading of Al 2024 near threshold, *Fatigue 84* (EMAS, Warley), p. 1913.
(18) FOTH, J., MARISSEN, R., NOWACK, H., and LÜTJERING, G. (1984) Fatigue crack initiation and microcrack propagation in notched and unnotched Al 2024-T3 specimens, Proc. ICAS 1984, p. 791.
(19) KUNG, C. Y. and FINE, M. E. (1979) Fatigue crack initiation and microcrack growth in Al 2024-T4 and 2124-T4 Al alloys, *Met. Trans, 10A*, 603–610.
(20) MORRIS, W. L. and BUCK, O. (1977) Crack closure load measurements for microcracks developed during fatigue of Al 2219-T851, *Met. Trans, 8A*, 597–601.
(21) MORRIS, W. L. (1977) Crack closure load development for surface microcracks in Al 2219-T851, *Met. Trans, 8A*, 1079–1086.
(22) MORRIS, W. L. (1977) A comparison of microcrack closure load development for stage I and II cracking events for Al 7075-T651, *Met. Trans, 8A*, 1087–1093.
(23) MORRIS, W. L. (1979) Microcrack closure phenomena for Al 2219-T851, *Met. Trans, 10A*, 5–11.
(24) JAMES, M. R. and MORRIS, W. L. (1983) Effect of fracture surface roughness on growth of short fatigue cracks, *Met. Trans, 14A*, 153–155.
(25) TAYLOR, D. and KNOTT, J. F. (1981) Fatigue crack propagation behaviour of short cracks, the effect of microstructure, *Fatigue Engng Mater. Structures*, **4**, 147–155.
(26) MORRIS, W. L. and JAMES, M. R. (1984) Investigation of the growth threshold for short cracks, Proc. Fat. Thresholds, Philadelphia, p. 479.
(27) SURESH, S. (1983) Crack deflection: implication for the growth of long and short fatigue cracks, *Met. Trans, 14A*, 2375–2385.
(28) SURESH, S. (1985) Fatigue crack deflection and fracture surface contact: micromechanical models, *Met. Trans, 16A*, 249–260.
(29) LANKFORD, J. (1985) The influence of microstructure on the growth of small fatigue cracks, *Fatigue Fracture Engng Mater. Structures*, **8**, 161–175.
(30) HIROSE, S. and FINE, M. E. (1983) Fatigue crack initiation and microcrack propagation in X7091 type Al p/m alloys, *Met. Trans, 14A*, 1189–1197.
(31) ALLEN, R. J. and SINCLAIR, J. C. (1982) The behavior of short cracks. *Fatigue Engng Mater. Structures*, **5**, 343–347.

(32) BLOM, A. F. and HOLM, D. K. (1985) An experimental and numerical study of crack closure, *Engng Fracture Mech.*, **22**, 997–1011.

(33) TANAKA, K. NAKAI, Y., and YAMASHITA, M. (1981) Fatigue growth threshold of small cracks, *Int. J. Fracture*, **17**, 519–533.

(34) NEWMAN, J. C. (1982) A non-linear FM approach to the growth of small cracks, Proc. AGARD 55th SMP meeting, Toronto.

(35) HOBSON, P. D. (1982) The formulation of a crack growth equation for short cracks, *Fatigue Engng Mater. Structures*, **5**, 323–327.

(36) COOPER, C. V. and FINE, M. E. (1984) Coffin–Manson relationship for fatigue crack initiation, *Scripta Met.*, **18**, 593–596.

(37) DE LOS RIOS, E. R., MOHAMED, H. J. and MILLER, K. J. (1985) A micro-mechanics analysis for short fatigue crack growth, *Fatigue Fracture Engng Mater. Structures*, **8**, 49–63.

(38) KITAGAWA, H. and TAKAHASHI, S. (1976) Applicability of fracture mechanics to very small cracks or the cracks in the early stages, Proc. ICM-2, ASM, p. 627.

(39) USAMI, S. and SHIDA, S. (1979) Elastic–plastic analysis of the fatigue limit for a material with small flaws, *Fatigue Engng Mater. Structures*, **1**, 471–481.

(40) MORRIS, W. L., JAMES, M. R. and BUCK, O. (1981) Growth rate models for short surface cracks in Al 2219-T851, *Met. Trans*, **12A**, 57–64.

(41) MORRIS, W. L., JAMES, M. R., and ZUREK, A. K. (1985) The extent of crack tip plasticity for short fatigue cracks, *Scripta Met.*, **19**, 149–153.

(42) ZUREK, A. K., JAMES, M. R., and MORRIS, W. L. (1983) The effect of grain size on the fatigue crack growth of short cracks, *Met. Trans*, **14A**, 1697–1705.

(43) TAYLOR, D. (1984) The effect of crack length on fatigue threshold, *Fatigue Engng Mater. Structures*, **7**, 267–277.

(44) LANKFORD, J. (1982) The growth of small fatigue cracks in 7075-T6 Al, *Fatigue Engng Mater. Structures*, **5**, 233–248.

(45) CHAN, K. S. and LANKFORD, J. (1983) A crack tip strain model for growth of small fatigue cracks, *Scripta Met.*, **17**, 529–532.

(46) LANKFORD, J. and DAVIDSON, D. L. (1983) Near threshold crack tip strain and crack opening for large and small fatigue cracks, Proc. Threshold Conf., Philadelphia.

(47) CHANANI, G. R., TELESMAN, I., BRETZ, P. E., and SCARICH, G. V. (1982) Methodology for the evaluation of fatigue crack growth resistance of Al alloys, Northrop Corporation Technical Report.

(48) STICKLER, R. and WEISS, B. (1982) Review of the application of ultrasonic fatigue test methods for the determination of crack growth and threshold behavior of metallic materials, *Proc. Int. Conf. on ultrasonic fatigue* (Edited by J. Wells) (AIME), p. 135.

(49) ASTM working document E647 – *Standard test method for measurement of fatigue crack growth rates*, 1985.

(50) BLOM, A. F., HADRBOLETZ, A., and WEISS, B. (1983) Effect of crack closure on near-threshold crack growth behavior in a high-strength Al-alloy up to ultrasonic frequencies, *Proc. ICM-4*.

(51) ASTM E740-80 – *Standard practice for fracture testing with surface crack tension specimens*, ASTM 1983-3-03.01).

(52) BLOM, A. F., HADRBOLETZ, A. and WEISS, B. unpublished results.

(53) LUKAS, P., KUNZ, L., KNESL, Z., WEISS, B., and STICKLER, R. (1985) Fatigue crack propagation rate and the crack tip plastic strain amplitude in polycrystalline Cu, *Mater. Sci. Engng*, **70**, 91–100.

(54) WEISS, B., MÜLLNER, H., STICKLER, R. LUKAS, P., and KUNZ, L. (1984) Influence of frequency on fatigue limit and fatigue crack growth behaviour of polycrystalline Cu, *Proc. ICF-6*, Vol. 3, p. 1783.

(55) ZAIKEN, E. and RITCHIE, R. O. (1985) On the development of crack closure and the threshold condition for short and long fatigue cracks in Al 7150, *Met. Trans*, **16A**, 1467–1477.

(56) HOLM, D. K. and BLOM, A. F. (1984) Short cracks and crack closure in Al 2024-T3, *Proc. ICAS* (ICAS-84-3.7.1), p. 783.

MATERIAL AND SPECIMEN PREPARATION EFFECTS

R. W. Suhr*

The Effect of Surface Finish on High Cycle Fatigue of a Low Alloy Steel

REFERENCE Suhr, R.W., **The Effect of Surface Finish on a High Cycle Fatigue of a Low Alloy Steel**, *The Behaviour of Short Fatigue Cracks*, EGF Pub 1 (Edited by K. J. Miller and E. R. de los Rios) 1986, Mechanical Engineering Publications, London, pp. 69–86.

ABSTRACT As fatigue cracks initiate predominantly at the free surface of a material, the condition of the surface can be assumed to be critical with regard to fatigue strength. Two features of a mechanically prepared surface which are considered to be major factors affecting fatigue strength are the surface roughness and the residual stress in the surface layer. It has been noted that little of the general body of published data on the effect of surface finish on fatigue has separated or, in many cases, recognized the additional effects of residual stress introduced by the finishing process which would interfere with the evaluation of surface irregularities. To understand fully the effects of surface irregularities they must be examined in isolation from residual stress effects using stress free specimens. This paper describes such an investigation on a low alloy steel to examine the relationship between fatigue limit and depth of shallow surface defects which were either inherent in the material or the result of the surface finishing process.

Introduction

As fatigue cracks initiate predominantly at the free surface of a material, the condition of the surface can be assumed to be critical with regard to fatigue crack initiation. Two features of a mechanically prepared surface which are considered to be major factors affecting fatigue strength are the surface roughness and the residual stress in the surface layer.

A measure of the surface roughness introduced by a particular machining process can be obtained from a profile scan of the surface and is often expressed as the centre line average or R_a value. The R_a value might vary from 0.2 microns for a good ground finish to 8 microns for a rough turned finish. Where the effects of surface finish on fatigue have been examined in carbon steels, reductions in fatigue limit varying between 10 and 25 per cent have been observed when comparing rough turned or rough ground with a fine ground or polished surface (1)−(7) and the effect of surface roughness has been shown (8) to be even more pronounced in high strength steels.

It may be that the most significant parameter categorizing the quality of a machined surface from the fatigue standpoint is maximum depth of the surface irregularities. Siebel and Gaier (9), for instance, compared fatigue strength with maximum depth of surface irregularities (R) and found a critical depth (R_o) below which there was no change in fatigue strength and above which there was a linear fall in fatigue strength with log R.

It is noted that little of the general body of data on the effect of surface finish on fatigue has separated or, in many cases, recognized the additional effects of

* GEC Turbine Generators Ltd, Willans Works, Newbold Road, Rugby CV21 2NH.

residual stress introduced by the machining process which would interfere with the evaluation of surface irregularities. To understand fully the effects of surface irregularities, they must be examined in isolation from residual stress effects on stress-free specimens.

In this paper, the effect of surface finish on the high cycle fatigue life of a low alloy steel forging is examined using a variety of surface finishes.

Material

The material employed was taken from a low alloy steel forging of composition (% wt) 0.29C, 0.21Si, 0.55Mn, 0.01S, 0.005P, 1.99Ni, 1.30Cr, 0.57Mo, 0.09V. This had been steam quenched from 850°C, and tempered at 610°C to give the mechanical properties of 844 MPa tensile strength, 710 MPa 0.2 per cent proof stress, 18.3 per cent elongation, 63 per cent reduction in area, and a hardness of 280 Hv30. The microstructure of the material (shown in Fig. 1) consisted of a tempered bainite structure with a grain size of ASTM 8. All test specimens used in this investigation were given a further heat treatment of 590°C for four hours, applied after specimen manufacture, to remove residual machining stresses from the surfaces. This treatment, conducted in vacuum, had no effect on the tensile properties of the material. Residual stress measurements showed residual stress levels of between +17 MPa and −22 MPa, which are of the same order as the accuracy of the X-ray diffraction method employed, and indicated full stress relief had been achieved.

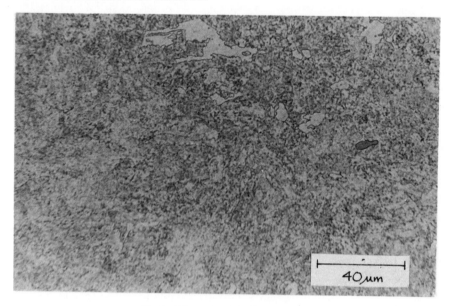

Fig 1   Microstructure of the material

Experimental details and results

The fatigue programme was conducted on test specimens in both the bending and push–pull modes in air at room temperature using Amsler Vibrophore machines under constant amplitude loading conditions.

Bending tests

The bending tests were carried out under four point loading using a specimen designed, as shown in Fig. 2(a), to give a uniform stress along a central 32 mm section of surface and to ensure that crack initiation occurred away from the edge with no stress concentration feature apart from that produced by the surface finishing process or the presence of surface inclusions. The design allowed the specimen surface to be easily machined in either the longitudinal or transverse direction.

In order to produce a range of surface roughness, test specimens were given a variety of ground finishes and in one case a series of shallow grooves were introduced into the surface by careful machining. The following types of surface finish were examined:

(A)   fine longitudinal ground;
(B)   rough longitudinal ground;
(C)   fine transverse ground;
(D)   rough transverse ground;
(E)   0.05 mm deep (90 degree included angle) transverse grooves;
(F)   transverse emery finish.

The results of the four point bending tests conducted with an R value of 0.1 are plotted in Fig. 3(a) for specimens taken axially from the forging and in Fig. 3(b) for specimens taken radially.

Push–pull tests

In order to examine the effect of mean stress from zero to 770 MPa, push–pull fatigue tests were conducted on specimens of the type shown in Fig. 2(b) with the following surface finishes (figure illustrations of results are as given in brackets):

(a)   fine circumferentially ground (Fig. 4(a));
(b)   0.05 mm deep circumferential grooves (Fig. 4(b));
(c)   0.50 mm deep circumferential grooves (Fig. 4(c));
(d)   rough circumferentially ground (Fig. 5(a));
(e)   longitudinally polished with diamond paste (Fig. 5(a)).

The results of the push–pull fatigue tests are given in Figs 4 and 5(a) for specimens taken axially from the forging, and in Figs 5(b) and 6 for specimens taken radially.

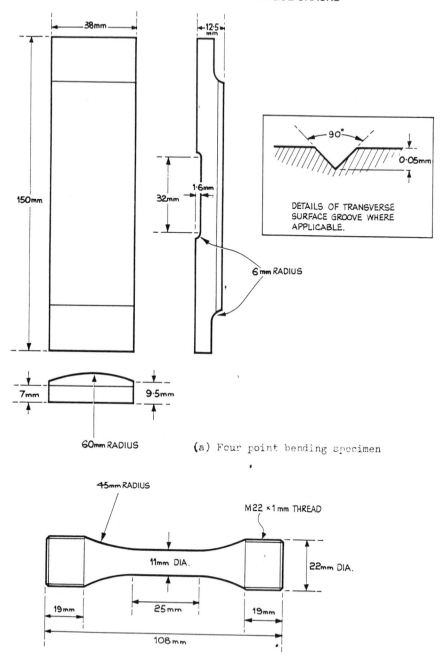

(a) Four point bending specimen

(b) Push-pull specimen

Fig 2   Details of fatigue test specimens

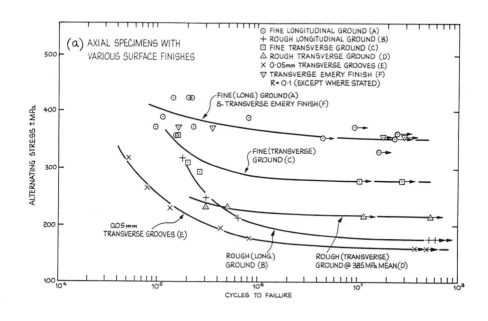

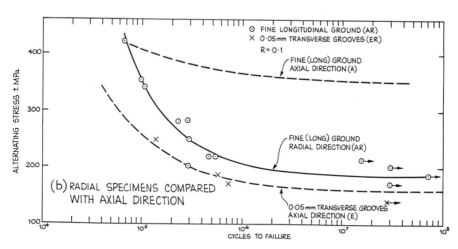

Fig 3   Bending fatigue results

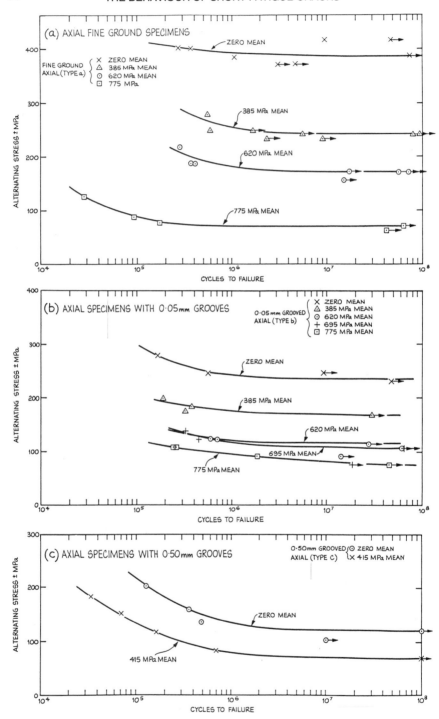

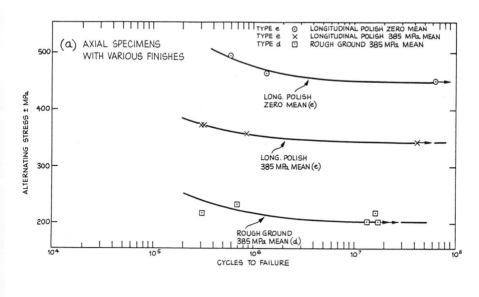

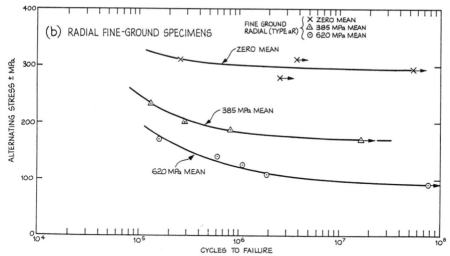

Fig 5   Push–pull fatigue results

Fig 4   Push–pull fatigue results

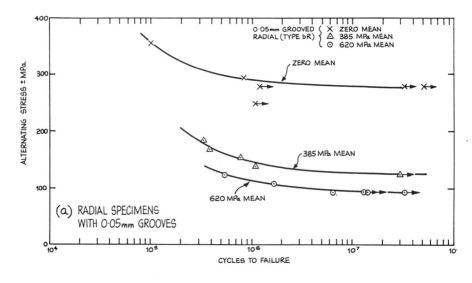

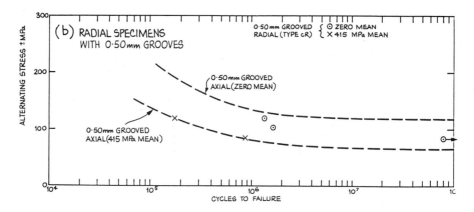

Fig 6   Push–pull fatigue results

Fig 7   Surface features introduced by machining   ▶
(a) Fine transverse ground surface (Type C) showing side flow of material
(b) 0.05 mm deep groove (Type E) showing tear marks on the flanks and a smooth finish at the base which contains a fatigue crack
(c) Profile of a 0.05 mm groove containing a fatigue crack
(d) Fatigue crack initiation from a surface groove introduced by transverse grinding (Type D)

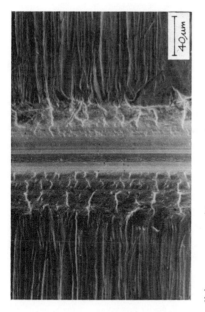

(a)

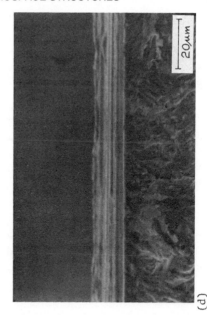

(b)

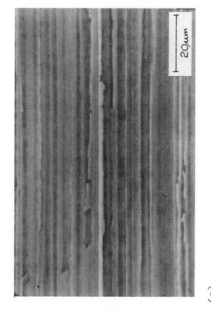

(c)

(d)

Metallographic examination

Surface features

An examination of the various surfaces generated was made using a scanning electron microscope with the following observations.

(a) *Ground finishes.* In all cases there was a considerable amount of plastic deformation or side flow of the material transverse to the direction of the grinding marks. This is shown typically in Fig. 7(a) for a fine ground finish. No significant difference was observed between the equivalent finishes in the bending and push–pull specimens or between the equivalent transverse and longitudinal grinding in the bend specimens.

(b) *Emery or polished finishes.* The transverse finish given to the bend specimens was similar in appearance to the fine ground finish except that there was little evidence of plastic deformation or side flow. Longitudinal polishing on push–pull specimens resulted in an extremely smooth surface which showed few surface irregularities even at high magnifications on the SEM.

(c) *Grooved finishes.* Examination of the 0.05 mm deep transverse grooves, which were machined in the surface of both bending and push–pull specimens, generally showed a fairly smooth finish at the base of the groove with tear marks along the flanks of the grooves. A typical groove is shown in Fig. 7(b) which indicates a fatigue crack running along the base of the groove after testing. Figure 7(c) shows a section through the groove.

Fracture features

Examination of the fracture faces after fatigue testing revealed that fatigue cracks usually initiated from some form of surface irregularity which was either the result of the surface finishing process or inherent in the material. These irregularities could be categorized as follows.

(a) *Machining.* Where specimens contained grinding or emery marks transverse to the direction of loading, fatigue cracks almost invariably initiated at the root of one of the grooves introduced by the process (Fig. 7(d)).

(b) *Inclusions from grinding.* Rough longitudinal grinding resulted in particles of the grinding wheel becoming detached and either embedding in the surface or leaving sharp notches in the surface. These inclusions or notches provided the usual initiation point for fatigue cracks in the rough longitudinal specimens and Fig. 8(a) is a typical example. Similar defects provided the occasional initiation site in fine longitudinal ground specimens.

Fig 8   Fracture features of fatigue specimens   ▶

(a) **Fatigue crack initiation from an alumina/silica inclusion embedded in the surface during grinding**
(b) **Fatigue crack initiation from a MnS inclusion at the surface when stressed in the radial direction**
(c) **Fatigue crack initiation from a closed pore at a surface with a fine longitudinal ground finish**
(d) **A pore on a fracture face showing the featureless type of surface associated with this type of defect**

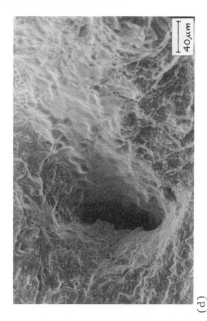

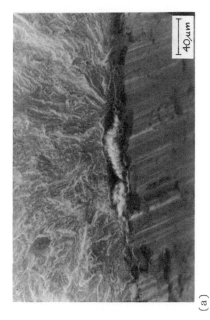

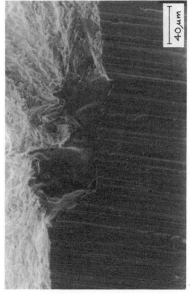

(c)   *Inherent defects*. MnS inclusions, inherent to the material, were elongated in the axial direction of the forging. Crack initiation was almost exclusively from these inclusions lying parallel and very close to the surface at 90 degrees to the direction of loading in specimens taken in the radial direction, but occasional instances occurred in specimens taken in the axial direction where an inclusion intersected the surface. A typical example of failure is shown in Fig. 8(b) and and evaluation of the MnS inclusion size found at crack initiation sites indicated lengths of up to 1.0 mm and depths of up to 0.038 mm.

Another inherent defect, from which occasional fatigue cracks were found to initiate in longitudinally ground or polished specimens, was a featureless type of area shown typically in Fig. 8(c). Analysis of the surface of these defects on the SEM showed no variation in chemical composition from the forging composition and it was considered that this type of defect might be a closed or partially closed pore. Evidence for a pore was found on one fracture face remote from the initiation site (Fig. 8(d)).

Discussion

Surface roughness

In general terms an increase in surface roughness as indicated by a Talysurf scan is accompanied by a decrease in fatigue limit in both bending and push–pull fatigue as expected. For example, a rough longitudinal ground finish with a recorded peak to valley height four times greater than a fine longitudinal ground finish has a fatigue limit which is 50 per cent less. From an examination of the fracture faces of failed specimens, however, it was apparent that for ground, emery and polished surfaces, fatigue initiation occurred from transverse surface grooves, inclusions or inherent defects which were all greater in depth than the maximum peak to valley height measured by a Talysurf scan. Tables 1 and 2 give the depth of these defects at the fatigue crack initiation point

Table 1   Surface measurements on bending fatigue specimens

Type	Surface finish	Depth of defect at crack initiation site (microns)
A	Fine longitudinal ground	8.4
B	Rough longitudinal ground	43.1
C	Fine transverse ground	6.3
D	Rough transverse ground	21.3
E	0.05 mm grooves	50.0
F	Transverse emery finish	4.3
AR	Fine longitudinal ground (Radial)	38.1
ER	0.05 mm grooves (Radial)	90.0

and it was clear that a profile scan made prior to testing failed to reveal the surface irregularities from which fatigue cracks could initiate.

The depths of defects at the crack initiation sites given for the various surface finishes in Tables 1 and 2 have been plotted against the corresponding fatigue limits for R values of 0.1 and -1 in Fig. 9(a) and (b). It is well known from specimens containing long cracks that a threshold condition exists below which the crack will not grow. This threshold is represented by the standard fracture mechanics equation

$$\Delta K_o = C \Delta \sigma_{(a)o} a^{1/2} \tag{1}$$

where ΔK_o is the threshold stress intensity, $\Delta \sigma_{(a)o}$ the limiting fatigue stress range for a test piece (component) with a crack of depth a, and C a geometry constant. For the material employed in this investigation, ΔK_o has been measured as 6.0 MPa$\sqrt{m}$ when $R = 0.1$ and 8.8 MPa$\sqrt{m}$ when $R = -1$. Lines representing equation (1) with the appropriate values of ΔK_o have been constructed in Fig. 9, C being taken as $\sqrt{(1.25\pi)}$ (10) and it becomes apparent that small defects will grow at reversing stresses below those given by equation (1). Such deviations have been reported by Kitagawa et al. (11) for short cracks of similar length and to allow for such deviations El Haddad et al. (12) have proposed a modified relationship of the form

$$\Delta K_o = C \Delta \sigma_{(a)o} (a + a_o)^{1/2} \tag{2}$$

where a_o is given by

$$a_0 = \frac{1}{1.25\pi} \left(\frac{\Delta K_o}{\Delta \sigma_o}\right)^2 \tag{3}$$

Equation (3) is for the condition $\Delta \sigma_{(a)o} = \Delta \sigma_o$ which is a limiting value where the maximum fatigue limit is achieved and no further reduction in crack depth has any effect. Applying this modified relationship, a regression analysis has been used to obtain a least squares fit for the data with defect depths of 0.05 mm

Table 2   Surface measurements on push–pull fatigue specimens

Type	Surface finish	Depth of defect at crack initiation site (microns)
a	Fine circumferential ground	17.8
b	0.05 mm grooves	50.0
c	0.50 mm grooves	500.0
d	Rough circumferential ground	30.5
e	Longitudinal polish	5.1
aR	Fine circumferential ground (Radial)	40.0
bR	0.05 mm grooves (Radial)	90.0

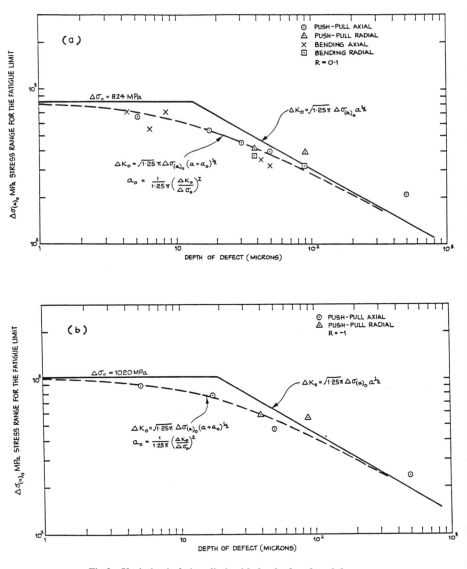

Fig 9   Variation in fatigue limit with depth of surface defect

or less. Mean curves have been constructed in Fig. 9 for which limiting values for the stress range $\Delta\sigma_0$ of 824 and 1020 MPa have been obtained for R values of 0.1 and −1, respectively. The agreement between the data and the modified threshold curve suggests that this form of equation gives a good representation of the data for stress free specimens and that defects below 0.05 mm might be regarded as equivalent to short cracks. In this material it would appear that the fatigue strength is limited by surface defects that are inherent in the material.

Under these circumstances no increase in $\Delta\sigma_{(a)o}$ could be achieved by an improvement in surface finish although it is recognized that smaller defects could lead towards the limiting value $\Delta\sigma_o$.

As already stated the data presented has been obtained on stress free specimens and therefore residual stresses arising from the surface finishing process do not have to be taken into account. Clearly, if there is a residual surface stress it may act as a superimposed mean stress varying in magnitude with depth and should be considered when analysing short crack growth behaviour. It is not apparent from the published work on this subject, however, that this aspect has been taken into account. For instance, the work already referred to by El Haddad *et al.* (**12**) does not consider the influence of possible shallow machining stresses in the experimental work conducted to correlate threshold and short crack results. Such machining stresses would have introduced an effective mean stress although the tests were carried out with an *R* value of -1.

Orientation

Fatigue specimens taken from the radial direction in the forging were tested in both the bending and push–pull mode of loading. The results of the bending tests given in Fig. 3(b) show that for a fine ground finish the fatigue limit in the radial direction is almost 50 per cent lower than in the axial direction for an *R* value of 0.1. The same comparison in push–pull loading, which can be made from the data presented in Fig. 10(a) and (b), shows that the radial tests give fatigue limits progressively lower by 24, 29, and 45 per cent at increasing mean stress levels of 0, 385, and 617 MPa, respectively. These reductions in fatigue limit are attributed solely to the effect of MnS inclusions from which fatigue cracks invariably initiated in the radial specimens.

Radial specimens containing shallow machined grooves of 0.05 mm in depth were also tested in the bending and push–pull modes (Figs 3(b) and 10(b), respectively). These results show that the MnS inclusions, which are typically 0.04 mm in depth and of a similar size to the groove depth, act as extensions to the groove to give an effective 0.09 mm defect. Tables 1 and 2 give the details of the radial fatigue tests and these are included in Fig. 9, showing good agreement with the corresponding axial tests.

Mean stress

The effect of tensile mean stress on the push–pull fatigue limit for various finishes is summarized in Fig. 10 for the axial and radial directions. The fatigue limits have been plotted against the mean stress levels and mean curves have been constructed for each type of surface finish. In general the relationship between fatigue limit and mean stress tends to be linear for mean stresses up to 620 MPa, but neither a modified Goodman nor a Gerber type relationship can be used to represent the data over the full range of tensile mean stress.

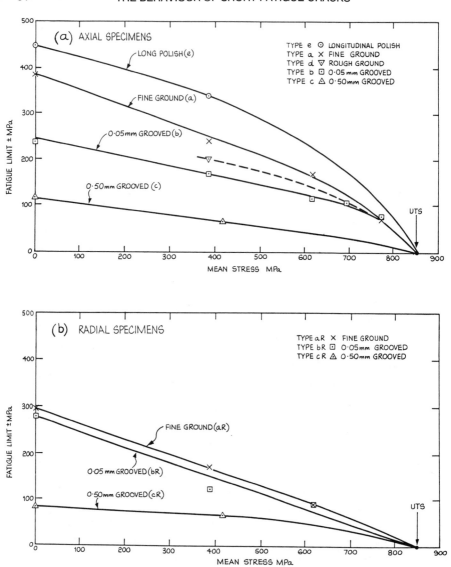

Fig 10   Effect of mean stress on fatigue limit tested in the push–pull mode

Fatigue limits are often regarded as approximating to half the tensile strength but Fig. 10 indicates that such a generalization only applies to polished specimens at zero mean stress. For specimens with significant surface irregularities or an orientation which aligns inclusion stringers at 90 degrees to the applied stress, the fatigue limit may be drastically reduced.

Conclusions

From an investigation into the effect of surface finish and orientation on the fatigue strength of a low alloy steel forging the following conclusions have been formed.

(1) Fatigue crack initiation was usually from surface irregularities either introduced by the finishing process, such as surface scratches, or inherent defects in the material such as inclusions and closed pores.

(2) The surface irregularities providing the fatigue crack initiation sites were not usually detected by either surface roughness scanning methods or random metallographic sectioning techniques and were only identified from examination of fracture faces.

(3) The fatigue limit decreased with increasing depth of defect at the crack initiation site and surface grooves or inclusions about 0.05 mm in depth reduced the fatigue limit of a fine ground surface by 50 per cent when the defects were aligned transverse to the applied stress. Hence radial specimens had lower fatigue strengths than axial specimens due to the presence of MnS inclusions elongated in the axial direction.

(4) Defects providing the crack initiation sites of approximately 0.05 mm in depth or less were found to grow below the threshold represented by the standard fracture mechanics equation

$$\Delta K_o = C \Delta \sigma_{(a)o} a^{1/2}$$

Using a modified equation of the form $\Delta K_o = C \Delta \sigma_{(a)o} (a + a_o)^{1/2}$ curves were constructed which suggested that this form of equation gives a good representation of the threshold for the defects observed. Values for a_o of 0.013 mm and 0.019 mm were obtained for R values of 0.1 and -1, respectively.

Acknowledgements

The author acknowledges with thanks the Directors of GEC Turbine Generators Ltd for permission to publish this paper and also Mr J. T. W. Smith who conducted the fatigue tests.

References

(1) MOORE, H. F. and KOMMER, J. B. (1927) *The fatigue of metals* (McGraw-Hill, New York), p. 201.

(2) KAWAMOTOE, M. and NISHOIKA, K. (1954) Effect of surface roughness on the fatigue strength of steel, Proceedings of the 4th Japan National Congress on Applied Mechanics, p. 211.

(3) CLEDWYN-DAVIES, D. N. (1955) Effect of grinding on the fatigue strength of steels, *Proc. Instn. mech. Engrs*, **169**, 83–91.

(4) TARASOV, L. P. and GROVER, H. J. (1950) Effects of grinding and other finishing processes on the fatigue strength of hardened steel, *Proc. Am. Soc. Test. Mater.*, **50**, 668–687.

(5) TARASOV, L. P., HYLER, W. S. and LETNER, H. R. (1958) Effects of grinding direction and of abrasive tumbling on the fatigue limit of hardened steel, *Proc. Am. Soc. Test. Mater.*, **58**, 528–539.

(6) VITOVEC, F. H. and BINDER, H. F. (1956) Effects of specimen preparation on fatigue, Wright Air Development Department, Tech. Rep., **56**, 289–305.

(7) HOUDREMONT, E. and MAILANDER, R. (1929) Bending fatigue tests on steels, *Stahl and Eisen*, **49**, 833–892.

(8) HANLEY, B. C. and DOLAN, T. J. (1953) Surface Finish, *Metals engineering – design* (Soc. Mech. Engrs), p. 100.

(9) SIEBEL, E. and GAIER, M. (1956) Influence of surface roughness on the fatigue strength of steels and non-ferrous alloys, *A. Ver. Dtsch. Ing.*, **98**, (in German); Translation: *Engineers Digest* (1957), **18**, 109–125.

(10) BROWN, W. F. and STRAWLEY, J. E. (1966) Plane strain crack toughness testing of high strength metallic materials, *ASTM STP 410*.

(11) KITAGAWA, H. and TAKAHASHI, S. (1976) Applicability of fracture mechanics to very small cracks, *Int. Conf. Mech. Behaviour of Materials (ICM2)*, (American Society of Metals), p. 627.

(12) EL HADDAD, M. H., TOPPER, T. H. and SMITH, K. N. (1979) Prediction of non-propagating cracks, *Engng Fracture Mech.*, **11**, 573–580.

V. M. Radhakrishnan* and Y. Mutoh†

On Fatigue Crack Growth in Stage I

REFERENCE Radhakrishnan, V. M. and Mutoh, Y., **On Fatigue Crack Growth in Stage I**, *The Behaviour of Short Fatigue Cracks*, EGF Pub. 1 (Edited by K. J. Miller and E. R. de los Rios) 1986, Mechanical Engineering Publications, London, pp. 87–99.

ABSTRACT Investigations have been carried out to study the effect of grain size on stage I crack growth of two types of cracks; cracks starting from smooth surfaces and cracks starting from pre-existing notches. It has been observed that increase in grain size increases the threshold stress intensity factor but reduces the fatigue strength. Analysis indicates that in both types of cracks, the crack driving parameter is $\sigma\sqrt{a}$ and the resistance parameter is $\sigma_{ys}\sqrt{d}$. Based on this approach the non-propagation of both types of cracks is discussed.

Introduction

Fatigue crack growth in stage I is dependent on the grain size and the yield strength of the material both for cracks starting from smooth surfaces and cracks starting from pre-existing notches. In the case of smooth-surfaced materials it is well known that a fine-grained material will give a better fatigue resistance than a coarse-grained one. However, it has been observed by many researchers **(1)–(3)** that, in general, the threshold stress intensity factor, ΔK_{th}, increases with increase in grain size. This implies that materials of larger grain size will be better suited to resist crack initiation and subsequent slow growth if a pre-existing crack or a sharp notch happens to be present in the structure. The practical implications of these opposing microstructural requirements for good resistance to fatigue crack propagation in smooth specimens (high fatigue limit) and a good resistance to fatigue crack propagation in previously fatigue cracked specimens (high ΔK_{th}) must be clearly understood in selecting a material for a given application **(4)(5)**. In between these two extreme cases of a smooth specimen and a specimen having a sharp long crack, actual structures will have blunt notches or very small cracks – mechanical or metallurgical – and, in such cases, fatigue crack nucleation and further propagation will very much depend on the notch field and the microstructural properties of the material **(6)**.

Experimental

The materials investigated are (a) type 304 stainless steel, and (b) a low carbon steel. The chemical compositions are given in Table 1. The heat-treatment and the corresponding grain size and yield strength are given in Table 2. Two types of specimen, as shown in Fig. 1, were used for the stainless steel, a CT specimen of thickness 12.7 mm for the study of crack propagation and a round bar

* Metallurgy Department, IIT, Madras-36, India.
† Mechanical Engineering Department, Technological University of Nagaoka, Nagaoka, Japan.

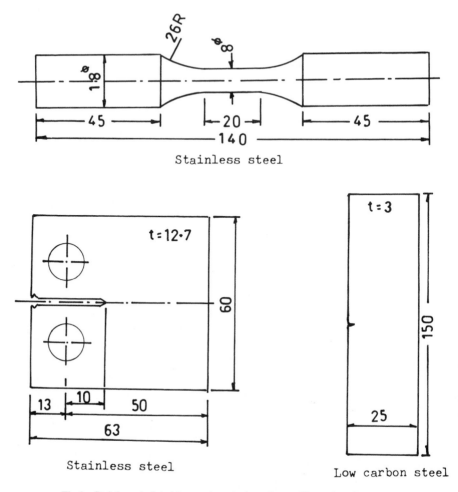

Fig 1   Stainless steel and low carbon steel specimens (dimensions in mm)

bar specimen for the determination of fatigue strength. In the case of the low
carbon steel a plate specimen of thickness 3 mm with an edge notch (SEN type)
was used. A servo-hydraulic MTS testing machine operating at a frequency of
40 Hz was used to estabish the fatigue threshold at a stress ratio $R = 0.05$ and
the fatigue limit at a stress ratio $R = -1$ of the stainless steel. A Vibrophore
fatigue testing machine was used to establish both the fatigue threshold and the
fatigue limit of the low carbon steel at $R = 0.1$. For both materials an optical
travelling microscope ($\times 30$) was used to measure the crack length. The
accuracy of measurement was 0.01 mm. In the establishment of ΔK_{th} a load
shedding method was employed and the crack was allowed to grow a distance
corresponding to two to three times the plane stress maximum plastic zone size
of the previous loading in order to avoid the residual stress effect due to load

Table 1   Chemical composition (wt%)

	C	Si	Mn	P	S	Ni	Cr
Stainless steel	0.06	0.74	1.24	0.027	0.005	8.45	18.10
Low carbon steel	0.09	0.12	0.2	0.03	0.03	—	—

Table 2   Mechanical properties

Material	Heat treatment	Grain size (μm)	σ_{ys} (MPa)
Stainless steel	S1 1223 K	34	399
	($\frac{1}{2}$ hr)		
	S2 1323 K	74	308
	S3 1473 K	86	288
	S4 1523 K	110	258
Low carbon steel	LC1 1173 K	18	240
	LC2 1223 K	28	210
	LC3 1273 K	44	190

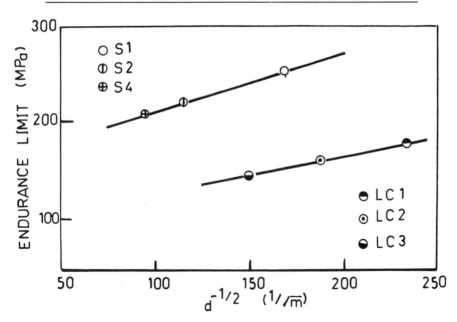

Fig 2   Relation between endurance limit and grain size (S = Stainless steel; LC = Low carbon steel)

reduction (**7**). At least two specimens were tested under similar loading conditions.

Results and discussion

Figures 2 and 3 show the relation between the endurance limit ($\Delta\sigma_e$) and the grain size, and that of the threshold stress intensity factor (ΔK_{th}) and the grain size (d), respectively, according to the relations

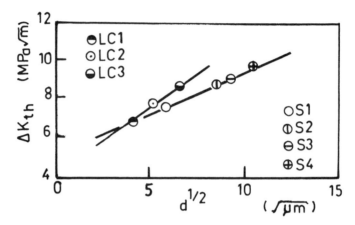

Fig 3 Relation between threshold stress intensity factor and grain size (S = Stainless steel; LC = Low carbon steel)

$$\Delta\sigma_e = 167 + 0.4/\sqrt{d} \qquad \text{Stainless steel}$$
$$ = 102 + 0.31/\sqrt{d} \quad \text{Low carbon steel} \qquad (1)$$

and

$$\Delta K_{th} = 6.2 + 320\sqrt{d} \quad \text{Stainless steel}$$
$$\phantom{\Delta K_{th}} = 4.1 + 667\sqrt{d} \quad \text{Low carbon steel} \qquad (2)$$

where $\Delta\sigma_e$ is in MPa, d in metres, and ΔK_{th} in MPa$\sqrt{m}$. Within the range of grain size investigated (where it has been found that the Hall–Petch relation is valid as shown in Fig. 4) the fatigue strength decreases and the threshold stress intensity factor increases with increase in grain size. Figure 5 shows the relation between the endurance limit $\Delta\sigma_e$ and the yield strength σ_{ys} given as

$$\Delta\sigma_e = 0.5\sigma_{ys} + 51 \qquad (3)$$

with both $\Delta\sigma_e$ and σ_{ys} in MPa.

Figures 6(a) and (b) show the relation between σ_{max} and the crack length, a, for different threshold levels of stainless steel and low carbon steel, respectively. It can be observed that below the value of crack length a_I the σ_{max} for crack initiation remains more or less constant at $\Delta\sigma_e$. The value of a_I is dependent on the grain size and is given by

$$a_I = nd \qquad (4)$$

where $n = 15$ in the case of low carbon steel and 4 in the case of stainless steel.

In some of the low carbon steel specimens, sharp notches with a depth equal to 2 mm and a root radius in the range 0.1–0.2 mm were introduced and the crack growth from the root of the notch was studied as explained earlier. The stress intensity factor for the crack at the notch tip was calculated using the relation (**8**)

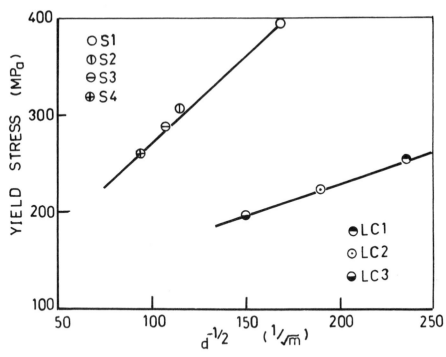

Fig 4   Hall–Petch relation (S = Stainless steel; LC = Low carbon steel

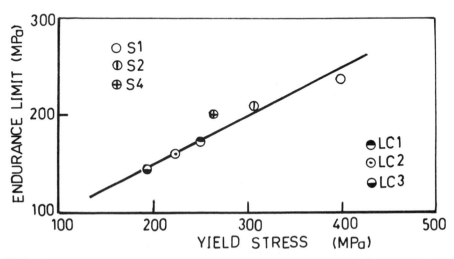

Fig 5   Dependence of endurance limit on the yield strength (S = Stainless steel; LC = Low carbon steel)

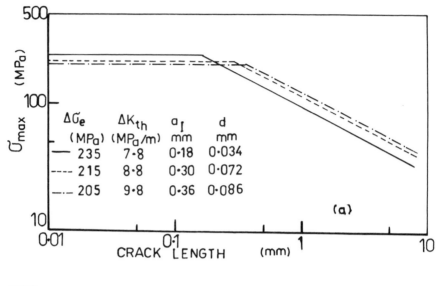

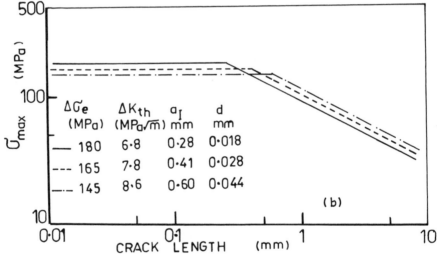

Fig 6   Kitagawa–Takahasi (16) diagrams
(a) Stainless steel type 304
(b) Low carbon steel

$$K = \{1 + 4.762\sqrt{(D/\rho)}\}^{1/2} K_{SEN} \tag{5}$$

where D is the depth of the notch and ρ is the root radius, and K_{SEN} is

$$K_{SEN} = \sigma\sqrt{(\pi a)}(1.12 - 0.23\alpha + 10.55\alpha^2 - 21.72\alpha^3 + 30.39\alpha^4) \tag{6}$$

where $\alpha = a/W$ and a is the fatigue crack length. Figure 7 shows the crack growth from notches of different included angles β and also from a sharp

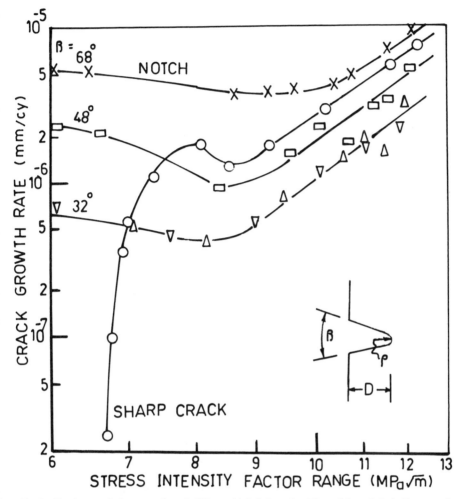

Fig 7   Crack growth from notches of different included angles. The solid symbols indicate crack
growth from a pre-existing crack. The material is low carbon steel

pre-existing crack. It can be seen that in the case of comparatively sharp
notches the crack propagation rate is somewhat lower than that in the case of
blunt notches. On the limit, when the notch is so sharp that it can be simulated
to a crack, the crack growth rate becomes zero until the stress intensity factor is
increased to ΔK_{th}. However, when the cracks start propagating and become
stage II cracks, all the curves tend to merge together.

A model for crack nucleation

A model for crack nucleation at the endurance limit from a smooth surface has
been proposed (9), the essence of which is as follows. Consider a smooth
specimen subjected to a stress range τ_{max}–zero. When the maximum shear

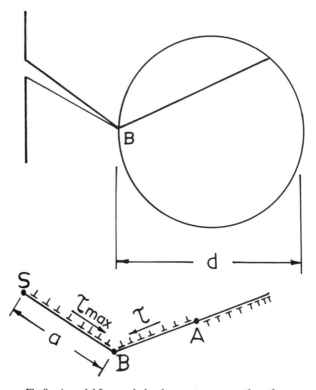

Fig 8   A model for crack development on a smooth surface

stress τ_{max} is equal to or greater than the yield stress in shear τ_{ys} of any grain oriented for easy slip, persistent slip bands will form which will transform themselves into a small crack. This small crack will encounter the next grain as shown in Fig. 8. The small surface crack can be treated as a row of dislocations emitted from an F-R source, S, and blocked by a grain boundary at B. If the stress concentration is of sufficient magnitude, it will activate the source at A in the next grain. A–B is the plane over which these activated dislocations will move. Assuming that the dislocation source in the next grain is activated when the shear stress at the distance r^* from the boundary B reaches a critical value τ^*, one obtains

$$\tau^* = (\tau_{max} - \tau_i)\left(\frac{a}{r^*}\right)^{1/2} \tag{7}$$

where τ_i is the internal stress and a the length of SB. On the other hand, from a similar assumption the yield shear stress τ_{ys} is given as

$$\tau^* = (\tau_{ys} - \tau_i)\left(\frac{d/2}{r^*}\right)^{1/2} \tag{8}$$

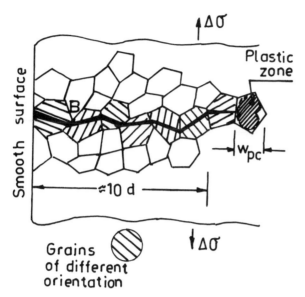

Fig 9   Crack growth in stage I and stage II. The crack encounters grains of different orientation on its path

where d is the grain diameter. Combining equations (7) and (8) we get

$$(\tau_{max} - \tau_i)\sqrt{a} = \frac{1}{\sqrt{2}}(\tau_{ys} - \tau_i)\sqrt{d} \qquad (9)$$

Assuming τ_{max} to correspond to the endurance limit, surface cracks of depth approximately equal to the grain size can form at the endurance stress which may or may not propagate further. The yield stress τ_{ys} of that grain which is oriented for easy slip represents a minimum value for the grain that will lie on the crack path. The above relation also shows that in general, if the bulk yield strength of the material increases, the endurance limit will also increase, as has been shown in the present investigation and indicated in Fig. 5 and equation (3) of this paper.

If the conditions are favourable the crack, which is of an initial length equal to the grain size, will grow further. In stage I the orientation of the crack will be in the direction of maximum shear. But after a certain growth, the mode will change to mode I and the direction will be normal to the applied stress. A schematic representation of this situation is shown in Fig. 9. At the point where the crack enters mode I, corresponding to stage II of the crack propagation, it is postulated that the plastic zone will be of the order of the grain size. Further increase in crack length will enlarge the plastic zone to encompass several grains and as a result the bulk yield strength will control the size of the enclave. Thus at the beginning of stage II we can write the cyclic plastic zone size w_{pc} as

$$w_{pc} = d \qquad (10)$$

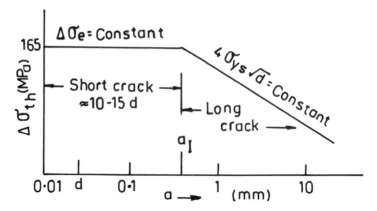

Fig 10   Schematic representation of regions of short and long cracks

Even for a crack emanating from a very sharp notch a similar assumption has been made by Yoder *et al.* (**3**) when the crack moves from stage I to stage II, so that

$$\Delta K_T = 5.5\sigma_{ys}\sqrt{d} \tag{11}$$

where ΔK_T corresponds to the SIF range at the transition.

The threshold stress intensity factor ΔK_{th} can be taken to be of the same order as ΔK_T. Liu and Liu (**10**) have shown that ΔK_{th} is of the order of $0.7\,\Delta K_T$. The value of a_I up to which the threshold stress will be approximately equal to the endurance limit (Fig. 10) can be given as

$$\Delta\sigma_e\sqrt{(\pi a_I)} = 0.7\,\Delta K_T$$
$$= 0.7 \times 5.5\sigma_{ys}\sqrt{d} \tag{12}$$

Taking $\Delta\sigma_e \approx (1/\sqrt{2})\sigma_{ys}$, as indicated in equation (9), we get

$$a_I \approx 10d \tag{13}$$

It has been noticed by Taylor and Knott (**11**) that a_I is structure dependent and is of the order indicated by the above relation.

Non-propagating cracks

It has been observed that for both types of cracks, the crack driving parameter is $\Delta\sigma\sqrt{a}$ and the resistance parameter is $\sigma_{ys}\sqrt{d}$, as given in equations (9) and (11). In stage I, as the crack propagates, these two parameters just balance each other and the important stresses encountered are the applied shear stress τ and the yield stress in shear τ_{ys} of the grain just ahead of the crack tip. Hence the crack growth is structure sensitive and will be governed by the orientation of the grain which it will encounter on its path. As schematically shown in Fig. 9, when the surface crack starts growing from position B the yield stress τ_{ys} of each grain will control the resistance parameter $\tau_{ys}\sqrt{d}$. If $\sqrt{(a/d)} > \tau_{ys}/\sqrt{2}\,\tau_{max}$ the crack

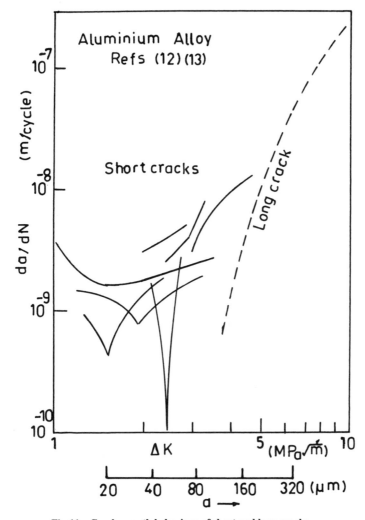

Fig 11   Crack growth behaviour of short and long cracks

will grow: otherwise there will be a retardation and sometimes non-propagation of the crack. Thus crack growth may be smoothly increasing (if all grains are favourably oriented) or zig-zag in character, with partial or complete arrest in some cases. Such behaviour is illustrated in Fig. 11, for an aluminium alloy; this data is taken from references (**12**) and (**13**). Thus the condition for non-propagation is that the resistance parameter is higher than the crack driving parameter and the length of such non-propagating cracks cannot be greater than a_I. This fact is also borne out in Fig. 11 where the crack lengths encountered are not more than 300 μm. Similarly an examination of the data reported in (**14**) indicated that the length of non-propagating cracks ranged from 50 μm to 250 μm.

For cracks starting from pre-existing notches or cracks the growth behaviour is very much determined by the stress and strain distribution ahead of the notch. A critical analysis on the prediction of non-propagating cracks for this condition, based on notch stresses, is given by El Haddad *et al.* (**15**). The plastic zone, w_{pn}, that forms ahead of the notch of length, $2D$, is a function of the stress concentration factor, K_t, and the root radius, ρ, of the notch and can be given by

$$w_{pn} = \rho f\left(\frac{K_t \sigma}{\sigma_{ys}}\right) \tag{14}$$

Whilst the crack is within the notch zone, it will propagate. Once it develops beyond the influence of the notch, its propagation depends on whether the crack driving parameter $\Delta\sigma\sqrt{\{\pi(D + a)\}}$ is greater than $4\sigma_{ys}\sqrt{d}$, i.e., the plastic zone size, w_{pc}, due to the total crack length $(D + a)$ needs to be larger than (or at least equal to) the grain size, d, for continued crack growth.

The influence of grain size on smooth surface cracks appears to be straightforward in that refining the grain size will, in general, increase the yield strength and the fatigue strength. However, in the case of notch generated cracks, as well as increasing the yield stress, the effect of grain refinement will also be to decrease the notch zone w_{pn} for a given notch, as indicated by equation (14). In this case, for a fine grain size to be advantageous requires the decrease in w_{pn} to be more than that in d, so that a crack cannot initiate from the tip of the notch.

Equation (14) also indicates that sharper notches will develop relatively smaller plastic zones. As a result it can be expected that under a given stress intensity condition the growth rate of cracks from sharp notches will be slower than that from relatively blunt notches. This trend can be seen from the experimental results, shown in Fig. 7. If the notch is very sharp and approaches a crack in shape, then the growth rate will be zero till the stress intensity is raised to the ΔK_{th} value. However, the growth rate curves for stage II cracks, whether from the notch or from the pre-existing crack, tend to merge, as can be seen in Fig. 7.

Concluding remarks

From the experimental investigations it is clear that for single phase materials and others for which a Hall–Petch type dependence on grain size is appropriate, increasing the yield stress will increase the endurance limit and decrease the threshold stress intensity factor. For a stage I crack generated from a smooth surface the resistance which the crack encounters as it grows can be given by the resistance parameter $\tau_{ys}\sqrt{d}$, where τ_{ys} is the yield strength of individual grains on the path of the crack. Since τ_{ys} of each grain depends on its orientation, the resistance will also change, as a result of which the crack growth may be smooth, or zig-zag, with partial or complete arrest in some cases. The maximum length of a non-propagating crack may not exceed more than 10–15 grain diameters. Increase in bulk yield strength will in turn increase the yield strength of the individual grains and, hence, the resistance to stage I crack growth will also increase.

In the case of cracks starting from notches, the growth rate will be relatively high when the crack lies inside the notch zone. Increase in yield stress will reduce the notch zone size and the crack growth rate inside the notch zone. Once the crack leaves the notch effect, its further propagation depends on the plastic zone size of the total crack under the applied stress range. Increasing the yield strength of the material thus appears to be beneficial to both types of stage I cracks.

Acknowledgement

The authors are thankful to the university authorities at the Technological University of Nagaoka, Japan, and to the Director, IIT, Madras, India, for their kind permission to publish this paper.

References

(1) MASOUNAVE, J. and BAILON, J. P. (1976) Effect of grain size on the threshold stress intensity factor in fatigue of a ferrite steel, *Scripta Met*, **10**, 165–170.
(2) USAMI, S. and SHIDA, S. (1979) Elastic–plastic analysis of fatigue limit for a material with small flaws, *Fatigue Engng Mater. Structures*, **1**, 471–481.
(3) YODER, G. R., COOLEY, L. A., and CROOKER, T. W. (1981) A critical analysis of grain size and yield strength dependence of near threshold fatigue crack growth in steels, NRC Memorandum, Report 4576, NRC, Washington DC.
(4) BENSON, J. P. and EDMONDS, D. V. (1978) Effect of microstructure on fatigue in threshold region in low alloy steel, *Metal Sci.*, **12**, 223–232.
(5) RITCHIE, R. O. (1977) Influence of microstructure on near threshold fatigue crack propagation in ultra high strength steel, *Metal Sci.*, **11**, 368–381.
(6) MILLER, K. J. (1982) The short crack problem, *Fatigue Engng Mater. Structures*, **5**, 223–232.
(7) MUTOH, Y. and RADHAKRISHNAN, V. M. (1986) Effect of yield stress and grain size on threshold and fatigue limit, *J. Engng Mater. Technol.*, in press.
(8) CAMERON, A. D. and SMITH, R. A. (1981) Upper and lower bounds for the lengths of non-propagating cracks, *Int. J. Fatigue*, **3**, 9–15.
(9) MUTOH, Y. AND RADHAKRISHNAN, V. M. (1981) An analysis of grain size and yield stress effects on stress at fatigue limit and threshold stress intensity factor, *J. Engng mater. Technol.*, **103**, 229–233.
(10) LIU, H. and LIU, D. (1984) A quantitative analysis of structure sensitive fatigue crack growth in steels, *Scripta Met.*, **18**, 7–12.
(11) TAYLOR, D. and KNOTT, J. F. (1981) Fatigue crack propagation behaviour of short cracks: The effect of microstructure, *Fatigue Engng mater. Structure*, **4**, 147–155.
(12) CHAN, K. S. and LANKFORD, J. (1983) A crack tip strain model for the growth of small cracks, *Scripta Met.*, **17**, 529–532.
(13) LANKFORD, J. (1982) The growth of small fatigue cracks in 7075-T6, *Fatigue Engng Mater. Structures*, **5**, 233–248.
(14) KUNIO, T. and YAMADA, K. (1979) Microstructural aspects of the threshold condition for non-propagating cracks in martensitic–ferritic structures, *ASTM STP 675*, pp. 342–360.
(15) EL HADDAD, M. H., TOPPER, T. H., and SMITH, K. N. (1979) Prediction of non-propagating cracks, *Engng Fracture Mech.*, **11**, 573–584.
(16) KITAGAWA, H. and TAKAHASHI, S. (1976) Applicability of fracture mechanics to very small cracks in the early stage, *Proc. ICM 2*, Boston, pp. 627–631.

R. K. Bolingbroke* and J. E. King*

A Comparison of Long and Short Fatigue Crack Growth in a High Strength Aluminium Alloy

REFERENCE   Bolingbroke, R. K. and King, J. E., **A Comparison of Long and Short Fatigue Crack Growth in a High Strength Aluminium Alloy**, *The Behaviour of Short Fatigue Cracks* (Edited by K. J. Miller and E. R. de los Rios) 1986, Mechanical Engineering Publications, London, pp. 101–114.

ABSTRACT   The behaviour of 'microstructurally short' fatigue cracks (20–100 μm in length) in the high strength aluminium alloy, 7010, has been investigated, in an under-aged and an over-aged condition, at the same strength level, to determine the influence of slip and precipitate distribution.

The results of the short crack propagation tests, at 20°C and $R = 0.1$, are compared with conventional (long) fatigue crack propagation and threshold results under the same conditions, and also with long crack tests carried out at constant maximum applied load but increasing mean stress.

In conventional long crack tests, better crack propagation resistance at low ΔK, and a higher threshold, is associated with the under-aged microstructure. The short cracks are found to propagate at significantly higher rates than the long cracks, at $R = 0.1$, below ΔK values of approximately 4 MPa$\sqrt{}$m. At ΔK levels above 7 MPa$\sqrt{}$m, and crack lengths of around 100 μm, the data from the short crack tests merge with the conventional da/dN vs ΔK data measured at $R = 0.1$. The under-aged microstructure again shows somewhat better behaviour in the short crack tests. When the short crack data are compared with high R ratio long crack results, from the constant maximum load tests, the average crack propagation rates are quite similar.

The differences between the long and short crack propagation behaviour at $R = 0.1$ are attributed partly to the absence of roughness-induced closure effects when the cracks are very short, and also to the effects of the high maximum applied stresses, in short crack tests, on the crack-tip plasticity. This argument is also used to explain the degree of correlation shown between short-crack behaviour at $R = 0.1$ and long crack growth rates at $R > 0.8$.

Introduction

There have been a large number of investigations into the fatigue behaviour of precipitation-hardened aluminium alloys. In general the results of these studies are in agreement in two important areas: those of crack initiation and near-threshold crack propagation.

(1)   *Crack initiation*. Peak-aged and over-aged alloys show better resistance to crack initiation in slip bands than under-aged alloys **(1)(2)**.

(2)   *Near-threshold crack propagation*. By contrast, under-aged micro-structures have better long crack propagation resistance at low stress intensities, and also higher threshold values (ΔK_{th}) than over-aged micro-structures **(1)–(6)**.

* Department of Metallurgy and Materials Science, University of Nottingham, University Park, Nottingham NG7 2RD, UK.

Both of these effects have been explained in terms of differences in slip distribution caused by the nature of the dislocation/precipitate interactions in under- and over-aged structures. In the under-aged condition, dislocations are assumed to cut precipitates producing local slip band softening. This makes cross-slip unlikely and so leads to the concentration of deformation into a small number of intense slip bands. In over-aged structures the necessity for dislocations to by-pass particles promotes cross-slip and, hence, a more homogeneous strain distribution.

When these effects occur in the surface of a material, leading up to crack initiation in a slip band, peak- and over-aged materials generate a large number of fine slip bands, whereas the intense slip bands formed in under-aged alloys give rise to large slip offsets at the surface, which in turn lead to easy crack initiation.

For the crack propagation behaviour it is suggested that slip occurring in the plastic zone at the crack tip is more reversible in under-aged structures (1)(3)(5) and this leads to slower crack propagation rates. Hornbogen and Zum Gahr (5) propose that, for a crack growing along a single slip band at the crack tip, the crack propagation rate should be proportional to the number of dislocations emitted along a slip plane during the loading half of the fatigue cycle which do not return along the same plane during the unloading half. The number of dislocations which contribute to crack advance in each cycle is thus directly related to heterogeneity of the strain, i.e., the ease of cross-slip.

Another factor which is also likely to be important in near-threshold crack propagation, especially where propagation along slip bands in involved, is roughness-induced crack closure (7)(8). If slip homogeneity affects the fracture profile, then it will affect the closure contribution (K_{cl}) in the measured ΔK_{th} values, so that differences in K_{cl} with ageing treatment may also provide part of the explanation for the higher threshold values in under-aged microstructures.

The behaviour of 'microstructurally short' (9) fatigue cracks is sometimes thought to be more closely related to crack initiation and fatigue or endurance limits, than to long crack propagation (10). The object of this work, therefore, was to investigate the effect of ageing treatment on short crack behaviour, a particularly interesting area because of the opposite effects of ageing condition on initiation and on long crack propagation resistance. The alloy used for the investigation was a high strength aluminium alloy, 7010, in an under-aged and an over-aged condition with the same yield strength.

Experimental

The material, aluminium alloy 7010, of the composition given in Table 1, was in the form of rolled plate. Fatigue specimens, in the form of square section bars, were cut with their top faces parallel to the surface of the plate, and their lengths aligned with the rolling direction.

The two heat treatments employed involved the following stages: (i) solution

Table 1 Composition (wt%)

Zn	Mg	Cu	Cr	Zr	Si	Fe	Mn	Al
6.2	2.5	1.7	0.05	0.14	0.07	0.11	0.10	balance

treat, 470°C 1 hr, water quench, (ii) pre-age, 90°C, 8 hrs, (iii) age: either 170°C 1 hr (UA), or 170°C 18 hrs (OA). The ageing times were chosen from an ageing curve of yield stress against time at 170°C, to give an under-aged (UA) and an over-aged (OA) condition with the same yield strength.

Long crack propagation tests were performed on single-edge-notch test pieces, 20 mm × 20 mm × 100 mm, containing central notches, 5 mm deep. The da/dN vs ΔK data were determined using a Mayes servo-controlled electro-hydraulic testing machine operating at 50 Hz, in air at 20°C. Specimens were loaded in four point bending and crack length was monitored continuously using a d.c. potential drop technique. Two types of threshold test were performed, conventional tests at a load ratio (R) of 0.1, in which threshold was approached by a load shedding technique, with load reductions of $\simeq 2$ per cent close to threshold, and constant maximum load tests. In this second type of test, ΔK was reduced towards threshold by increasing the minimum load whilst keeping the maximum load constant. The effect of this is to increase the R ratio as ΔK falls. For these tests, the initial loads were chosen such that threshold was reached at R values of about 0.8.

Unnotched bars, in the same orientation as the notched bars, were used for the short crack tests, with dimensions 12.5 mm × 12.5 mm × 70mm. The surface of these specimens were ground and electropolished, after heat treatment, to ensure a smooth, stress-free surface, and then lightly etched in Keller's reagent. The four point bend loading produced a region on the top surface of the specimen, 10 mm × 12.5 mm, which experienced a constant, maximum value of bending moment. The cracks initiated and grew within this region. Tests were run at a frequency of 10 Hz. (The short crack tests were run at a lower frequency than the threshold tests because of the high loads and short replication intervals. Running the threshold tests at 10 Hz would have produced unacceptably long test durations; however, very little effect of frequency between 10 and 50 Hz is anticipated at 20°C.) Crack growth was monitored using a replication technique (11), with replicas being taken every 4000 cycles. The maximum surface stress during these tests was 440 MPa (0.9 σ_y), with $R = 0.1$.

Stress intensity values for the short crack tests were calculated using a calibration for semi-elliptical cracks in pure bending (12) assuming a ratio of half surface crack length to crack depth of 0.85. This value was determined from measurement of crack shape for a number of cracks present at the end of tests on the smooth specimens. All the cracks measured were found to be of similar shape in both heat-treatment conditions.

Results

Yield strength

The two conditions, UA and OA, were produced with very similar yield strength levels, between 490 and 495 MPa.

Microstructure

The optical micrographs in Fig. 1(a) and (b) illustrate the pancake grain structure common to both ageing conditions. Grain size ranges in the three sections are: L 100 μm–1 mm, LT 60 μm–200 μm, and ST 10 μm–40 μm. The size of the $MgZn_2$ precipitates (η and η') after the two treatments can be seen in Fig. 2, along with occasional $ZrAl_3$ dispersoid particles.

Fatigue crack growth.

Figure 3 shows da/dN vs ΔK for the OA condition at $R = 0.1$. The additional crack depth scale applies to the 'short crack' results from the smooth specimen tests (solid triangles). The long crack data come down to a threshold value of about 2.7 MPa$\sqrt{m}$. The short crack points are for a single, naturally initiated crack. The dip in the crack growth curve for the short crack at ≈ 2 MPa$\sqrt{m}$ or ≈ 10 μm crack depth, indicated by an arrowed point, is a position at which the crack tip was held up at an obstacle over many cycles, thus giving a growth rate too low to be plotted within the axes of this figure. For crack depths greater than about 100 μm the long and short crack curves merge. Smaller cracks than this grow in an erratic manner, but at average rates above those of the long cracks, and at ΔK values below the long crack threshold at $R = 0.1$.

In Fig. 4 a comparison is made between the OA and the UA conditions. The UA material has the higher long crack threshold of 3.9 MPa$\sqrt{m}$, although growth rates for both conditions are very similar in the Paris regime, above $\Delta K \approx 6$ MPa$\sqrt{m}$. The short crack data for a single crack in the UA material shows the same type of behaviour as the OA, although average growth rates are slightly lower below ≈ 6 MPa$\sqrt{m}$. This is shown more clearly in Fig. 5, where the scatter bands for data from three cracks in OA specimens and two in UA tests are drawn, much of the data overlaps, but the OA material shows consistently higher maximum and average growth rates.

Figure 4 also shows the long crack thresholds from the constant maximum load tests. These thresholds are at R ratios of slightly above 0.8. The high R thresholds are reduced to 1.4 MPa$\sqrt{m}$ for OA and 2 MPa$\sqrt{m}$ for UA material.

Fractography

The fracture surface morphology in the near-threshold regime for long crack tests in the four combinations of ageing treatment and R ratio is shown in Fig. 6. The crack propagation is of a very similar faceted, crystallographic type in each case.

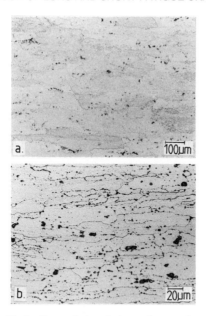

Fig 2   Transmission electron micrographs:
(a) under-aged
(b) over-aged

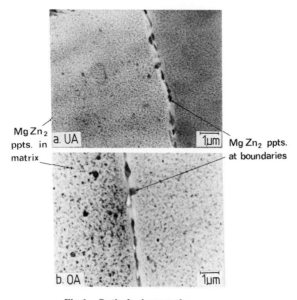

Fig 1   Optical micrographs:
(a) top surface
(b) short-transverse (S-T) section

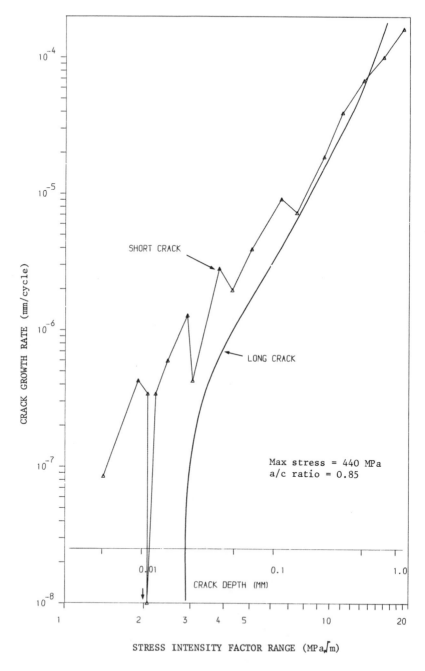

Fig 3 Crack growth rate characteristics. Long and short crack results for over-aged material, air, 20°C, $R = 0.1$

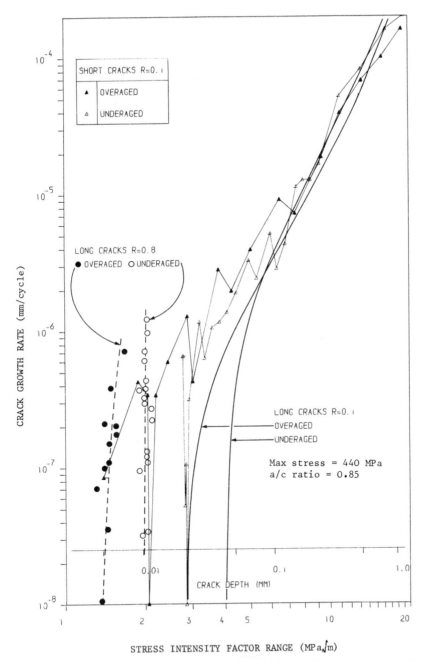

Fig 4   Crack growth rate characteristics in air at 20°C. Long cracks are shown by full lines
($R = 0.1$) and dashed lines ($R = 0.8$).
Short cracks are shown by triangular points

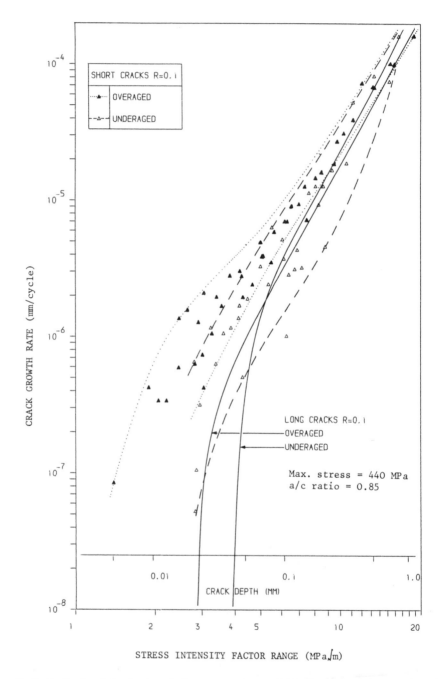

Fig 5   Scatter bands for short crack data representing results from several cracks: $R = 0.1$. Air: 20°C

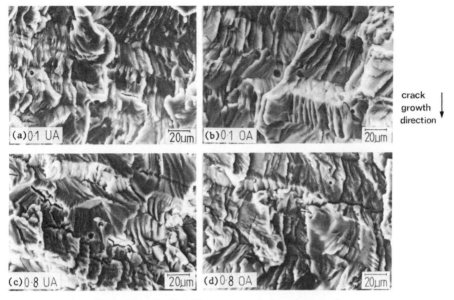

crack
growth
direction

Fig 6   Near-threshold fracture morphology, long cracks:
(a) $R = 0.1$ under-aged (b) $R = 0.1$ over-aged (c) $R = 0.8$ under-aged (d) $R = 0.8$ over-aged

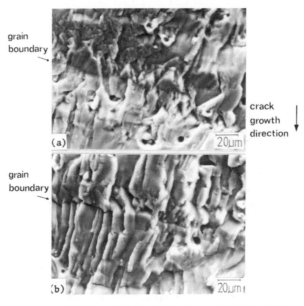

grain
boundary

crack
growth
direction

grain
boundary

Fig 7   Short crack fracture surfaces (a) $R = 0.1$ under-aged (b) $R = 0.1$ over-aged

Short crack fracture surfaces are illustrated in Fig. 7. Again, the OA and UA fracture paths are very similar, and closely resemble the near-threshold growth in Fig. 6.

Discussion

Behaviour of long, through-thickness cracks

At a load ratio of 0.1 the long crack behaviour shows the normal trend (1)–(6)(14) of higher thresholds and lower near-threshold crack propagation rates in UA material than in OA material.

The threshold values themselves, 3.9 MPa$\sqrt{}$m (UA) and 2.7 MPa$\sqrt{}$m (OA), are in good agreement with those from the work of Lankford (13) (about 3.8 MPa$\sqrt{}$m) and Suresh et al. (14) (3.7 MPa$\sqrt{}$m) for under-aged 7075, a similar high strength aluminium alloy, and Suresh et al.'s (14) value of 2.6 MPa$\sqrt{}$m for over-aged 7075.

The faceted nature of the near-threshold fracture surfaces in both conditions (Fig. 6) suggests that there will be a significant asymmetrical shear component in crack-tip deformation. This will introduce a degree of mismatch between the two fracture faces and so there will be a roughness-induced closure contribution to the threshold value at $R = 0.1$.

Long crack behaviour at high R will be free of roughness-induced closure effects, because of the larger crack openings involved (14)(15), so, since the mechanism of crack propagation is the same at $R = 0.1$ and $R = 0.8$ (Fig. 6), the R = 0.8 threshold values can be taken as intrinsic material thresholds. These will be equal to the effective crack tip stress intensity ranges (ΔK_{eff}) at threshold in the $R = 0.1$ tests. The measured values of 2 MPa$\sqrt{}$m (UA) and 1.4 MPa$\sqrt{}$m (OA) are again in good agreement with those of Suresh et al. (4) in 7075, of 1.7 MPa$\sqrt{}$m (UA) and 1.2 MPa$\sqrt{}$m (OA).

The superior near-threshold crack propagation resistance of the UA condition observed at $R = 0.1$ persists, in the absence of roughness induced closure, at $R = 0.8$. This suggests that it is due to a difference in the material's inherent resistance to fatigue, perhaps due to increased slip reversibility, rather than a mechanical effect such as a difference in closure behaviour. Suresh et al.'s (14) proposal that the improved resistance of UA material is due to increased crack deflection does not seem to apply in the present case where the fracture mode in both UA and OA material is very similar, although increased crack deflection might also be a slip reversibility effect.

Comparison of long and short crack behaviour

The propagation rates of short, semi-elliptical cracks in this alloy are greater than those of long, through-thickness cracks at equivalent linear elastic ΔK values and $R = 0.1$, despite a similar fracture mode (Figs 6 and 7). This observation is in agreement with the early work of Pearson (16) and more

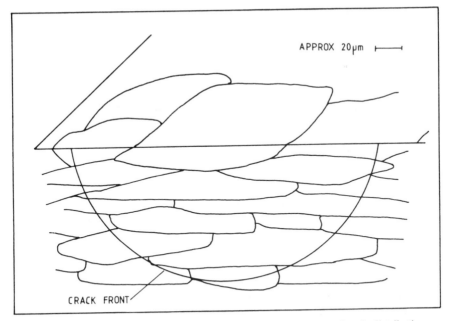

APPROX 20µm

CRACK FRONT

Fig 8   Schematic diagram showing relationship between crack front and grain distribution

recent studies of Morris and co-workers (**17**)(**18**) and of Lankford (**13**), on aluminium alloys.

There has been some discussion about the improper use of linear elastic ΔK to plot short crack data. It is used here, alongside a crack depth scale for the semi-elliptical cracks, for the important reason tht it shows when the cracks in the smooth specimen tests start to behave like long, through-thickness cracks. This occurs for crack depths greater than about 100 μm for both OA and UA conditions. The grain size in the S–T orientation (Fig. 1(b)) is between 10 and 40 μm, so at this stage the crack front will be sampling somewhere between 5 and 20 grains (Fig. 8) and an averaging of crack growth behaviour results.

Once the data from the two types of test have merged (Figs 3 and 4), the agreement between the growth rates form the semi-elliptical cracks and the through-thickness cracks is very good, indicating that, in the range of ΔK involved ($\approx$ 7 to 20 MPa$\sqrt{m}$) there is no crack shape effect on fatigue crack propagation behaviour, if an appropriate K calibration is used. This confirms the work of Pickard, Brown and Hicks (**19**) on crack shape effects in Ni-base and titanium alloys.

It has been suggested that some of the difference between the behaviour of long and short cracks at low ΔK is due to the absence of roughness-induced closure effects when the crack is only of the order of one or two grains in length (**9**)(**20**). It is therefore appropriate to compare short crack data with closure-free long crack results, i.e., ΔK_{eff} values. This comparison can be seen in Fig. 4, where long crack, high R thresholds are plotted in addition to long and short crack results at $R = 0.1$. Very few of the short crack data appear below the long crack threshold at $R = 0.8$, however, this may simply be due to the difficulty in obtaining short crack results below a ΔK of ≈ 1.5 MPa$\sqrt{m}$ as the inclusions at which the cracks initiated in the present tests were usually several microns in diameter. Further work at lower stress levels and longer test durations is needed to obtain such data. Nevertheless, when compared with closure-free long crack data, very little of the short crack growth seems anomalously fast, suggesting that the absence of roughness induced closure for short cracks does contribute to their high average growth rates.

Another reason for comparing short crack results with high R long crack data is the suggestion, put forward by Brown et al. (20) that the apparent inability of short cracks to close, even at low R, lies in the extent of crack tip plasticity generated by the high top surface stresses in the short crack test-pieces. They argue that a large maximum plastic zone size is produced, due to the high maximum stress, combined with small reverse plastic zone becaue of the lack of gross yielding on unloading. Thus, it is as if the crack is growing at a high value of R, compared with a simple linear elastic analysis. This should also lead to larger crack openings for short cracks than predicted from linear elastic fracture mechanics, as has been reported by Chan and Lankford (21).

The short crack in the smooth bend specimen is growing, effectively, at high R. In the Paris regime long crack data are relatively insensitive to R ratio, because of the declining importance of closure effects as symmetrical, continuum deformation takes over at the crack tip and the fracture surfaces become flatter. Thus the short crack data would be expected to merge with the $R = 0.1$ long crack data in the Paris regime, as the experimental results show. There may also be an additional effect in the smooth bend specimen. As the crack lengthens the effective value of R, at the deepest point of the crack, may decrease as the fibre stress in the specimen falls on moving away from the top surface. This would also cause the short crack data to move from the $R = 0.8$ long crack curve towards the $R = 0.1$ long crack results as crack depth increases. This might produce a slight difference in the crack lengths at which long and short crack data merge, depending on whether the short crack tests are carried out in bend or in tension.

Aspects of short crack growth

The irregular nature of the short crack growth appears to be closely associated with the grain size and, therefore, interactions between the crack tip and grain boundaries. This is seen clearly in Fig. 3 for the OA material, where dips in growth rate occur for the short crack at depths of ≈ 10, 30, and 50 μm, corresponding to the grain size of between 10 and 40 μm in the S–T orientation (Fig. 1(b)). The resulting changes in orientation of the fracture path at each grain boundary can be seen on the fracture surface in Fig. 7(b). This is in agreement with many previous observations (9)(13)(17)(18) and has been interpreted as being due to the difficulty of reinitiating slip in a new orientation in the next grain.

Such observations might lead one to expect that the crack shape for the short cracks would be related to the grain shape, such that cracks growing through long, thin pancake grains would be shallow cracks with large surface length to depth ratios. However, the cracks observed here were almost semicircular, over a wide range of crack lengths, as were those observed by Lankford in 7075 (13) at crack depths as small as 10 μm. The consistent crack shape, maintained from within the first grain through to the long crack regime where the crack front spans numerous grains, suggests that it is appropriate to use a single controlling parameter to describe crack propagation over the whole range of crack length.

The UA condition shows better short crack propagation resistance than the OA condition, in common with the long crack results and in contrast to reports of crack initiation resistance (1)(2). This reinforces the view that it is an inherent microstructural property, such as slip reversibility in the plastic zone, that causes the effect.

Conclusions

(1) Short fatigue cracks (depths <100 μm) in under- and over-aged 7010 propagate faster than long through-thickness cracks at the same apparent applied ΔK, and $R = 0.1$.

(2) Short crack growth rates at $R = 0.1$ are bounded by the long crack data measured at high R, over the range of crack length and ΔK it has been possible to obtain in these tests, but lie above the long crack growth rates at $R = 0.1$.

(3) Long and short fatigue cracks at low ΔK produce similar, faceted fracture surfaces, indicating that the same crack propagation mechanism is operative, despite differences in propagation rate.

(4) Under-aged 7010 shows significantly better crack propagation resistance than over-aged 7010 in the near threshold regime for long cracks. Higher threshold values are obtained in under-aged 7010, even when roughness-induced closure effects are eliminated. Short crack propagation resistance also appears to be slightly better in the under-aged condition, with consistently higher average and maximum growth rates in over-aged material in the short crack regime.

(5) The discontinuous nature of short crack growth is associated with the grain size of the material.

(6) There is no crack shape effect on fatigue crack propagation in under- and over-aged 7010 between ΔK values of 7 and 20 MPa$\sqrt{m}$. Semi-elliptical and through-thickness cracks propagate at the same rates in this range, with the same fracture mode.

Acknowledgements

The authors wish to thank Professor J. S. L. Leach for provision of laboratory facilities, Dr R. N. Wilson and Dr C. W. Brown for helpful discussions, and Mr K. Dinsdale and Mr F. Garlick for assistance with electron microscopy and mechanical testing. This work has been carried out with the support of the procurement Executive, Ministry of Defence.

References

(1) GYSLER, A., LINDIGKEIT, J., and LUTJERING, G. (1979) Correlation between microstructure and fatigue fracture, *Proceedings of ICSMA5*, Aachen, p. 1113.

(2) MUGHRABI, H. (1983) Deformation of multi-phase and particle-containing materials, *Proceedings of 4th Riso Int. Symp. on Met. and Mat. Sci.* Denmark, p. 65.

(3) LINDIGKEIT, J., GYSLER, A. and LUTJERING, G. (1981) The effect of microstructure on the fatigue crack propagation behaviour of an Al–Zn–Mg–Cu alloy, *Metal. Trans*, **12A**, 1613–1619.

(4) GARRETT, G. G. and KNOTT, J. F. (1975) Crystallographic fatigue crack growth in Al alloys, *Acta Met.*, **23**, 841–848.

(5) HORNBOGEN, E. and ZUM GAHR, K. H. (1976) Microstructure and fatigue crack growth in a γ-Fe–Ni aluminium alloy, *Acta Met.*, **24**, 581–592.

(6) ANTOLOVICH, S. D. and JAYARAMAN, N. (1980) Fatigue: environment and temperature effects, *Proceedings of 27th Sagamore Army Materials Research Conf.*, p. 119.

(7) WALKER, N. and BEEVERS, C. J. (1979) A fatigue crack closure mechanism in titanium, *Fatigue Engng Mater. Structures*, **1**, 135–148.

(8) RITCHIE, R. O. and SURESH, S. (1982) Some considerations on fatigue crack closure at threshold stress intensity due to fracture surface morphology, *Met. Trans*, **13A**, 937–940.

(9) RITCHIE, R. O. and SURESH, S. (1982) Behaviour of short cracks in airframe components, *Proceedings of 55th AGARD Struct. and Mats. Panel meeting*, Canada.

(10) MILLER, K. J. (1982) The short crack problem, *Fatigue Engng Mater. Structures*, **5**, 223–232.

(11) BROWN, C. W. and SMITH, G. C. (1982) A two stage plastic replication technique for monitoring fatigue crack initiation and early fatigue crack growth, *Advances in crack length measurement* (EMAS), p. 41.

(12) ROOKE, D. P. and CARTWRIGHT, D. J. (1976) *Compendium of stress intensity factors* (HMSO, London).

(13) LANKFORD, J. (1982) The growth of small fatigue cracks in 7075-T6 aluminium, *Fatigue Engng Mater. Structures*, **5**, 233–248.

(14) SURESH, S., VASUDEVAN, A. K., and BRETZ, P. E. (1984) Mechanisms of slow fatigue crack growth in aluminium alloys: The role of microstructure and environment, *Met. Trans.*, **15A**, 369–379.

(15) VENABLES, R. A., HICKS, M. A., and KING, J. E. (1984) Influence of stress ratio on fatigue threshold and structure sensitive crack growth in Ni base superalloys, *Fatigue crack growth threshold concepts* (ASM), p. 341.

(16) PEARSON, S. (1975) Initiation of fatigue cracks in commercial aluminium alloys and the subsequent propagation of very short cracks, *Engng Fracture Mech*, **7**, 235–247.

(17) MORRIS, W. L., JAMES, M. R., and BUCK, O. (1981) Growth rate models for short surface cracks in Al 2219–T851, *Met. Trans*, **12A**, 57–64.

(18) ZUREK, A. K., JAMES, M. R., and MORRIS, W. L. (1983) The effect of grain size on fatigue growth of short cracks, *Met. Trans*, **14A**, 1697–1705.

(19) PICKARD, A. C., BROWN, C. W., and HICKS, M. A. (1983) The development of advanced specimen testing and analysis techniques applied to fracture mechanics lifting of gas turbine components, *Advances in life prediction methods* (ASME), p. 173.

(20) BROWN, C. W., KING, J. E., and HICKS, M. A. (1984) The effects of microstructure on long and short fatigue crack growth in Ni-base superalloys, *Met. Sci.*, **18**, 374–380.

(21) CHAN, K. S. and LANKFORD, J. (1983) A crack tip strain model for the growth of small fatigue cracks, *Scripta Met.*, **17**, 529–532.

A. Fathulla,* B. Weiss,* and R. Stickler*

Short Fatigue Cracks in Technical pm-Mo Alloys

REFERENCE Fathulla, A., Weiss, B., and Stickler, R., **Short Fatigue Cracks in Technical pm-Mo Alloys**, *The Behaviour of Short Fatigue Cracks*, EGF Pub. 1 (Edited by K. J. Miller and E. R. de los Rios) 1986, Mechanical Engineering Publications, London, pp. 115–132.

ABSTRACT The nucleation and growth behaviour of fatigue microcracks in pm-Mo and pm-Mo-Ti-Zr alloys was investigated for room temperature cyclic loading at amplitudes slightly above the fatigue limit, and at a stress ratio of $R = -1$. In deformed Mo and recrystallized multiphase Mo-Ti-Zr-alloys the microcracks nucleated at intracrystalline sites and propagated in a transcrystalline manner. The advance of these cracks was considerably impeded by grain boundaries. In recrystallized pure Mo crack nucleation occurred in cyclic deformed regions in the vicinity of grain boundaries, followed by predominantly intercrystalline propagation. A transition from short-crack to long-crack growth behaviour – indicated by a pronounced change of the crack path and the extent of plastic deformation – could be noticed at a characteristic crack length independent of grain size. Based on these observations it is possible to differentiate several stages in fatigue crack growth behaviour, that is: (i) short-crack growth characterized by a marked reduction in growth rate when encountering microstructural features; (ii) a transition interval (with the lower bound related to the effective threshold stress intensity) over which with increasing crack length these interactions fade out while an increase in crack closure can be noticed; (iii) long crack growth which prevails beyond a certain crack length and which can be described by LEFM on the basis of the long-crack threshold stress intensity.

Introduction

The nucleation and growth behaviour of fatigue cracks is a field of considerable practical and theoretical significance. However, as pointed out in several recent surveys (e.g., (**1**)(**2**)(**3**)) discrepancies exist between the growth behaviour of the initially short fatigue cracks (SC), of length comparable with microstructural features, and of long fatigue cracks (LC) grown to macroscopic dimensions. Studies of nucleation and growth behaviour of SCs have been the subject of a considerable number of publications which revealed the complexity of the SC problem. One of the most striking features of SCs is their apparently irregular growth behaviour which may pose considerable problems for a conservative prediction of fatigue life, in particular under loading conditions near the fatigue limit (**1**)(**2**). Microcracks and SCs have been reported to grow at a much faster rate than LCs when subjected to cyclic loading at the same stress intensity range, in fact SCs may grow below the threshold stress intensity range of LCs. Consequently, several investigators have questioned the validity of LEFM concepts when applied to describe SC growth (**3**)(**4**).

The differences between SC and LC behaviour have been attributed to an inadequate continuum mechanics characterization of the crack tip stress and

* University of Vienna, Vienna, Austria.

strain fields associated with SCs, to effects of mixed mode crack propagation, to interactions with microstructural features, to differences in crack shape and extension mechanisms, and, in particular, to varying contributions of crack closure with increasing crack length (5)–(8).

In view of the lack of a general model for the description of SC behaviour, Schijve (2) has urged that information on SC propagation in technical alloys should be collected. The present investigation was planned in line with Schijve's stipulation to provide quantitative data on damage accumulation, microcrack nucleation, SC growth, and transition to LC growth for some technical alloys. Technically pure pm-Mo and multiphase pm-Mo-Ti-Zr alloys were selected as specimen materials because of their relatively low plasticity under room temperature test conditions, and their stable dislocation configuration and chemical stability in air at temperatures near ambient. These features should minimize some of the effects reported to influence the SC behaviour. Previous results on Mo materials (10)(11) have indicated significant effects of microstructure and pretreatment on SC growth.

For a more detailed investigation similar Mo-base materials with different microstructures were selected to study SC initiation and propagation at cyclic amplitudes slightly above the fatigue limit. Special emphasis was placed on the investigation of the transition from SC to LC growth behaviour. In addition, fatigue threshold and effective threshold values were determined. The measurements were carried out in a fatigue test system equipped with a high-resolution light microscope and video-recording systems for *in situ* observations of fatigue damage accumulation, SC propagation, and LC threshold measurements. In addition, SEM-replica techniques were used to provide information on details of the microcrack initiation processes.

Specimen material

Pure Mo and two Mo-Ti-Zr alloys produced by standard powder-metallurgical processing techniques were selected. The materials were provided either in the form of hot-rolled sheet (5 mm thickness) or as swaged bars (12 mm diameter), heat treated after hot-working to obtain either stress-relieved or recrystallized microstructures. Alloy designation, composition, and heat treatment of the specimen materials are listed in Table 1.

Information on microstructure, hardness, tensile properties, and values of the dynamic Young's modulus is compiled in Table 2. In the single phase pm-Mo, occasionally very fine particles could be detected. The Mo-Ti-Zr materials contained globular Zr-rich intracrystalline particles and a fine Ti-Mo rich precipitate along grain boundaries and on intracrystalline sites. The size of the globular particles was significantly different between the two Mo-Ti-Zr alloys.

Table 1 Material and specimen preparation data

Alloy designation	Shape	Chemical composition							Heat treatment (°C/h)	Remarks
		O (ppm)	N (ppm)	C (ppm)	Fe (ppm)	H_2 (ppm)	Ti (%)	Zr (%)		
Mo–SR	Bar (12 mm dia.)	22	5	5	20	5	—	—	Swaged and stress relieved (850/1)	This investigation
Mo–R	Bar (12 mm dia.)	20	5	8	22	5	—	—	Swaged and annealed (1400/2)	This investigation
Mo–Ti–Zr–1/R	Bar (12 mm dia.)	35	5	205	20	5	0.49	0.78	Swaged and annealed (1700/2)	This investigation
Mo–Ti–Zr–2/R	Sheet (5 mm thick)	215	5	310	22	5	0.47	0.077	Rolled and annealed (1800/1)	(10)

Table 2 Material properties

Alloy designation	Microstructure	Grain size (μm)	Tensile properties $\sigma_{0.2}$ (MPa)	σ_m (MPa)	R.A. (%)	Hardness HV_{10}	Dyn. Young's modulus (GPa)	Fatigue limit, σ_{FL} (MPa) for $N = 2 \times 10^8$, 20 kHz, 20°C
Mo–SR (bar)	Elongated grains with equiaxed sub-grains (10 μm dia.)	300 × 60	590	590	33	228	330.5	420
Mo–R (bar)	Equiaxed recrystallized	100	318	440	4	179	331.3	360
Mo–Ti–Zr–1/R (bar)	Equiaxed recrystallized*	40	390	562	19	193	319.0	350
Mo–Ti–Zr–2/R (sheet)	Elongated recrystallized†	45 / 20	863	914	1	305		Not determined

* Small globular intracrystalline particles (Zr-rich) and fine intracrystalline and grain boundary particles (Mo + Ti rich).
† Coarse globular intracrystalline particles (Zr-rich) and fine intracrystalline and grain boundary particles (Mo + Ti rich).

Experimental procedures

The room-temperature tensile properties were determined at a strain rate of 0.0016/s. Standard specimens with a gauge length of 20 mm and a diameter of 3 mm were used. The values of the dynamic Young's modulus were obtained at room temperature by a resonance test procedure (12).

The fatigue properties were evaluated in a resonance test system operated at a frequency of approximately 20 kHz at zero mean stress ($R = -1$) at room temperature. Details of this test method have been published (13). In the present investigation fatigue tests at constant total-strain amplitudes were carried out between 10^6 and 10^9 loading cycles (N). From the total-strain amplitudes (assuming negligible plastic strain) the cyclic stress amplitudes were calculated by use of the appropriate values of the dynamic Young's modulus.

The geometry of the test specimens machined from the bars is shown in Fig. 1, plate-shaped specimens have been machined from the sheet metals with a cross-section of 5 mm × 20 mm, as reported earlier (10). It was the objective of this study to use identical specimens for the SC and the LC growth investigations. Thus, specimens with a rectangular cross-section in the gauge length were prepared from the cylindrical bars. A mild notch was machined in the mid-section of the gauge length in order to limit the area of maximum fatigue damage to a size readily scanned by a travelling microscope. For all experiments the gauge section of specimens was electropolished in order to eliminate residual stresses and microstructural damage resulting from the machining operation. The stress concentration factor of the notch was calculated to be 1.63. For calibration measurements miniature strain gauges were applied to the root of the notch, the experimentally determined stress concentration factor was found to be 1.67, in good agreement with the calculated value. All cyclic amplitudes cited in this paper refer to the measured surface strain in the root of the shallow notch. For LC measurements semi-elliptical surface notches were introduced by electro-discharge machining, these specimens were also used for the determination of closure effects by a modified compliance test technique (14).

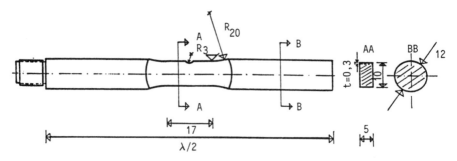

Fig 1 Test specimen for fatigue life and fatigue crack growth studies, Mo and Mo-alloys (dimensions in mm) ($\lambda/2$ = half wavelength)

The crack initiation and growth studies were performed in the same test system as described for the fatigue tests, equipped with a high-resolution travelling light microscope for the in situ observation of damage accumulation, microcrack nucleation, and fatigue crack growth up to a total length of approximately 1000 μm. The observations were recorded on video-tape synchronously with pertinent experimental information for a quantitative post-test evaluation on replay.

During the early stages of damage accumulation, of SC initiation and SC growth some specimens were subjected to a loading sequence consisting of $N = 10^4$ stress cycles (approximately 0.5 s) separated by short pauses (e.g., 1 s) during which the total notch region could be scanned and recorded by the light microscope/video system. In addition, when characteristic changes in surface topography could be noticed, plastic replicas were prepared of the entire notch surface to investigate with higher resolution the features of microcrack nucleation and initial growth. For the SEM examinations these replicas were Au-coated.

The fatigue crack growth behaviour was studied in the range of growth rates between the threshold (assumed to occur near 10^{-13} m/cycle) and 10^{-9} m/cycle for crack lengths from 2 to 1000 μm. In the present investigation the length (c) of a semi-elliptical crack on the specimen surface was measured with an accuracy of better than 2 μm. All growth experiments were carried out at cyclic stress amplitudes only slightly above the fatigue limit (experimentally determined for $N = 2 \times 10^8$) of the respective specimen materials, i.e., $\sigma_c/\sigma_{FL} = 1.07$. Crack growth ($dc/dN - c$) curves were constructed from the tangents to the measured $c - N$ curves at closely spaced points.

After completion of the crack growth experiments the electropolished region of the notch of all specimens was directly examined in the SEM. In some instances the electropolished regions were lightly etched after cyclic loading to reveal microstructure and damage zones. (Etchant: 15 g $K_3Fe(CN)_6$, 2 g NaOH dissolved in 100 ml H_2O). Subsequently some of the specimens were broken in tension to provide fracture surfaces for microfractographic evaluation and for the determination of the shape-factor of the semi-elliptical cracks at various stages of growth.

Results

Fatigue life and fatigue limit

The results of the fatigue experiments plotted as $S - N$ diagrams indicated the existence of a fatigue limit for N exceeding 2×10^8; the corresponding stress amplitudes are recorded as a fatigue failure limit in Table 2. The values of the stress amplitudes refer to the measured surface strain in the root of the shallow edge-notch of the specimens. Lack of sufficient specimen material did not permit an accurate determination of the fatigue limit of the sheet material (**10**).

THE BEHAVIOUR OF SHORT FATIGUE CRACKS

Short-crack initiation and growth

Light micrographs of characteristic stages during the in situ observations of damage accumulation in a stress-relieved Mo specimen (Mo-SR) are shown in Fig. 2. As a result of the cyclic loading 7 per cent above the fatigue limit black dots were observed within individual elongated grains early in fatigue life (Fig. 2(a)). SEM-studies revealed that these dots consisted of small extrusions and intrusions associated with the subgrain structure in the stress relieved matrix. Occasionally, small particles could be detected within these deformed regions. the density of these dots increased with N until a microcrack could be observed to nucleate in one of these markings (Fig. 2(b)). This microcrack increased in length and continued to grow as a characteristic SC in crystallographic directions. Interactions with subgrain boundaries and grain boundaries can be deduced from the changes in growth direction. In this specimen the initiation of only two cracks (crack 1 and crack 2) was observed. Crack 1 apparently became non-propagating while crack 2 continued to grow and eventually linked up with crack 1.

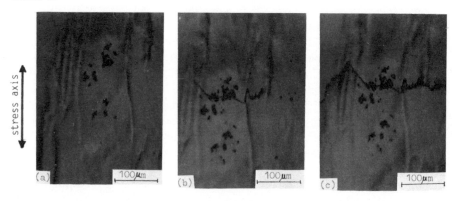

Fig 2   Damage accumulation, microcrack nucleation and SC growth in stress relieved Mo–SR, tested at $\sigma_c = 450$ MPa ($\sigma_c/\sigma_{FL} = 1.07$).
(a) $N = 6 \times 10^6$, first indication of damage, light micrograph (Point A in Fig. 3(a))
(b) $N = 5 \times 10^7$, short fatigue crack, light micrograph
(c) $N = 5, 8 \times 10^7$, section of the same fatigue crack grown to a length of approximately 350 μm, light micrograph

The growth of these two SCs is plotted in the $c - N$ diagram of Fig. 3(a). The step-wise growth is a result of the retarding effects of grain boundaries, the blocking action of grain boundaries can extend for a large number of loading cycles. For example, crack 1 was held up at a grain boundary for almost 4×10^7 cycles until the slowly growing crack 2 linked-up with it. The growth data are plotted in the $dc/dN - c$ diagram in Fig. 3(b), clearly revealing a pseudo-threshold behaviour of the SC advance. It can be seen that the step-wise growth rate extends for both cracks up to a length of approximately $c = 100$ μm. this step-wise growth behaviour is considered typical for SC.

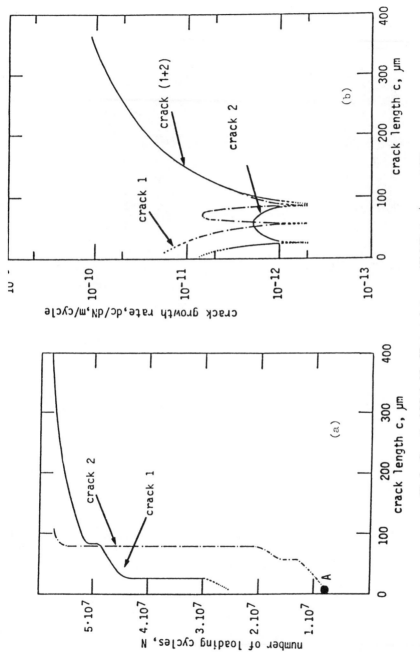

Fig 3   Growth of SCs in Mo–SR, tested at σ_c = 450 MPa (σ_c/σ_{FL} = 1.07).
(a) $c - N$ curves
(b) crack growth rate as function of crack length

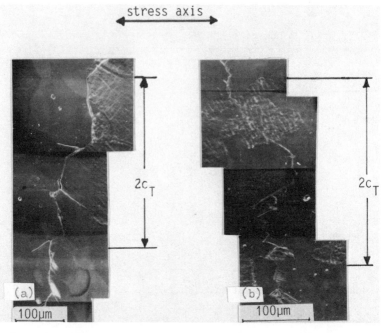

Fig 4   **Comparison of crack path in recrystallized Mo and Mo-Ti-Zr specimens, SEM-micrographs of specimen surfaces.**
(a)   **Mo–R tested at $\sigma_c = 385$ MPa ($\sigma_c/\sigma_{FL} = 1.07$) $N = 4.8 \times 10^7$ (see point A in Fig. 5)**
(b)   **Mo-Ti-Zr–1/R tested at $\sigma_c = 375$ MPa ($\sigma_c/\sigma_{FL} = 1.07$) $N = 2.8 \times 10^7$ (see Fig. 6 and point D in Fig. 7(a))**

Beyond a certain SC length the cracks assume a continuous growth behaviour characteristic of LCs. The transition from SC to LC can be deduced from micrographs such as shown in Fig. 4. The growth of the former is characterized by a crack advance along a single crystallographic plane through entire grains (typical for Stage I of fatigue crack growth); the latter exhibits an irregular transcrystalline crack path essentially normal to the stress axis (typical for Stage II of fatigue crack growth). The experiments revealed that the transition from SC to LC behaviour occurred at a characteristic crack length indicated in Fig. 4 as c_T and listed as 'transition crack length' in Table 4.

The accumulation of the fatigue damage in the recrystallized Mo (Mo–R) differs significantly from that observed in the stress-relieved material (Mo–SR). As shown in Fig. 4(a), isolated grains became gradually filled with slip markings, while neighbouring grains remained apparently unchanged. A fatigue crack nucleated along one of the boundaries between a deformed and an apparently undeformed grain; further crack growth followed, predominantly, grain boundaries without any resolvable interactions. At a crack length of approximately 300 μm indications of local plastic deformation could be seen in the vicinity of the advancing crack tip. Furthermore, a sizeable residual crack

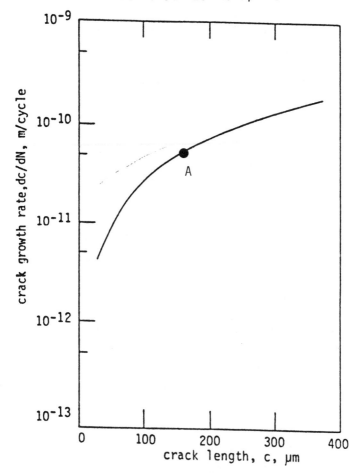

Fig 5 Crack growth rate as function of crack length for recrystallized Mo–R, tested at $\sigma_c = 385$ MPa ($\sigma_c/\sigma_{FL} = 1.07$)

opening became apparent in the unloaded specimen. The crack growth curve, Fig. 5, shows no decelerations.

In the recrystallized alloy Mo-Ti-Zr–1/R containing the finer dispersion of second phase particles the damage accumulation in the initial stages appeared similar to that in the recrystallized Mo specimen. Isolated surface grains become covered with slip markings (Fig. 6(a)). These markings, however, followed crystallographic directions and gave rise to 'embryo' microcracks (Fig. 6(b)), which joined up to form a characteristic SC (Fig. 6(c)). (The corresponding stages of these micrographs are marked on the crack growth curve in Fig. 7(a)). This SC continued to grow during further cyclic loading along single crystallographic directions through entire grains (Stage I). With increasing length the crack changed to a zig-zag crystallographic growth within

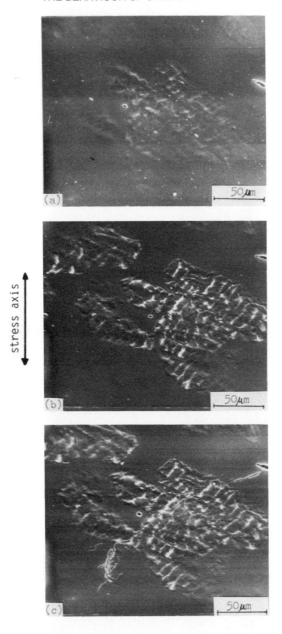

stress axis

Fig 6 Damage accumulation, microcrack nucleation and SC growth in recrystallized Mo-Ti-Zr–
1/R, tested at σ_c = 375 MPa (σ_c/σ_{FL} = 1.07) (a), (b), and (c) SEM micrographs of replicas; (d)
SEM micrograph of specimen surface.
(a) $N = 6 \times 10^6$ (point A in Fig. 7(a))
(b) $N = 2 \times 10^7$ (point B in Fig 7(a))
(c) $N = 2.3 \times 10^7$ (point C in Fig. 7(a))

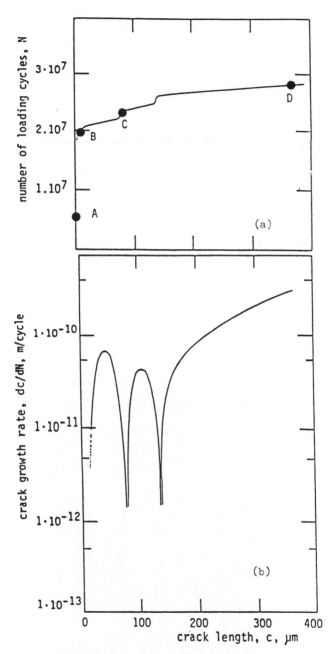

Fig 7   Growth behaviour of SCs in Mo-Ti-Zr–1/R, tested at $\sigma_c = 375$ MPa ($\sigma_c/\sigma_{FL} = 1.07$)
(points A to D refer to micrographs in Figs 6 and 4(b), respectively).
(a) crack length as function of number of loading cycles
(b) crack growth rate as function of crack length

grains ('crystallographic stage II'), and finally following a transcrystalline direction normal to the stress axis ('non-crystallographic stage II'), as shown in the composite micrographs in Fig. 4(b). Slip markings which indicate plastic deformation were observed at the transition to Stage II crack growth. This change from SC to LC behaviour occurred at a crack length of approximately 300 μm. The growth curve in Fig. 7(a) shows the retarding effects of grain boundaries during the SC growth stage, less pronounced, however, than in the stress relieved Mo, (Mo–SR). In addition, the SC in the Mo-Ti-Zr alloy exhibits higher maxima of growth rate between the transient crack arrests at grain boundaries (Fig. 7(b)), although in both cases the tests were carried out at the same relative stress level (σ_c/σ_{FL}). The SC growth behaviour in this Mo-Ti-Zr alloy is very similar to that reported previously (10) for alloy Mo-Ti-Zr–2/R which has a different microstructure and a significantly higher tensile strength. However, the nucleation of SCs in the latter alloy occurred at large globular Zr-rich particles (up to 20 μm diameter) (Fig. 8), in contrast to the initiation within slip lines observed in the alloy Mo-Ti-Zr–1/R devoid of such large particles.

To reveal the shapes of the cracks at various stages of growth some of the specimens were broken in tension after fatigue. Fractographic examinations

Fig 8   Fatigue damage and crack initiation at globular particle (Zr-rich) in Mo-Ti-Zr–2/R, SEM micrograph

Table 3   Mechanical properties

Alloy	Yield strength $\sigma_{0.2}$ (MPa)	Fatigue limit (N = 2×10^8) (MPa), ±5%	Threshold stress intensity (MPa$\sqrt{m}$), ±5%	Effective threshold stress intensity (MPa$\sqrt{m}$), ±10%
Mo–SR	590	420	10.3	7.0
Mo–R	318	360	8.5	6.6
Mo-Ti-Zr–1/R	390	350	10.1	6.3

showed that crystallographic microcracks are confined to individual surface grains, with a shape corresponding to the crack plane outlined by grain boundaries. Short cracks in the initial stages exhibited a roughly semi-circular crack front; on further SC growth the crack front changes to a semi-elliptical shape.

LC growth behaviour

The results of LC growth measurements are summarized in Table 3, in which the values of yield strength and fatigue limit are listed together with the measured threshold stress intensities and the effective threshold stress intensities calculated on the basis of the closure data.

Discussion and conclusions

The objective of this investigation was to present a detailed description of the microprocesses leading to fatigue failure. Specifically, quantitative data about the various aspects of nucleation and growth of 'short' cracks, under cyclic loading near the fatigue limit were to be determined for technical pm-Mo and multiphase Mo-Ti-Zr alloys. These alloys were selected because of their low plasticity at the test temperature, their stable dislocation substructure, and their chemical resistance to air environment near ambient temperatures, features which were thought to be essential for a minimization of the effects of plasticity at the crack tip and of corrosion products along the crack interfaces. In view of the inter-relationship between SC behaviour and the fatigue limit both properties were investigated by the use of identical specimen material, identical specimen geometry, and test procedures. The crack growth studies were carried out under cyclic loading only 7 per cent above the high-cycle fatigue failure stress limit, since earlier tests showed that in Mo alloys the fatigue limit constitutes the critical stress amplitude above which microcracks continue to grow. The experiments were designed to reveal details of the major events during the total fatigue life of a specimen, that is: (i) fatigue damage accumulation prior to microcrack formation; (ii) microcrack nucleation (length less than one grain diameter); (iii) 'short' crack development and propagation (up to a length of typically between 3 and 10 grain diameters); (iv) transition from 'short' to 'long' crack growth behaviour.

The test results reveal that under the chosen comparable test conditions the

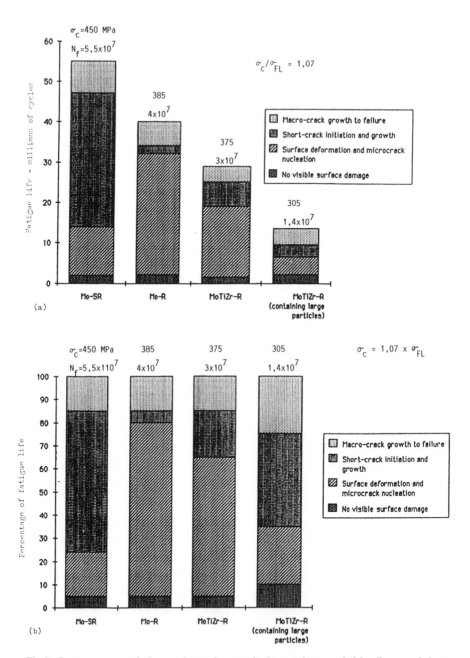

Fig 9 Damage accumulation and crack growth in specimens of Mo-alloys cycled at
$\sigma_c/\sigma_{FL} = 1.07$.
(a) plotted as function of the number of loading cycles, N
(b) plotted as function of normalized fatigue life, N/N_f

characteristics of fatigue damage (e.g., damage sites, nature of damage, nucleation sites of microcracks, growth paths of 'short' cracks) as well as the fractions of fatigue life related to the various phases of fatigue, vary considerably with composition, microstructure, second phase particles, and pre-treatment of the material.

The observations of fatigue damage accumulation and fatigue crack growth behaviour are summarized in Fig. 9. The relationship of surface deformation, SC initiation, and growth, as well as the transition to LC growth, are shown in the bar-graph of Fig. 9(a) as a function of the absolute number of loading cycles. In Fig. 9(b) the data are normalized by the number of cycles to failure. Both diagrams show that in all the materials, changes in surface topography could be noticed very early in fatigue life (i.e., 5–10 per cent of N_f), in spite of the low loading amplitudes. The numbers of cycles at which SC nucleation could be detected depend sensitively on the alloy and microstructure. It is possible to observe SCs very early in the stress relieved Mo; also a large fraction of fatigue life is spent on SC growth. In the recrystallized Mo, on the other hand, the deformation phase extends up to 80 per cent of the fatigue life, leaving only a short period to be divided amongst SC nucleation, SC growth, and the transition to LC growth.

The detrimental effect of large dispersoid particles is indicated by the early nucleation of SC in the high strength Mo-Ti-Zr alloy, while in a similar alloy without large particles a considerable amount of plastic deformation precedes SC formation. In all alloys investigated the period of LC growth leading to fatigue failure consumes only approximately 20 per cent of the life time, slightly more for the Mo-Ti-Zr alloy containing the large globular particles.

A significant aspect of 'short' crack behaviour is that we find a characteristic crack length associated with the transition from 'short' to 'long' crack growth. This transition is visible in a change of the crack path and/or in a marked increase in the plastic deformation associated with the advancing crack tip. The transition length appears to be a grain-size-independent parameter. The values of the measured transition length are summarized in Table 4 and compared to the grain size of the material. For the specimens with an elongated grain structure only the dimension in the growth direction (i.e., the width of the elongated grains) is considered. One interesting point is that the transition length, $2c_T$, corresponds to approximately three times the grain diameter for

Table 4   Data on transition lengths between short and long crack growth

Material	Grain size (μm)	$2c_T$ (μm)	$2c_T/grain\ dia.$	$\sigma_c\ (MPa)$	σ_c/σ_{FL}	Mode fracture
Mo–1/SR	60*	180	3	450	1.07	Transcrystalline
Mo–1/R	100	320	3.2	385	1.07	Grain boundary
Mo-Ti-Zr–1/R	40	328	10.9	375	1.07	Transcrystalline
Mo-Ti-Zr–2/R	20*	260	13	305		Transcrystalline

* In crack growth direction.

the Mo specimens, while for the two Mo-Ti-Zr alloys it exceeds 10 grain diameters, in spite of significant variations in microstructure, grain size, fatigue strength, and SC initiation.

As pointed out in previous publications (10)(11)(14) the transition length appears to be the characteristic critical upper bound of semi-elliptical SCs which do not affect the fatigue limit. Cracks longer than this transition length cause a gradual reduction in fatigue strength. With increasing length, crack growth occurs at consecutively lower stress amplitudes until crack advance coincides with the stress amplitudes predicted by LEFM analysis based on LC threshold considerations.

The transition length appears also to indicate the crack length beyond which a LEFM description of the crack growth process becomes applicable. However, in view of the microscopic nature of the cracks at the transition length we may assume only a negligible closure contribution; thus, we infer that the SC growth behaviour is characterized by the effective stress intensity value. A crack of length corresponding to the beginning of the SC regime should then resume growth at stress amplitudes computed from the effective threshold stress intensity. In fact, our results presented in a Kitagawa-type diagram show that the transition length corresponds closely to the crack length given by the intersection of the sloping line calculated for the value of the effective threshold stress intensity and the fatigue limit, as shown in Fig. 10. The differences between the numerical values in Table 3 and in Fig. 10 may be attributed to differences in the sensitivity of the various test methods. An interesting point realized by Fig. 10 is that the lines corresponding to the effective threshold stress intensities for pure-Mo and for the dilute Mo-Ti-Zr alloy are close together, while the fatigue limits as well as the LC threshold data differ. It follows that the values of the critical transition crack length vary to a lesser degree than the values of the crack length beyond which crack growth obeys the LC threshold relationship.

We may conclude that the growth behaviour of SCs beyond the transition length can be predicted by LEFM, provided the effective value of the threshold stress intensity is applied. Only for considerably longer cracks (in the order of 1 mm) can the growth behaviour be characterized by the conventional LC threshold stress intensity.

The observations show that the growth of SCs shorter than the transition length occurs by a glide-plane decohesion mechanism, details of which are not at present understood, but for which it may be assumed that conventional LEFM concepts are in principle not applicable. In this context the observations that residual crack opening can only be observed for cracks exceeding the transition length, at a length where measurable plastic deformation associated with the crack tip can be revealed, may be of significance.

The information accumulated in the present investigation is considered to have significant implications for advanced models of fatigue life prediction and engineering fail-safe design concepts. It should also be of relevance to guidelines for optimization of alloy processing and selection.

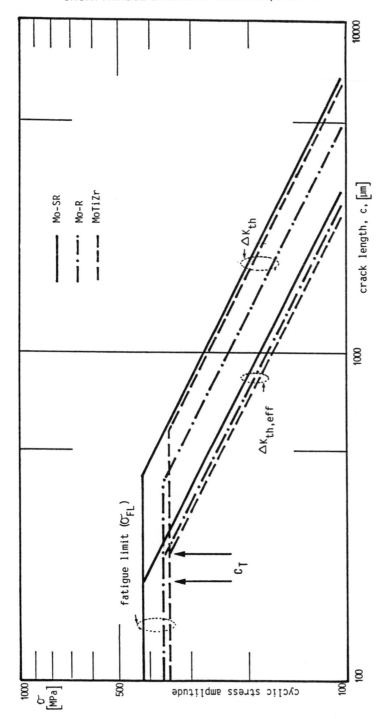

Fig 10 Effect of crack length on stress amplitude for crack growth in Mo-alloys

Acknowledgements

This investigation was supported in part by the Hochschuljubiläumsstiftung der Gemeinde Wien, Vienna, and the Metallwerk Plansee GmbH, Reutte. The authors thank in particular Mr J. Femböck of the Metallwerk Plansee GmbH for technical support of this program and for providing specimen materials with pertinent data.

References

(1) SURESH, S. and RITCHIE, R. O. (1984) Propagation of short fatigue cracks, *Int. Met. Rev.*, **29**.
(2) SCHIJVE, J. (1984) The practical and theoretical significance of small cracks, *Fatigue 84* (Edited by Beevers, J.), (EMAS, Warley), Vol. II, p. 751.
(3) MILLER, K. J. (1984) Initiation and growth rates of short fatigue cracks, *Proc. Eshelby Memorial Conf.*, Sheffield, IUTAM Conference (Cambridge University Press), pp. 473–500.
(4) KLESNIL, M., POLAK, J., and LISKUTA, P. (1984) Short crack growth close to the fatigue limit in LC-steel, *Scripta Met.*, **18**, 1231.
(5) HEUBAUM, F. and FINE, M. E. (1984) Short fatigue crack growth behavior in a HSLA-steel, *Scripta Met.*, **18**, 1235.
(6) McEVILY, A. J. and MINAKAWA, K. (1984) Crack closure and growth of short cracks and long cracks, *Scripta Met.*, **18**, 71.
(7) MINAKAWA, K. (1984) On the development of crack closure with crack advance in a steel, *Scripta Met.*, **18**, 1371.
(8) JAMES, M . N. and SMITH, G. C. (1984) Short crack behavior in steels, *Proc. ICF-6* (Edited by P. R. Rao *et al.*), New Delhi, p. 2117.
(9) RITCHIE, R. O. (1984) *Fatigue crack growth threshold concepts* (Edited by S. Suresh *et al.*), (AIME, New York), p. 555.
(10) FATHULLA, A., WEISS, B., and STICKLER, R. (1984) Initiation and propagation of short cracks, *Proc. International Spring Meeting, French Metals Society* (Edited by P. Rabbe), p. 182.
(11) FATHULLA, A., WEISS, B., STICKLER, R., and FEMBÖCK, J. (1984) The initiation and propagation of short cracks in pm-Mo and Mo-alloys, *Proc. 11th Plansee Seminar* (Edited by H. Ortner), p. 45.
(12) HESSLER, W. (1982) PhD thesis, University of Vienna.
(13) STICKLER, R. and WEISS, B. (1982) Review of the application of ultrasonic fatigue test methods for the determination of crack growth and threshold behaviour of metallic materials, *Proc. Inter. Conf. on Ultrasonic Fatigue* (Edited by J. Wells *et al.*) (AIME, New York), p. 135.
(14) BLOM, A. F., HEDLUND, A., ZHAO, W., FATHULLA, A., WEISS, B., and STICKLER, R. (1986) Short fatigue crack growth in Al 2024 and Al 7475, The Behaviour of Short Fatigue Cracks, EGF Pub. 1 (Edited by K. J. Miller and E. R. de los Rios), (Mechanical Engineering Publications, London), 37–65 (these proceedings).

F. Soniak* and L. Remy*

Fatigue Growth of Long and Short Cracks in a Powder Metallurgy Nickel Base Superalloy

REFERENCE   Soniak, F. and Remy, L., **Fatigue Growth of Long and Short Cracks in a Powder Metallurgy Nickel Base Superalloy**, *The Behaviour of Short Fatigue Cracks*, EGF Pub. 1 (Edited by K. J. Miller and E. R. de los Rios) 1986, Mechanical Engineering Publications, London, pp. 133–142.

ABSTRACT   Some aspects of the differences in fatigue crack growth behaviour and threshold data, for long and short cracks, are presented in this paper. Tests were carried out at room temperature on a H.I.P. + forged Astroloy, which has a necklace structure. For long cracks, closure is found to depend upon the stress ratio, but is independent of crack length and the maximum stress intensity factor. Short cracks were found to grow faster than long cracks. The influence of crack length on fatigue crack growth rate has been rationalized using the effective range of the stress intensity factor, and a unique intrinsic crack growth law is proposed for long and short cracks.

Introduction

Powder metallurgy superalloys are currently being developed for turbine disc applications in advanced turbine engines. However, defects are inherent to this manufacturing process and they can initiate short cracks under fatigue cycling. An accurate prediction of fatigue life is needed for aeroengine components. Generally, life calculations are made by the integration of the Paris law (1): $da/dN = C(\Delta K)^m$, from the largest size of initial defect to the crack length at final fracture. But the growth rate of short cracks (2)(3), and of long cracks at high stress ratios (4)(5) is faster than that of long cracks, say about 5 mm, at low stress ratios, which are generally used in laboratories to characterize fatigue crack growth behaviour. It follows that the use of the 'defect tolerant' approach with such laws leads to non-conservative predictions of the fatigue lives of components. This work reports on the influence of the stress ratio R and crack length upon fatigue crack growth rate in a powder metallurgy nickel base alloy, at room temperature. The aim of this work was to find a common parameter for all the crack configurations, which could rationalize any difference in fatigue crack growth behaviour and lead to conservative fatigue life predictions.

   Fatigue crack growth rate results are reported here for long cracks, using CT and SEN specimens, together with crack closure measurements for various stress ratios. Short crack growth rate data are then presented for artificial short through-cracks. Differences in behaviour between short and long cracks are discussed using crack closure measurements.

* Centre des Matériaux de l'Ecole des Mines de Paris, UA CNRS 866, B.P. 87–91003 Evry Cédex, France.

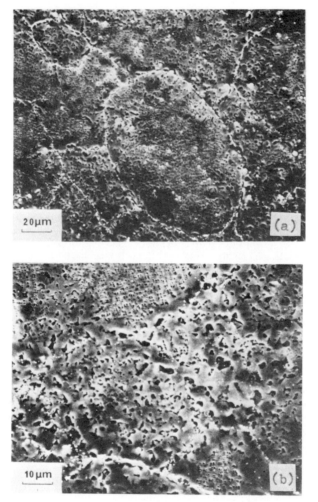

**Fig 1   Microstructure of the coarse grain (a) and of the necklace structure (b) of the H.I.P. +
forged Astroloy**

Material and experimental procedure

The H.I.P. + forged Astroloy studied here has a necklace structure which is
shown in Fig. 1, composed of coarse grains about 50 μm, and fine grains about
2 μm in diameter. The nominal percentage composition by weight is 0.022C,
0.003S, 14.8Cr, 5.04Mo, <0.01Cu, 16.9Co, 3.52Ti, 0.044Zr, 3.98Al, and
0.02B. The 0.2 per cent proof strength and ultimate tensile strength at room
temperature are 1110 and 1510 MN. m^{-2}, respectively.

Long and short fatigue crack propagation studies were carried out at room temperature in laboratory air. The fatigue tests were conducted on a servo-hydraulic fatigue testing machine using sine wave loading in the frequency range 30–60 Hz.

The growth rates of long fatigue cracks were measured on compact tension specimens 40 mm in width and 8 mm in thickness. A load shedding procedure was adopted for the threshold measurement on these specimens, using a 5–10 per cent decrease in load after a growth of 0.5 mm, down to about $5 . 10^{-11}$ m/cycle. Four stress ratios were investigated 0.1, 0.5, 0.7, and 0.9. The growth rate measured under the ΔK decreasing procedure were found to be in agreement with those measured under constant load at increasing fatigue crack growth rates.

Tests were also carried out down to the threshold under a stress ratio of 0.1 on single edge notched specimens 18 mm wide and 4 mm thick (Fig. 2(a)). Cracks were grown about 2 mm ahead of the notch in SEN specimens using the same load shedding procedure as in CT specimens. These SEN specimens were then machined down to about 12 mm in width and 2 mm in thickness (Fig. 2(b)). These machined SEN specimens contain a rectangular profile through-crack of 2mm in width and about 0.25–0.4 mm in depth. No heat treatment was given in order to prevent any modification of the crack surface. The plasticity left after all previous operations was kept to a minimum due to the threshold procedure adopted and the very cautious machining of specimens. The tests on these machined specimens were conducted at constant load, or using an increasing load procedure with steps of 5 per cent if crack growth did not occur after 10^6 cycles.

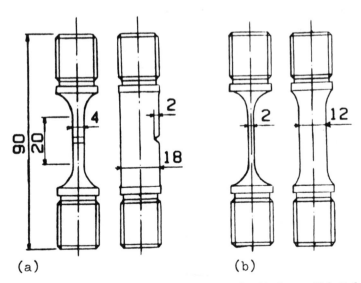

Fig 2 Single edged notched specimen geometry (in mm). (a) initially, (b) after machining to leave only a short crack

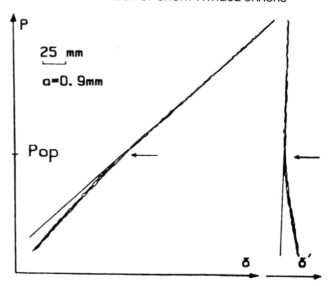

Fig 3 Experimental load–displacement curve

Crack growth was monitored using optical measurements on both sides of the specimen, also by a potential drop technique, and finally by a clip gauge extensometer located on the front part of the specimen. Load–displacement $(P-\delta)$ curves were recorded periodically at low frequency (0.5 Hz). The crack opening load P_{op} was determined by the upper break of the loading $P-\delta$ curve, which indicated that the crack was fully open (**6**). The opening load was more accurately measured using a corrected displacement $\delta' = \alpha P - \delta$ where α is an adjustable constant which was given by an electronic processor. One of the experimental curves is shown in Fig. 3 for a crack length of 0.9 mm. The value of P_{op} was quite difficult to assess but the evolution of the load displacement curve was so regular that the accuracy level was estimated to be about 10 per cent. The stress intensity factor at crack opening K_{op} was deduced from the opening load using the calibration formula for the relevant specimen geometry.

Experimental results

Long cracks

The results of the crack propagation tests using long cracks are reported in Fig. 4. The fatigue crack growth rates are reported on Fig. 4(a), as a function of the stress intensity range ΔK, for two specimen geometries (CT and SEN specimens), and four stress ratios (0.1, 0.5, 0.7, and 0.9). The threshold stress intensity range ΔK_{th} was conventionally defined for a crack growth rate of 10^{-10} m/cycle. The ΔK_{th} value decreases from about 10.5 MPa$\sqrt{m}$ to about 3 MPa$\sqrt{m}$ when the stress ratio increases from 0.1 to 0.9. The same fatigue crack growth data are reported in Fig. 4(b) versus the effective stress intensity range

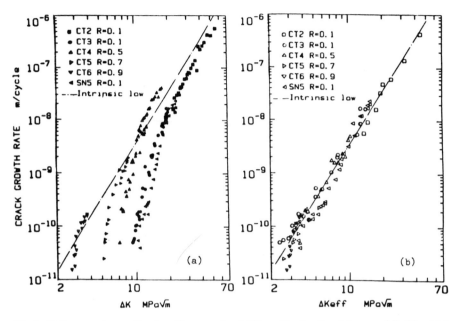

Fig 4   Fatigue crack growth rates of long cracks at different load ratios. (a) versus ΔK; (b) versus ΔK_{eff}

ΔK_{eff}, defined as $K_{\mathrm{max}} - K_{\mathrm{op}}$. All the results fit a single straight line within a small degree of scatter from the low rate region to the high rate region, i.e., in the range $5 . 10^{-11}$–10^{-6} m/cycle. The stress ratio dependency of FCGR has completely disappeared when using ΔK_{eff}, by accounting for the crack closure phenomenon. The value of the opening stress intensity factor K_{op} has been plotted versus the maximum stress intensity factor K_{max} and the normalized crack size a/W, respectively, in Fig. 5(a) and (b). For each stress-ratio, 0.1, 0.5, and 0.7 the K_{op} value seems to be a constant for these long cracks. By increasing the stress ratio from 0.1 to 0.7 the K_{op} value increases from 8.3 to 14.9 MPa $\sqrt{\mathrm{m}}$.

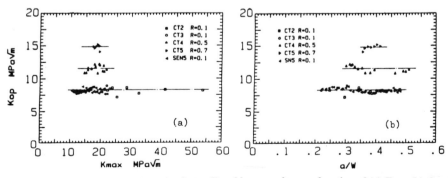

Fig 5   Crack opening stress intensity factor K_{op} of long cracks as a function of (a) K_{max}, (b) the normalized crack size, a/W

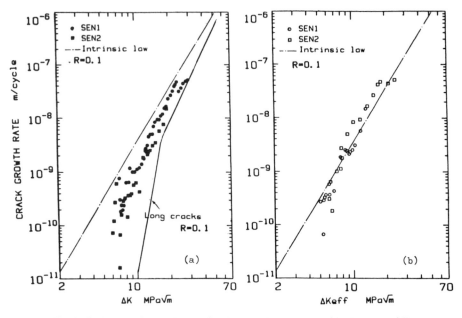

Fig 6   Fatigue crack growth rates for short cracks. (a) versus ΔK; (b) versus ΔK_{eff}

Short cracks

Fatigue crack growth rates of short cracks are reported as a function of the stress intensity range ΔK in Fig. 6(a) for a stress-ratio of 0.1. It can be seen that short cracks are able to propagate below the long crack threshold stress intensity and at a crack growth rate more than thirty times that of long cracks. The results are plotted versus the effective stress intensity range ΔK_{eff} in Fig. 6(b). As for long cracks all the results are along a single straight line which is similar to that for long cracks. The K_{op} values are reported versus ΔK and the crack size a, in Fig. 7(a) and (b) respectively. The value of K_{op} increases with crack size and merges with the long crack data at a value of 8.3 MPa$\sqrt{m}$ at crack sizes ranging from 2 to 3 mm. The K_{op} value has been found to increase also with increasing ΔK values up to a value of 20 MPa$\sqrt{m}$.

Discussion

From the present results it is clear that the range of the stress intensity factor alone cannot account for stress ratio and crack size effects on fatigue crack growth rate, as seen by previous authors (7)–(12). Many authors since Pearson (2) have studied the growth of natural cracks in smooth specimens. Assumptions must be made to calculate the fatigue crack growth rate and also the stress intensity range ΔK (13). For example, many studies use observations of cracks

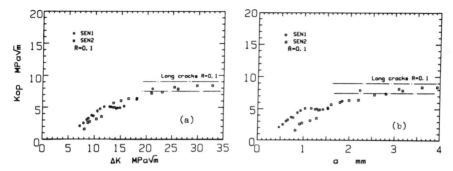

Fig 7   Crack opening stress intensity factor K_{op} of short cracks as a function of (a) ΔK, (b) the crack length, a

on the specimen surface and the assessment of crack shape and depth relies upon a calibration curve which is established from broken specimens. Such data are not easily generated for the smaller crack sizes, and the accuracy of the calibration curve becomes questionable. Natural cracks initiate generally at material inhomogeneities (10)–(14) such as inclusions, pores, and second phase particles. When the crack is much larger than the size of the initiation defect it has an equilibrium shape and the calibration curve should apply. However this becomes questionable at short crack lengths where defects are at or near the outer surface where cracks may not be of equilibrium shape (15). Furthermore short crack behaviour near the initiating defect can be influenced by local stress concentrations (at pores or inclusions) and residual stress patterns inherited from the manufacture process due to inclusions, second phase particles, etc. Finally in the whole range of crack sizes, natural cracks involve a three-dimensional fracture mechanics problem and the variation of the stress intensity factor along the crack front has to be taken into account.

On the other hand, the artificial short cracks as described here, may be accurately described as two dimension through-cracks. The assumptions made, concerning bidimensionality and a uniform stress intensity factor along the crack front are realistic in this case. The artificial short cracks, as used in the present work, are a convenient way to study the crack length dependency of fatigue crack growth behaviour, even though natural cracks are more representative of situations occurring in real components.

As seen in Fig. 5(b) the opening stress intensity factor K_{op} is equal to about 8.3 MPa√m for a stress ratio of 0.1 for long cracks (>3 mm) and for two specimens geometries (CT and SEN). In Fig. 7(b) the K_{op} value of short cracks increases from zero for crack sizes ranging from 0.2 to 0.5 mm, and merges with the long crack K_{op} value for crack sizes ranging from 2 to 3 mm. We estimated the length of the crack faces which 'close' during unloading to the minimum load, from measurements of the compliance change associated with closure. We took care of any underload effect and we unloaded specimens to the

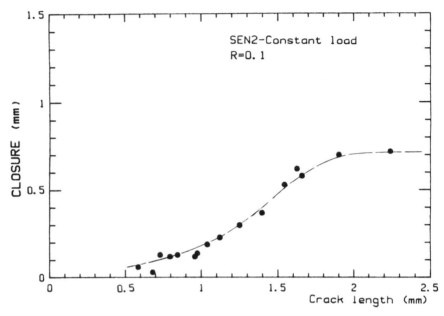

Fig 8   Length of closure in wake versus crack length

minimum load. So when the crack is closed, the measured compliance provides an apparent crack length from graphs of compliance against crack length. This has been done previously by James and Knott (**16**) and for cracks in this study (Fig. 8). In this figure the closed crack length in the wake is plotted against the crack length. It should be emphasized that these data refer to crack growth under increasing ΔK. As seen by James and Knott the length of the closure in the wake increases from about zero for a crack size of 0.5 mm and stabilizes at about 0.7 mm for a crack size of 1.5 mm when ΔK is increasing. As seen in Fig. 5(b), for long cracks the K_{op} value is constant whatever the crack size. For the CT specimens, the closure in the wake was observed to be identical to the real crack length ahead of the notch (**17**); this behaviour was observed for specimens tested under a load shedding procedure (i.e., at decreasing ΔK) as well as for specimens tested under constant load (i.e., at increasing ΔK). On the other hand, for the long cracks studied on SEN specimens (4 mm in thickness), the closure in the wake was always smaller than the crack length using a load shedding procedure. A constant K_{op} value was observed for long cracks for both CT and SEN specimens, using either a decreasing ΔK or an increasing ΔK procedure. However, this constant K_{op} behaviour corresponds to very different closure behaviours in the wake and so the opening stress intensity factor K_{op} is not simply connected to the closure in the wake for this material nor the slightly different closure stress intensity factor.

The crack length dependency of K_{op} and of the closure in the crack wake observed for an artificial short crack suggests that a minimum closure in the wake is required to establish a stabilized crack opening stress intensity factor. In the present case the length of material behind the crack tip responsible for closure would be about 0.5–1 mm. This conclusion is in pretty good agreement with the electro discharge machining experiments of Minakawa et al. (11) on CT specimens which showed that most of the closure behaviour was due to material 1 mm behind the crack tip.

The fatigue crack growth rate of long cracks and short cracks are plotted versus the effective stress intensity range in Figs 4(b) and 6(b), respectively. All the data fit onto a single straight line within a small degree of scatter. A single intrinsic law at room temperature for this H.I.P. + forged Astroloy is observed whatever the crack length ($R = 0.1$) or the stress ratio of the long crack. Previously we saw that the closure phenomenon is associated with closure in the crack wake but we saw also that the closure in the wake is not connected with the K_{op} value. Crack closure should be dependent upon surface roughness in the present alloy, but other mechanisms could also be operative ignoring any environmental influence.

Conclusions

(1)   The present investigation into the influence of stress ratio on long crack growth behaviour at room temperature in H.I.P. + forged Astroloy has shown that consideration of crack closure can normalize fatigue crack growth rate curves. This consideration leads to an intrinsic crack growth law of da/dN versus ΔK_{eff} for long cracks.

(2)   Differences in fatigue crack growth rates between long and short cracks, at the same stress ratio of 0.1, have been shown to be the consequence of the crack length dependency of crack closure behaviour at short crack lengths. A unique intrinsic crack growth law, da/dN versus ΔK_{eff}, has been observed for long and short cracks. This law should lead to conservative predictions of fatigue crack growth life.

Acknowledgements

The authors are indebted to Turboméca for provision of research facilities and the DRET for financial support.

References

(1)   PARIS, P. C., BURKE, J. J., REED, N. L., and WEISS, V. (1964) Fatigue, an interdisciplinary approach, *Proceedings of the 10th Sagamore Army Materials Research Conference* (Syracuse University Press), pp. 107–127.
(2)   PEARSON, S. (1975) Initiation of fatigue cracks in commercial aluminium alloys and the subsequent propagation of very short cracks, *Engng Fracture Mech.*, 7, 235–247.
(3)   LANKFORD, J. (1977) Initiation and early growth of fatigue cracks in high strength steel, *Engng Fracture Mech.*, 9, 617–624.

(4) KLESNIL, M. and LUKAS, P. (1972) Effect of stress cycle asymmetry on fatigue crack growth, *Mat. Sci. Engng*, **9**, 231–240.

(5) USAMI, S. (1982) Applications of threshold cyclic-plastic-zone-size criterion to some fatigue limit problem, *Fatigue Thresholds*, Edited by J. Bäcklund, A. F. Blom, and C. J. Beevers (EMAS), Vol. 1, pp. 205–238.

(6) ELBER, W. (1971) The significance of fatigue crack closure, *Damage tolerance in aircraft structures, ASTM STP 486* (American Society for Testing and Materials), pp. 230–242.

(7) KITAGAWA, H. (1982) Limitations in the applications of fatigue threshold ΔK_{th}, *Fatigue thresholds*, Edited by J. Bäcklund, A. F. Blom, and C. J. Beevers (EMAS), Vol. 2, pp. 1051–1068.

(8) NAKAI, Y., TANAKA, K., and NAKANISHI, T. (1981) The effects of stress ratio and grain size on near-threshold fatigue crack propagation in low carbon steel, *Engng Fracture Mech.*, **15**, 291–302.

(9) VENABLES, R. A., HICKS, M. A., and KING, J. E. (1984) Influence of stress ratio on fatigue thresholds and structure sensitive crack growth in Ni-base superalloys, *Fatigue crack growth threshold concepts*, Edited by D. Davidson and S. Suresh (T.M.S. AIME, Warrendale, Pennsylvania), pp. 341–357.

(10) SCHIJVE, J. (1982) Differences between the growth of small and large fatigue cracks in relation to threshold in *Fatigue thresholds*, Edited by J. Bäcklund, A. F. Blom, and C. J. Beevers (EMAS), Vol. 2, pp. 881–908.

(11) MINAKAWA, K., NEWMAN, J. C., Jr, and McEVILY, A. J. (1983) A critical study of the crack closure effect on near-threshold fatigue crack growth, *Fatigue Engng Mater. Structures*, **6**, 359–365.

(12) BREAT, J. L., MUDRY, F., and PINEAU, A. (1983) Short crack propagation and closure effects in A508 steel, *Fatigue Engng Mater. Structures*, **6**, 349–358.

(13) BROWN, C. W., KING, J. E., and HICKS, M. A. (1984) Effects of microstructure on long and short crack growth in nickel base superalloys, *Metal Sc.*, **18**, 374–380.

(14) CLEMENT, P., ANGELI, J. P., and PINEAU, A. (1984) Short crack behaviour in nodular cast iron, *Fatigue Engng Mater. Structures*, **7**, 251–265.

(15) FOTH, J., MARISSEN, R., NOWACK, M., and LUTJERING, G. (1984) A fracture mechanics based description of the propagation behaviour of small cracks at notches, *Proceedings of the 5th European Convention on Fracture*, Edited by L. Faria, pp. 135–144.

(16) JAMES, M. N., and KNOTT, J. F. (1985) An assessment of crack closure and the extent of the short crack regime in Q1N (HY80) steel, *Fatigue Fracture Engng Mater. Structures*, **8**, 177–191.

(17) SONIAK, F. and REMY, L. (1984) Centre des Matériaux, E.M.P., Evry, France, Unpublished results.

ENVIRONMENTAL EFFECTS

J. Mendez,* P. Violan,* and G. Gasc*

Initiation and Growth of Surface Microcracks in Polycrystalline Copper Cycled in Air and in Vacuum

REFERENCE Mendez, J., Violan, P., and Gasc, C. **Initiation and growth of surface microcracks in polycrystalline copper cycled in air and in vacuum**, *The Behaviour of Short Fatigue Cracks*, EGF Pub. 1(Edited by K. J. Miller and E. R. de los Rios) 1986, Mechanical Engineering Publications, London, pp. 145–161.

ABSTRACT Quantitative information has been obtained on the initiation and early growth of surface microcracks in polycrystalline copper fatigued in air and in vacuum at two different testing amplitudes (high cycle and low cycle fatigue). Histograms are presented that show the number of surface microcracks as a function of their length at different fractions of the fatigue life in air and in vacuum. Analysis of data such as microcrack density, mean crack length and major crack length, permit quantitative characterization of the effect of the atmospheric environment on the different stages of the fatigue failure process.

Notation

$\Delta\varepsilon_t$	Strain range
$\Delta\sigma$	Stress range $\sigma_{max} - \sigma_{min}$
ν	Test frequency
R	Load ratio $\sigma_{min}/\sigma_{max}$
N	Number of cycles
N_F	Number of cycles to failure
N_1	Number of cycles to the initiation of the first surface microcrack one grain boundary (g.b.) long
N_2	Number of cycles at which the first microcrack propagates out of its initiation site
δ	Surface microcrack density (number per mm^2)
δ_{max}	Surface microcrack density at failure
a	Surface crack length
$\bar{a}$	Mean surface crack length
a_{max}	Length of the major surface crack

* Laboratoire de Mécanique et Physique des Matériaux U.A. CNRS 863, 86034 Poitiers Cedex, France.

Introduction

It is well known from the early studies of Gough and Sopwith (1) that fatigue lives of most metals cycled in air at room temperature are considerably reduced when compared to tests performed in vacuum or in inert gases. However, it is still not clear how the gaseous environments affect the different stages in the process of fatigue failure.

For copper, Wadsworth and Hutchings (2) have shown that the reduction in fatigue life is essentially due to the effect of gaseous oxygen; water vapour increases the oxygen action to a certain degree, but it does not play a direct role in fatigue resistance. These results have been confirmed by Hunsche and Neumann in a more recent study (3).

The general opinion about gaseous environments is that they have little or no effect on crack nucleation and initial growth, but affect primarily crack propagation (2)(4)–(7). On the other hand, Thompson *et al.* (8) and Broom and Nicholson (9), for example, expressed the opinion that gaseous environments play a major role in the crack initiation stage.

Grinberg *et al.* (10) and Verkin and Grinberg (11) have related the influence of vacuum on fatigue failure to the fact that slip is more homogeneous in vacuum, which delays both the initiation and the propagation of fatigue cracks. However, more recently Wang *et al.* (12) and Mendez and Violan (13) have found that the surface slip features in air and in vacuum exhibit no significant differences when the comparisons are made at the same number of cycles, and that the homogeneous distribution of surface slip marks is only the result of the extended cycling in vacuum, and not its cause.

In the last ten years, by conducting numerous fatigue experiments on copper single crystals (see, for example (14)–(19)), significant progress has been made in the understanding of cyclic behaviour and of microcrack initiation processes in Persistent Slip Bands. Concerning the effect of environment, some authors have investigated the fatigue behaviour of copper single crystals in air and in high vacuum or ultra high vacuum (UHV). Wang and Mughrabi (12)(18)(20) have shown that, compared to air, fatigue life was 15 to 30 times greater in high vacuum. Up to the number of cycles to failure in air, the cyclic deformation behaviour in air and in vacuum were similar, however, continuing fatigue in vacuum led to a secondary cyclic hardening stage and to an homogeneous distribution of surface slips traces. Also the crack growth rates, particularly in Stage I, were observed to be lower in vacuum. Hunsche and Neumann (3), by conducting the fatigue tests in air or in oxygen, also demonstrated that the surface topography was not modified by the environmental reactions and that crack growth rates are lower in UHV than in air. Moreover these authors found evidence that the fatigue behaviour in vacuum was associated with rewelding.

On the other hand, only a few studies have been done on the effect of environment on polycrystalline copper. However, it is well known that intergranular crack initiation frequently occurs in low cycle fatigue tests (21) and sometimes in high cycle fatigue tests (22)(23).

The aim of the present work is to obtain quantitative information about initiation and growth of small surface intergranular microcracks in air and in vacuum in a fine-grained OFHC copper, in order to characterize the effect of the atmospheric environment on the fatigue damage process during different stages of the lifetime.

Experimental procedure

Fatigue specimens were machined from commercial polycrystalline OFHC copper bars of 22 mm diameter.

Two types of specimens were used: (a) specimens of square cross section with widths of about 6 mm and a gauge length of 16 mm (these specimens were used in total strain controlled tests); (b) specimens with cylindrical gauge lengths 6 mm long and 6 mm in diameter, which were used in load-controlled tests. The specimens were annealed under vacuum for 3 h at 460°C, giving a mean grain size of 30 μm; all the specimens were electropolished before annealing and once again just before the fatigue test. The tests were carried out in a total strain control mode at an amplitude of $\Delta\varepsilon_t/2 = 3.2 \times 10^{-3}$ and a frequency of $\nu = 0.1$ Hz, or in a load control mode using amplitudes varying between 86 and 115 MPa with a frequency $\nu = 37$ Hz and a load ratio $R = -0.95$. The tests were performed at room temperature in air or in a vacuum better than 10^{-3} Pa, using a servo-hydraulic testing machine equipped with a vacuum chamber.

Fatigue specimens were cycled up to failure, or with periodic interruptions to permit their examination by scanning electron microscopy (SEM). At each interruption a quantitative characterization of the fatigue damage was established by determining the number of surface microcracks per unit area (1 mm^2) and by making an estimation of their length.

The method used to evaluate the intergranular microcrack length in the case of fine grained specimens has been indicated elsewhere (24). The length of each intergranular microcrack has been estimated by counting the number of grain boundary segments related to the microcrack at the specimen surface. For specimens with a mean grain diameter of 30 μm, the mean length of the grain boundary segments is 0.012 mm. This method is illustrated in Fig. 1 where class 3 indicates three cracked boundaries.

From these measurements we have established histograms for each test condition (amplitude, environment) which give the number of microcracks, in each class of length, at various fractions of the fatigue life. The quantitative analysis of these histograms permits a characterization of the role of environment during the early stages of fatigue damage.

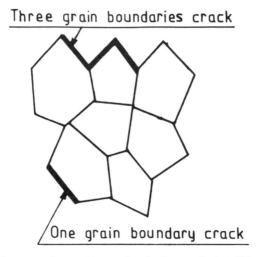

Three grain boundaries crack

One grain boundary crack

Fig 1 The fatigue damage at the specimen surface is characterized at different fractions of the fatigue life in air and in vacuum by the number of intergranular microcracks per square millimeter and by their length estimated as the number of cracked grain boundary segments

Results and discussion

Fatigue life and crack initiation mechanisms

The fatigue life, the nature of crack initiation sites, and the propagation modes in air and in vacuum for the fine-grained copper investigated here, are listed in Table 1 which classifies the three main test series previously described into class A, class B, and class C.

Table 1 Fatigue life, crack initiation sites, and propagation modes in air and in vacuum for different cyclic loading conditions

Test conditions		N_F cycles	Initiation site	Propagation mode	$\dfrac{N_{F,\text{vac}}}{N_{F,\text{air}}}$
Class A $\Delta\varepsilon_t/2 = 3.2 \times 10^{-3}$ $\nu = 0.1\,\text{Hz}$	Air	7.2×10^3	g.b.	Inter	
	Vac.	4.4×10^4	g.b.	Inter	6
Class B $\Delta\sigma/2 = 115\,\text{MPa}$ $R = -0.95$ $\nu = 37\,\text{Hz}$	Air	3.25×10^5	g.b.	Inter→Trans	
	Vac.	2.12×10^6	g.b.	Inter→Trans	6.5
Class C $\Delta\sigma/2 = 86\,\text{MPa}$ $R = -0.95$ $\nu = 37\,\text{Hz}$	Air	5×10^6	PSB	Trans	
	Vac.	53×10^6	g.b.	Inter (→Trans?)	>11 (≈ 20)

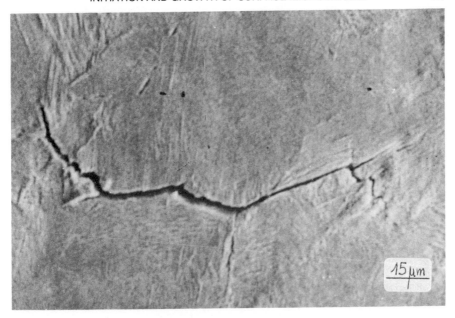

Fig 2   SEM photograph of a secondary surface crack on a specimen tested in air under load amplitude control, $\Delta\sigma/2$ = 115 MPa (test conditions B). Microcracks form early and grow at grain boundaries then start propagating transgranularly

The effect of the environment on fatigue life is significant even for high amplitudes, and especially so in the lowest stress amplitude range (class C) where the ratio $N_{F,vac}/N_{F,air}$ is much higher than 11. The test performed in vacuum at $\Delta\sigma/2$ = 86 MPa was stopped before failure, at 53×10^6 cycles; the examination of the free surface by SEM revealed a major crack only 10 grain boundary (g.b.) segments long ($a \approx 120~\mu$m), thus indicating an important residual life. From other results obtained in vacuum, at the same stress amplitude, but in coarse-grained specimens (d = 0.5 mm) (**25**), it can be assumed that in class C the residual life in vacuum can be about 30 to 50×10^6 cycles leading to $N_{F,vac}/N_{F,air} \approx 20$.

Let us consider now the mechanisms of crack initiation and propagation.

At the highest stress amplitude, class A tests, initiation and propagation of surface microcracks take place at grain boundaries in air as in vacuum. The deformation is uniformly distributed in the grains and few transgranular microcracks are initiated. At the intermediate stress level of 115 MPa, class B tests, microcracks also initiate and propagate early at grain boundaries, but in air this propagation changes to transgranular after a length of 4–6 g.b. segments. In this case, microcracks are also created in Persistent Slip Bands (PSB), in air as in vacuum, but they remain confined inside a grain and do not play any role in the initiation of the main cracks.

At the lowest stress level of 86 MPa, class C tests, crack initiation sites are different in air and in vacuum. In air our observations are in agreement with those of Thompson *et al.* (8) who have shown that, in high cycle fatigue, microcracks initiate in PSBs and propagate transgranularly. On the contrary we have shown that, in vacuum, the initiation and early growth stages always take place at grain boundaries, as for high cyclic loading amplitudes (13).

The objective of the present study was to determine the effect of the atmospheric environments on surface crack evolution and so it seemed appropriate to consider only the testing amplitudes leading to identical crack initiation and propagation mechanisms in air and in vacuum. To this effect the studies of microcrack surface features, for example, crack length and density, were only made for the first two test conditions, classes A and B. The interest of comparing quantitative data for these two different cyclic loading amplitudes, low cycle fatigue for the first tests and high cycle fatigue for the second tests, class B, lies in the fact that, in both cases, the same intergranular damage mechanisms are present, and, at the same time, the environmental effect quantified in terms of life ratios is of the same order. At 115 MPa, class B tests, the few transgranular microcracks initiated in air as in vacuum were not taken into account for our quantitative characterization of the fatigue damage. The test at 86 MPa will only be considered here to confirm the tendencies of the environment-induced surface features revealed by comparisons at the higher stress levels, classes A and B.

Histograms of microcrack lengths

Figure 3 shows the results of the measurements performed at different instants of fatigue life in air and in vacuum for the tests corresponding to the highest amplitude of cyclic strains. The histograms giving the number of intergranular microcracks per square millimeter in each class of length have been represented on a linear scale as a function of the number of cycles.

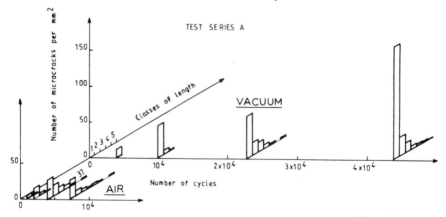

Fig 3 Histograms giving the number of intergranular microcracks per mm² in each class of length for different number of cycles in air and in vacuum. Test class A

The observations were performed using several specimens. First in vacuum a specimen was examined after $N = 1000$ cycles, but no microcrack was observed at this fraction of the fatigue life. The same specimen was cycled again until $N = 4000$ cycles, then $N = 10\,000$ cycles. A second specimen was cycled up to $N = 22\,700$ cycles and a third specimen was directly cycled up to failure ($N_{F,vac} = 43\,700$ cycles).

In air a first specimen was examined at $N = 1000$ cycles and at $N = 2000$ cycles. The second specimen was examined at 4000 cycles and the third completely cycled to failure ($N_{F,air} = 7200$ cycles) before being examined. The second specimen tested until $N = 4000$ cycles in air was continued in vacuum at the same amplitude, leading to a residual fatigue life in vacuum of 10 000 cycles. Therefore, the degree of fatigue damage in air at $N = 4000$ cycles is equivalent to that in vacuum at 34 000 cycles. This result is in agreement with the histograms of Fig. 3; in air at 4000 cycles the maximum crack length is already over 37 g.b. segments, whereas after 22 700 cycles in vacuum it is only seven g.b. segments long.

The histograms in Fig. 4 show the behaviour of small surface cracks in the case of intermediate stress levels, class B tests. These results have been obtained using only one specimen in vacuum and two specimens in air; the first one examined at 5×10^4 cycles then at 10^5 cycles whilst the second one was examined at 1.5×10^5 cycles then at 2.75×10^5 cycles.

A first analysis of Fig. 3 and Fig. 4 shows the following.

(1) the important accelerating effect of air on the processes of initiation and early growth of the small surface cracks. For example at the highest strain level tests, class A, at 10 000 cycles the largest crack in vacuum is only 3 g.b. segments long, whereas in air the specimen fractured at only 7200 cycles. At the intermediate stress level of 115 MPa, class B, the formation

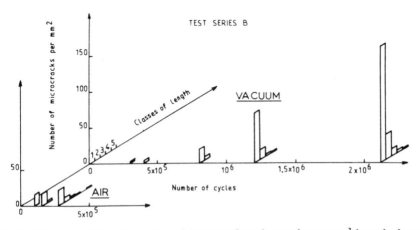

Fig 4   Histograms giving the number of intergranular microcracks per mm² in each class of length at different number of cycles in air and in vacuum. Test class B

of a small surface microcrack 2 g.b. segments long in vacuum requires a greater number of cycles than $N_{F,air}$.

(2) that the prolonged cycling in vacuum leads to the initiation of a much higher number of intergranular microcracks as compared to the fatigue behaviour in air (compare the heights of the histograms corresponding to the specimens cycled at failure in air and in vacuum).

(3) that for all the testing conditions, most of the surface microcracks remain small; their length on the fractured specimens being smaller than 5 or 6 g.b. segments.

Evolution of microcrack density and mean crack length

Figures 5 and 6 show the increase of the density δ of intergranular microcracks in air and in vacuum with cycling at the two highest stress–strain levels. The values of the different quantities calculated from Fig. 5 and 6 plots, are listed in Table 2.

The extrapolation of the curves δ–N to zero density gives an estimation of the number of cycles N_1 leading to the initiation of first surface microcracks: $N_{1,vac}$ is three times higher than $N_{1,air}$ in class A tests and about five times higher in class B tests.

From the curves δ–N the effect of the environment can also be characterized by differences obtained in the rate of initiation of new microcracks. In class A tests, Fig. 5 shows that δ increases linearly with the number of cycles in air, as in vacuum, with a slope, $\Delta\delta/\Delta N$, three times lower in vacuum than in air.

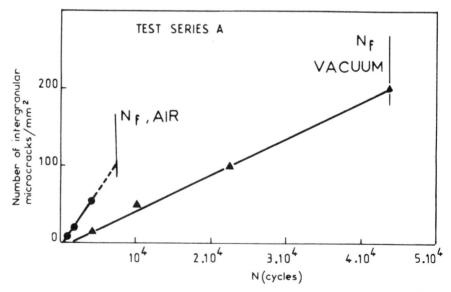

Fig 5   Evolution of the microcrack density as a function of the number of cycles in air and in vacuum for the test series A (high amplitudes)

Table 2   Number of cycles leading to the initiation of the first microcracks (N_1), rate of initiation of new microcracks, and maximum density

Test conditions		N_I cycles	N_I/N_F	$\Delta\delta/\Delta N$ microcracks per cycle	δ_{max} microcracks per mm^2
Class A $\Delta\varepsilon_t/2 = 3.2 \times 10^{-3}$	Air	500	0.07	1.5×10^{-2}	56
	Vac.	1400	0.03	4.8×10^3	200
Class B $\Delta\sigma/2 = 115\,\text{MPa}$	Air	5×10^4	0.15	1.3×10^{-4}	32
	Vac.	1.5×10^5	0.07	$\begin{cases} 0.45 \text{ then} \\ 1.5 \times 10^{-4} \end{cases}$	218

For class B tests in air a similar behaviour is found, but the results obtained in vacuum exhibit two distinct domains as indicated by two straight line plots. The slope $\Delta\delta/\Delta N$ is three times lower in vacuum than in air in the first part of the test, but reaches a value as high as in air in the latter part of the test.

Another important characteristic of the environmental effect is the difference between the maximum microcrack density values δ_{max} reached in air and in vacuum, both in low cycle and in high cycle fatigue ranges. One can see from Table 2 that δ_{max} is four times higher in vacuum than in air in class A tests and six times higher in class B tests. Moreover, the differences in δ_{max} because of the environment are still more striking for the lower stress levels of class C tests (see

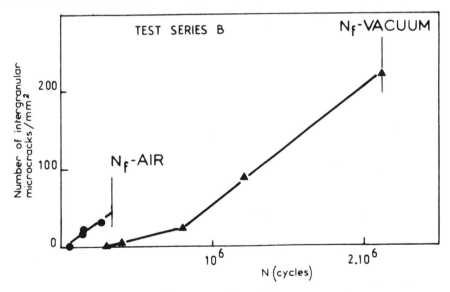

Fig 6   Evolution of the microcrack density as a function of the number of cycles in air and in vacuum for test series B (low amplitudes)

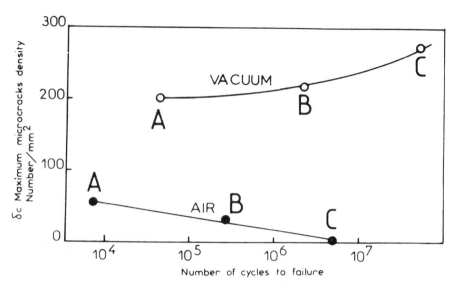

Fig 7   Plots of the maximum density of intergranular initiated microcracks in air and in vacuum as a function of the number of cycles to fracture for test series A, B, and C

Table 1). In this case we found a microcrack density of 270 microcracks per mm^2 in vacuum, higher than for class A and class B tests, although the measurements were made before the specimen fractured. In air, however, the microcrack density, in PSBs or grain boundaries, is very low, just a few microcracks per square millimeter. In Fig. 7, we have plotted δ_{max} against N_F in air and in vacuum from the curves of δ against N obtained for the test classes A, B, and C.

The evolution of δ_{max} observed in air is a similar result to that obtained by other authors on different materials (see, for instance, (**26**) and (**27**)). In air the fatigue damage process becomes more and more heterogeneous as the fatigue life increases. This behaviour is not observed in vacuum; on the contrary, Fig. 7 shows that in this environment δ_{max} tends to increase slightly with N_F. Thus it appears that in vacuum the coalescence of small cracks could play an important role in the formation of the main crack, whatever the cyclic loading amplitudes. This is not so in air, particularly at low stress amplitudes.

Figure 8 shows the evolution of the mean value of the crack length, $\bar{a}$, at the specimen surface, calculated from the measurements that produced histograms of the Figs 3 and 4. The extrapolation of the experimental curves $\bar{a}$ to the value $\bar{a} = 1$ g.b. segment gives an estimation of the number of cycles, N_2, beyond which $\bar{a}$ becomes higher than 1 g.b. segment; in other words it is considered that at N_2 the first microcracks start spreading out from their initiation sites into the neighbouring grains.

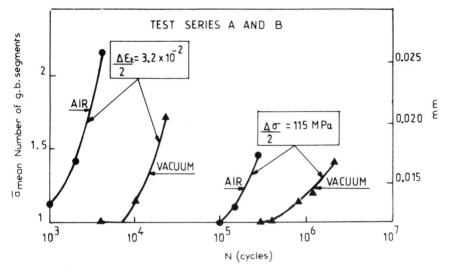

Fig 8 Evolution of the mean crack length $\bar{a}$ in air and in vacuum as a function of the number of cycles for test series A and B

The values of N_2 determined from the $\bar{a}$–N curves are given in Table 3. The effect of the environment on this stage of fatigue damage is more important for the high amplitude than for the low amplitude tests. Moreover, it is interesting to note that in the class B tests the effect of the environment is higher on N_2 than on the total fatigue life. However, in the class B tests our results are more in agreement with the general belief according to which the effect of environment concerns particularly crack propagation. The $\bar{a}$–N plots also clearly show that for all the test conditions of stress and environment, the majority of the intergranular cracks remain very small during the specimen life since the mean crack length stays lower than 30 μm.

Table 3 Number of cycles leading to the propagation of the first microcrack out of its initiation site (N_2)

Test classes	Environment	N_2 (cycles)	Ratio
A	Air Vac.	800 7000	8.75
B	Air Vac.	10^5 4.10^5	4

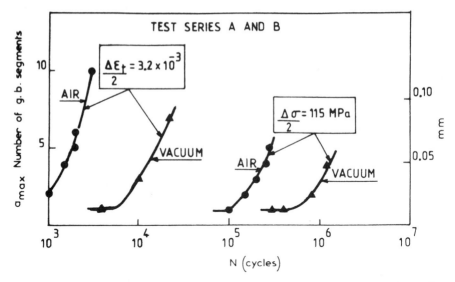

Fig 9 Evolution in the early stages of damage of the major crack length a_{max} versus the number of cycles in air and in vacuum. Test series A and B

The increase of the major crack length in the early stages of microcrack growth

The length of the largest surface crack observed at different number of cycles in air and in vacuum for the class A and B tests conditions has been plotted as a function of the number of cycles N in Fig. 9 and as a function of N/N_F in Fig. 10, with a linear scale. In these plots it has been taken into account that the crack only starts propagating at the number of cycles, N_2, previously defined. The fatigue behaviour investigated here concerns the very early stage of microcrack development in which the major crack length involves only a few boundary segments ($a_{max} \approx 100~\mu m$). In this range of small crack length the evolution of a_{max} versus N can be described by linear plots as it can be seen in Fig. 10.

An estimation of the growth rates at the specimen surface in this micro-propagation stage suggests the values given in Table 4. In this table are also

Table 4 Propagation rate in the first stages of microcrack development ($a_{max} \approx 100~\mu m$) and number of cycles of propagation in air and in vacuum

Test classes	Environment	da/dN (mm/cycle)	Ratio	$N_F - N_2$ (cycles)	Ratio
A	Air Vac.	$3.50~10^{-5}$ $5.45~10^{-6}$	6.4	6400 37000	5.8
B	Air Vac.	$2.40~10^{-7}$ $4.14~10^{-8}$	5.8	225000 1720000	7.6

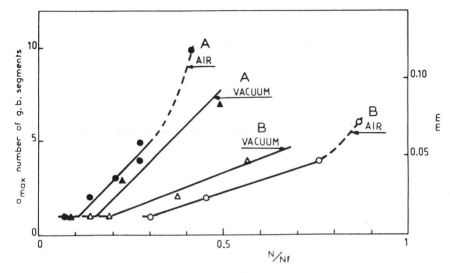

Fig 10   Evolution of the major crack length a_{max} plotted versus the fraction of the fatigue life. Test series A and B

listed the number of cycles ($N_F - N_2$) corresponding to the duration of the whole propagation stage. Our results on the effect of environment on the growth rates of very small surface cracks are in opposition to the usual opinion, according to which the lower the propagation rate, the higher the environmental effect. Indeed, it must be noted that the rates measured in class A tests are more than 100 times higher than in class B tests.

On the other hand, from the comparisons between the curves in Fig. 10 giving a_{max} as a function of the fraction of lifetime, and from the values of N_2 reported in Table 3, the following features were noted.

(1)   In air, the well known behaviour according to which crack initiation occurs earlier in life for low cycle than for high cycle fatigue is verified. The first microcracks start propagating at $N_2/N_F = 0.11$ during class A tests and at $N_2/N_f = 0.31$ for class B tests. Moreover, the early stages of the major crack development take a small fraction of the fatigue life in class A tests, but a very significant fraction of the fatigue life in class B tests. For example, for a major crack length of 4 g.b. segments, $N/N_F = 0.25$ for class A tests but 0.8 for class B tests.

(2)   In vacuum, a similar behaviour is found; however, from one test to another, the differences in the fatigue life fractions occupied by the initiation and the early development of the microcracks are less marked than in air. The first microcracks start propagating at $N_2/N_F = 0.16$ for class A tests and $N_2/N_F = 0.19$ for class B tests, and for a major crack

length of 4 g.b. segments the fatigue life fractions are, respectively, 0.3 and 0.6.

(3)　The relative positions of the air and vacuum curves are different for the test classes A and B. For the class A series (Low Cycle Fatigue) Fig. 10 shows that the curve a_{max} in air is at lower values of N/N_F than that for vacuum, while for the class B tests, the curve for air is at higher values of N/N_F.

From points (1) and (2) above one can see that the crack initiation occurs early in the fatigue life for high amplitude tests. That this is verified in vacuum as well as in air clearly shows that this behaviour is characteristic of low cycle fatigue, whatever the environment. However, the opinion that the crack initiation stage takes a larger fraction of the fatigue life for low amplitude than for high amplitude fatigue tests is verified in air more than in vacuum. Therefore this behaviour appears typical of environment-affected fatigue damage processes. In vacuum the fraction of the fatigue life associated with propagation remains important in high cycle fatigue tests.

The third point concerning the relative position of the curves a_{max}–(N/N_F) in air and in vacuum, see Fig. 10, shows that in low cycle fatigue the effect of the environment is important for the initiation and early growth of very small cracks ($a \approx 100$, μm) compared to the later stages of fatigue damage. However, in the case of high cycle fatigue, the environmental effect appears to be less in the early stages of fatigue damage than in the subsequent stages.

Let us summarize now the effect of the environment on the different stages of fatigue damage considered in this study. Concerning the class A tests, the effect of environment is especially marked on the stage characterized by the number of cycles $(N_2 - N_1)$ which leads from the formation of the first surface microcracks one grain boundary long (N_1) to the beginning of the stage of propagation (N_2); we found a ratio of 18 between the tests in vacuum and in air. On the contrary, the effect on N_1 is weak, only a ratio of 3. We have also found that the environmental effect is greater on the micropropagation stage, leading to a surface microcrack of a few grain boundary segments long, than in the later stages of propagation, see Table 4. Thus it is easy to understand that the environmental effect is less during the later crack propagation stages where the crack growth rates are higher.

More surprising are the results in class B tests (low amplitudes) where the effects of the environment become higher in the propagation stages following the formation of a microcrack 5 or 6 g.b. segments long: indeed, we have found a ratio of 5 between $(N_2 - N_1)$ values in vacuum and in air and a ratio of 5.8 between the characteristics of the micropropagation stage. Note that the global effect on fatigue life is 6, that the effect during all the propagation stage $(N_F - N_2)$ is 7.6, but that the effect on propagation in the stage leading from a microcrack of 4 g.b. segments long to specimen failure is as high as 13. The difference in this case, compared to the test series A, is the occurrence of a transition from an intergranular crack propagation mode to a transgranular one at microcrack lengths of about 5–6 g.b. segments. Thus, our results tend to prove that the effect of environment is more marked on transgranular propagation than on intergranular propagation. This behaviour is in good agreement with previous work by Violan *et al.* (**28**) on polycrystalline OFHC copper which gave a ratio of 17 between the mean crack growth rates in air and in vacuum.

Another result which appears very surprising at first sight is that the influence of the environment during the first stages of fatigue damage is higher for the high amplitude levels than for the lower amplitudes, although for both cases this stage corresponds to the same microcrack length (up to 5 or 6 g.b. segments). However, in interpreting these results, one must take into account the differences in testing frequencies (370 times lower in test series A). It is well known that the effect of an aggressive environment on fatigue crack propagation increases when the test frequency decreases (see, for example, (**29**) and (**30**).

Conclusions

The initiation and early growth of intergranular surface fatigue microcracks in air and in vacuum have been studied in a fine-grained OFHC copper under different loading conditions which leads to the following main conclusions.

(1) Prolonged cycling in vacuum, in both high cycle and low cycle fatigue, leads to the initiation of a greater number of surface microcracks than in air.
(2) Atmospheric environment has a significant effect on the initiation period leading to the formation of a microcrack one grain boundary segment long. However, this effect is less than the environmental effect on the total fatigue life.
(3) In the case of high cyclic amplitudes, environment mainly affects the subsequent fatigue damage stages, until the formation of a microcrack about 100 μm long. This behaviour appears to be related to the low frequencies used during the present tests.
(4) In low amplitude fatigue tests the effect of environment appears to be much more marked in the crack propagation stage following the formation of intergranular microcracks 5 or 6 grain boundary segments long. This behaviour is associated with a change in crack propagation mode from intergranular to transgranular.

References

(1) GOUGH, H. J. and SOPWITH, D. C. (1932) Atmospheric action as a factor in fatigue of metals, *J. Inst. Metals*, **49**, 92–112.
(2) WADSWORTH, N. J. and HUTCHINGS, J. (1958) The effect of atmospheric corrosion on metal fatigue, *Phil. Mag.*, **3**, 1154–1166.
(3) HUNSCHE, A. and NEUMANN, P. (1984) Crack nucleation in persistent slip bands,

ASTM Symposium *Fundamental Questions and Critical Experiments on Fatigue*, Dallas, October, 1984, to be published as ASTM Special Technical Publication.

(4) LAIRD, C. and SMITH, G. C. (1963) Initial stages of damage in high stress fatigue in some pure metals, *Phil. Mag.*, **8**, 1945–1963.

(5) ACHTER, M. R. (1967) Effect of environment on fatigue cracks, *Fatigue crack Propagation, ASTM STP 415* (American Society for Testing and Materials, Philadelphia), pp. 181–204.

(6) LAIRD, C. and DUQUETTE, D. J. (1972) Mechanisms of fatigue crack nucleation, *Corros. Fatigue*, **NACE-2**, 88–117.

(7) DUQUETTE, D. J. (1979) Environmental Effects I: General fatigue resistance and crack nucleation in metals and alloys, *Fatigue and microstructure* (American Society for Metals), pp. 335–363.

(8) THOMPSON, N., WADSWORTH, N. J., and LOUAT, N. (1956) The origin of fatigue fracture in copper, *Phil. Mag.*, **1**, 113–126.

(9) BROOM, T. and NICHOLSON, A. (1961) Atmospheric corrosion-fatigue of age-hardened aluminium alloys, *J. Inst. Metals*, **89**, 183–190.

(10) GRINBERG, N. M., ALEKSEYEV, A. I., and LYUBARSKI, I. M. (1972) Influence of vacuum on the various stages in the fatigue failure of copper, *Fiz. Metal. Metalloved*, **34**, 1259–1263.

(11) VERKIN, B. I. and GRINBERG, N. M. (1979) The effect of vacuum on the fatigue behaviour of metals and alloys, *Mater. Sci. Engng*, **41**, 149–181.

(12) WANG, R., MUGHRABI, H., McGOVERN, S., and RAPP, M. (1984) Fatigue of copper single crystals in vacuum and in air I: Persistent Slip Bands and dislocation microstructure, *Mater. Sci. Engng*, **65**, 219–233.

(13) MENDEZ, J. and VIOLAN, P. (1984) Modifications in fatigue damage processes induced by atmospheric environment in polycrystalline copper, ASTM Symposium *Fundamental questions and critical experiments on fatigue*, Dallas, October 1984, to be published as ASTM Special Technical publication.

(14) WINTER, A. T. (1974) A model for the fatigue of copper at low plastic strain amplitudes, *Phil. Mag.*, **30**, 719–738.

(15) MUGHRABI, H. (1978) The cyclic hardening and saturation behaviour of copper single crystals, *Mater. Sci. Engng*, **33**, 207–223.

(16) BASINSKI, Z. S. KORBEL, A. S., and BASINSKI, S. J. (1980) The temperature dependence of the saturation stress and dislocations substructure in fatigue copper single crystals, *Acta. Met.*, **28**, 191–208.

(17) CHENG, A. S. and LAIRD, C. (1981) Fatigue life behaviour of copper single crystals. Part I: observations of crack nucleation; Part II: model for crack nucleation in persistent slip bands, *Fatigue Engng Mater. Structures*, **4**, 331–342; 343–354.

(18) WANG, R. and MUGHRABI, H. (1984) Secondary cyclic hardening in fatigued copper monocrystals and polycrystals, *Mater. Sci. Engng*, **63**, 147–164.

(19) BASINSKI Z. S. and BASINSKI, S. J. (1985) Low amplitude fatigue of copper single crystals: II – Surface observations; III – PSB sections, *Acta Met.*, **33**, 1307–1318; 1319–1328.

(20) WANG, R. and MUGHRABI, H. (1984) Fatigue of copper single crystals in vacuum and in air. II: Fatigue crack propagation, *Mater. Sci. Engng*, **65**, 235–244.

(21) KIM, W. H. and LAIRD, C. (1978) Crack nucleation and Stage I propagation in high strain fatigue – I: Microscopic and interferometric observations; II: Mechanism, *Acta Met.*, **26**, 777–788; 789–800.

(22) FIGUEROA, J. C. and LAIRD, C. (1978) Crack initiation mechanisms in copper polycrystals cycled under constant strain amplitudes and in step tests, *Mater. Sci. Engng*, **60**, 45–58.

(23) MUGHRABI, H. (1983) A model of high-cycle fatigue crack initiation at grain boundaries by persistent slip bands, *Defects, fracture, and fatigue*, Edited by G. C. Sih and J. W. Provan (Martinus Nijhoff, The Hague) pp. 139–140.

(24) MENDEZ, J., VIOLAN, P., and GASC, C. (1984) Characterization of fatigue processes in a fine-grained copper tested in air and in vacuum, *Life assessment of dynamically loaded materials and structures*, Edited by L. Favia, Lisbon Vol. 1, pp. 515–522.

(25) MENDEZ, J. (1984) *Etude comparative des mécanismes d'amorçage des microfissures de fatigue sous air et sous vide dans le cuivre polycristallin. Influence d'une implantation ionique*, Thèse de Doctorat, Poitiers, France.
(26) KITAGAWA, H., TAKAHASHI, S., SUH, C. M., and MIYASHITA, S. (1979) Quantitative analysis of fatigue process-Microcracks and slip lines under cyclic strains, *Fatigue Mechanisms, ASTM STP 675*, Edited by J. T. Fong (American Society for Testing and Materials), pp. 420–449.
(27) SUH, C. M. YUUKI, R., and KITAGAWA, H. (1985) Fatigue microcrack in a low carbon steel, *Fatigue Fract. Engng Mater. Structures*, **8**, 193–203.
(28) VIOLAN, P., COUVRAT, P., and GASC, C. (1979) Influence of crystalline orientation on the environment affected fatigue crack propagation in copper, *Strength of metals and alloys*, Edited by P. Haasen and V. Gerold (Pergamon Press, New York) Vol. 5, pp. 1189–1194.
(29) COFFIN, L., Jr (1972) The effect of high vacuum on the low cycle fatigue law, *Met. Trans*, **3**, 1777–1788.
(30) BIGONNET, A., LOISON, D., NANDAR-IRANI, R., BOUCHET, B., KWON, J. H., and PETIT, J. (1983) Environmental and frequency effects on near-threshold fatigue crack propagation in a structural steel, *Fatigue crack growth threshold concepts*, Edited by Davidson and Suresh (The Metallurgical Society of AIME), pp. 99–114.

J. Petit* and A. Zeghloul*

On the Effect of Environment on Short Crack Growth Behaviour and Threshold

REFERENCE Petit, J. and Zeghloul, A., **On the Effect of Environment on Short Crack Growth Behaviour and Threshold**, *The Behaviour of Short Fatigue Cracks*, EGF Pub. 1 (Edited by K. J. Miller and E. R. de los Rios) 1986, Mechanical Engineering Publications, London, pp. 163–177.

ABSTRACT The growth in fatigue of physically short bidimensional cracks has been studied in air and in vacuum on the 7075 T651 high strength aluminium alloy in the low growth rate range and near threshold conditions. On the basis of crack closure measurements, and taking into account a large environmental influence, a rationalization of short and long cracks is proposed in terms of the effective stress intensity factor.

Introduction

Very few studies have been carried out on the influence of environment on the fatigue crack growth behaviour of short cracks apart from research specifically related to so called corrosive environments (**1**)(**2**). Recently Gerdes *et al.* (**3**) have shown the existence of initial propagation rates higher in air than in vacuum after initiation of short cracks in a Ti–8.6 Al alloy. Zeghloul and Petit (**4**) have shown that initial growth of short bidimensional cracks in 7075 T7351 Al alloy occurs at a very much lower stress intensity range (ΔK) level in an active environment of nitrogen containing traces of water vapour than in vacuum. Further, as inferred from long crack closure measurements (**5**), these authors have suggested that, for both environmental conditions, the crack growth rate of short bidimensional cracks could be rationalized in terms of the effective stress intensity factor range (Fig. 1(a) and (b)) which is in accordance with the observations of Breat and Pineau (**6**), and Tanaka and Nakai (**7**) working on steels.

However Lankford (**8**) has suggested the absence of any significant environmental influence on the growth of surface cracks in a 7075 T651 alloy on the basis of the observation of an overlapping of microcrack data in air and in purified nitrogen. But a comparative study of the long fatigue crack behaviour of several aluminium alloys in vacuum, ambient air and purified nitrogen (3 ppm H_2O and 1 ppm O_2) has clearly shown that the crack growth data near threshold in nitrogen is environmentally controlled with a threshold ΔK range lower than the one in vacuum and equal to or lower than the one determined in ambient air (**5**)(**9**) (Fig. 2).

To get a better understanding of the environmental influence on short crack

* Laboratoire de Mécanique et Physique des Matériaux-U.A. CNRS 863, 86034 Poitiers Cedex, France.

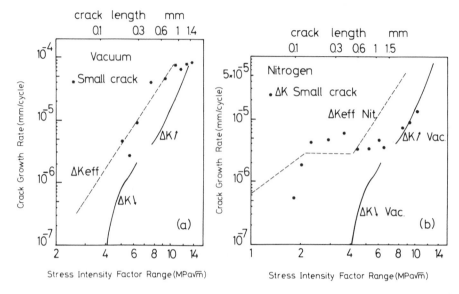

Fig 1 Crack growth rate versus ΔK for small cracks compared with long crack behaviour in terms of ΔK and ΔK_{eff}.

    (a) 7075 T7351 in vacuum

    (b) 7075 T7351 in vacuum and nitrogen (3 ppm H_2O)

behaviour, a study has been undertaken on the propagation of short bidimensional cracks. The present results obtained on the 7075 T651 alloy are discussed on the basis of crack closure measurements performed during tests conducted in ambient air, vacuum, and purified nitrogen.

Experimental conditions

The composition (% wt) of the 7075 T651 alloy studied was 6.0 Zn, 2.44 Mg, 1.52 Cu, 0.20 Cr, 0.16 Fe, 0.07 Si, 0.04 Mn, 0.04 Ti, Al (balance). Mechanical properties for this alloy are: yield stress 527 MPa, UTS 590 MPa, and elongation 11 per cent. The average grain size of the pancake structure is 40 × 150 × 600 μm. The microstructure of the peak aged condition is characterized by the presence of G.P. zones (about 10 Å in diameter), coherent dispersoïd plates (about 15 × 50 Å) and intermetallic constituent particles (~1 μm).

The specimens used were of 10 mm thick CT 75 type, machined in the LT direction. A long crack was first obtained at $a/W = 0.6$ by cycling at decreasing load amplitude down to threshold ($R = 0.1$, test frequency 35 Hz, in ambient air) so as to get a very small plastic zone at the crack tip. Then the plastic wake was removed by spark erosion all along the cracked surfaces so as to leave a short through-thickness crack of a length of about 0.1 mm. This technique was first proposed by Breat *et al.* (**9**) and McEvily *et al.* (**10**).

Tests were conducted in a chamber mounted on an electrohydraulic machine;

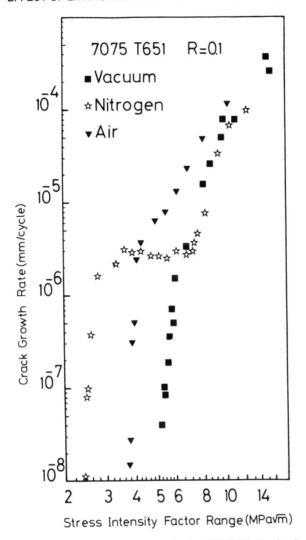

Fig 2   Crack growth rate versus ΔK for long cracks in the 7075 T651 alloy in air, vacuum, and nitrogen (3 ppm H$_2$O)

the environmental conditions were ambient air (~50 per cent RH) and vacuum (<$5 \cdot 10^{-4}$ Pa). A tension-tension sine wave was applied at 35 Hz with a load ratio of 0.1. Crack advance was optically monitored and crack closure was detected by means of a differential compliance technique with a gauge mounted across the notch mouth to measure the notch mouth opening δ with respect to the load P at a frequency of 0.2 Hz. Small variations of compliance were amplified using an improved differential technique (9)(11), giving $\delta' = G(\delta - \alpha_o P)$ where G is an electronic amplification factor and α_o the compliance of the

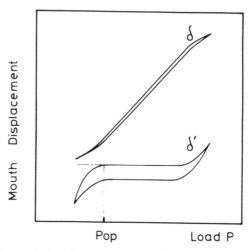

Fig 3 Determination of the load P_o corresponding to crack opening from δ vs P and differential δ' vs P diagrams

specimen with a crack fully open. From the δ' versus P diagrams the opening load P_o is defined as seen in Fig. 3.

Experimental results

Crack closure measurements were performed at different steps during machining of the crack wake (Fig. 4). The trend in the crack opening stress intensity factor K_{op} (corresponding to P_o) versus the remaining crack length Δa indicates a progressive decrease in K_{op} for values of Δa lower than 2 mm; closure becomes undetectable for a crack length of about 100 μm which is the length of the remaining short through-thickness crack. Such a length corresponds also to the limit of resolution of the compliance technique used for the detection of closure (**12**).

Further crack growth presented in Fig. 5 was performed in ambient air at decreasing load steps; the threshold level here obtained is significantly lower than the one previously determined for a long crack. Subsequent crack growth at increasing ΔK shows a progressive change in the short crack behaviour which reaches the long crack behaviour at a length about 2 mm.

The same test performed in vacuum (after machining of the wake of the crack grown in air) shows that initial growth occurs (Fig. 6) at a higher ΔK level than in air (Fig. 5). Crack arrest was obtained at the first decreasing load step (~1 million cycles) at a ΔK range of 3.6 MPa $\sqrt{m}$ instead of 5 MPa $\sqrt{m}$ for the long crack test. Further crack growth was performed under increasing ΔK conditions and, as in air, a progressive shift is observed leading to a behaviour similar to that observed for long cracks for a crack length of about 1 mm, which is shorter than the corresponding crack length in air.

The fluctuations observed in the short crack data can be related to the

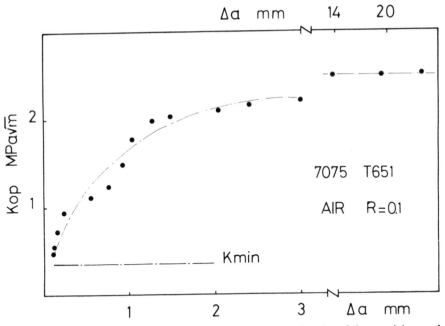

Fig 4   Variation of the opening stress intensity factor K_{op} as a function of the remaining crack
length Δa after machining of the crack wake on the 7075 T651 alloy

changes detected in the crack growth direction; the crack profile (Fig. 7) observed on one of the faces of the specimen exhibits evidence of crack deflection and crack branching at the high angle grain boundaries all along the crack path in vacuum starting at the level of the two microhardness imprints. So, the above fluctuations can be related to a microstructural influence on crack growth.

Crack closure measurements (Fig. 8) show a progressive increase in K_{op} during the short crack growth. The K_{op} development with respect to short crack length in air is similar to the one obtained from machining the plastic wake (the dash-dotted curve in Fig. 8 is from Fig. 4). Such a result indicates that the experimental procedure has little influence on closure measurements.

In vacuum, K_{op} increases also with respect to the crack length but more rapidly than in air. This result is consistent with the observation of a shorter 'short crack effect' in vacuum.

Figure 9 compares the short crack data expressed in terms of the effective stress intensity range ΔK_{eff} $(= K_{max} - K_{op})$ with results obtained for long cracks. A good correlation is observed in vacuum and, consequently, short and long crack behaviour appears to be rationalized in term of ΔK_{eff}. Compared to vacuum, a poorer agreement is observed in air between short and long cracks growth rates; however, the general trend is consistent with the observation made in vacuum.

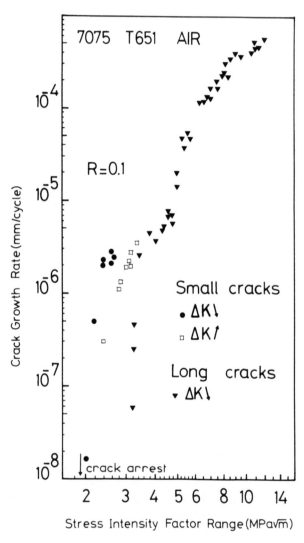

Fig 5 Crack growth rate versus ΔK in air for small cracks compared with long crack behaviour in the 7075 T651 alloy

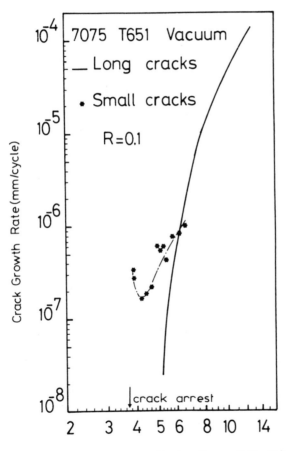

Fig 6   Crack growth rate versus ΔK in vacuum for small cracks compared with long crack behaviour in the 7075 T651 alloy

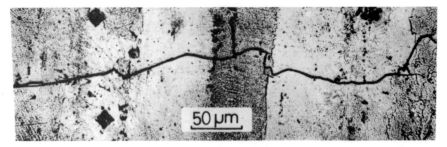

Fig 7   Micrographic observation of the crack profile in the 7075 T651 alloy tested in vacuum. The initial crack length corresponds to the microhardness indents (~ 0.17 mm)

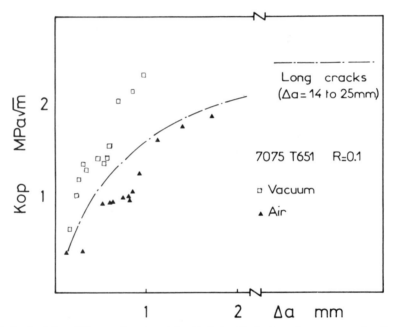

Fig 8   Variation of the opening stress intensity factor K_{op} in air and vacuum as a function of Δa the short crack length; 7075 T651 alloy

The results are consistent with dominant crack closure effects controlling the crack growth behaviour of bidimensional through-thickness short cracks and confirm observations made by different authors (6)(10)(13)(14). In addition, these results show a substantial influence of environment on crack growth rates and threshold ranges. The following discussion will be essentially focussed on the environmental influence.

Discussion

Figure 10 compares the long crack growth data expressed in terms of ΔK_{eff}

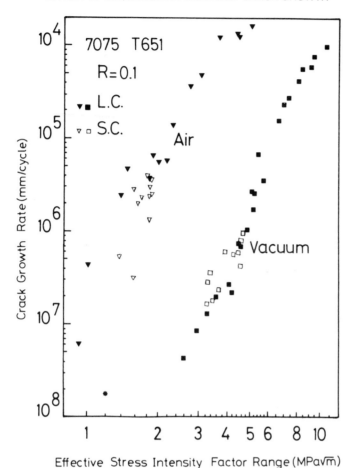

Fig 9   Crack growth rate versus ΔK_{eff} for small and long cracks in air and vacuum; 7075 T651 alloy

obtained on the 7075 T651 alloy tested at $R = 0.1$ and 35 Hz in vacuum, ambient air and purified nitrogen. This figure is quite complex and can be explained only on the basis of previously obtained results on different aluminium alloys and especially the 7075 alloy in the underaged (T351) and overaged conditions (T7351) (5).

In the absence of any environmental effect, in vacuum, the crack propagation mechanism is governed only by microstructural factors whose action in turn is governed by the loading conditions. The crack propagation is intergranular controlled by slip in one or many active planes. In the crack growth range where the Paris law is valid, i.e., in stage II (Fig. 10), the crack tip loading conditions permit at least two slip systems to be active (15) which in turn leads to a plane crack growth path affected only by the presence of large inter-metallic precipi-

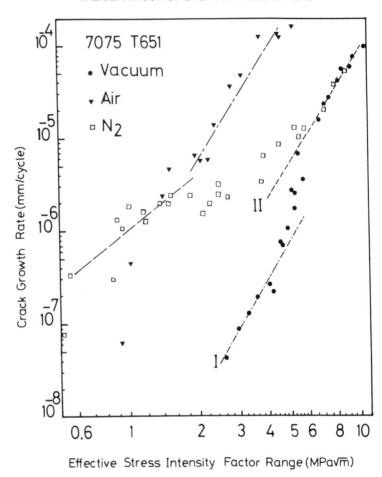

Fig 10 Crack growth rate versus ΔK_{eff} for long cracks on the 7075 T651 alloy in air, vacuum, and nitrogen (3 ppm H_2O)

tates. In the range which represents the transition from stage II to stage I, ($da/dN \sim 3 \cdot 10^{-6}$ mm/cycle), a crystallographic crack propagation mode is observed (Regime I in Fig. 10) corresponding to planar slip and where slip is localized to a single system. A similar process was analysed in the case of a Ni-based super-alloy with γ' hardening precipitates (**16**) and the above mentioned transition was attributed to the existence of deformation levels lower than that required for the saturation of slip bands near the crack tip. A similar mechanism may be present in the 7075-T651 alloy with coherent GP zones and dispersoid plates (15 × 50 Å), leading to low crack propagation rates and high thresholds ($\Delta K_{th} = 5.5$ MPa$\sqrt{m}$ and $\Delta K_{eff \cdot th} \sim 2.5$ MPa$\sqrt{m}$).

The crack propagation in vacuum can be represented by a law derived from Weertman's model (**17**)

$$\frac{da}{dN} = \frac{A(\Delta K_{eff})^4}{\mu \sigma^2 \gamma} \tag{1}$$

where A is a dimensionless constant, σ flow stress, μ shear modulus, and γ the energy required to create a unit free surface.

The environmental effect consists of two distinct processes which are water adsorption and oxidation. The adsorption of water vapour molecules on the freshly created surfaces at the crack tip leads to reduction in the atomic bond strength in the case of tensile and shear type deformation (Rehbinder effect (18)); this phenomenon results in a decrease of fatigue strength and consequently to a reduction in the value of energy, γ. When the time required for the creation of a single layer of adsorbed molecules at the crack tip is attained in each cycle, the adsorption effect is saturated and an acceleration of crack growth is observed without any change in crack propagation mechanism, which remains similar to that in vacuum, only the value of γ is diminished. Typically, this effect is observed in the 7075-T651 alloy in air at medium growth rates.

In the case where gas transport at the crack tip does not result in the formation of an adsorbed layer (even partial), no environmental effect is observed. Such a situation exists in high purity N_2 in the mid rate range where the crack growth behaviour in this environment is similar to the one in vacuum. When the conditions existing at the crack tip permit the creation of a partial adsorbed layer, local crack growth acceleration is observed as in the transition range in the N_2 environment. In this range the crack growth curve in N_2 shifts from a behaviour similar to that in vacuum (negligible adsorption effect) to a behaviour similar to that in air (saturated adsorbed layer): the degree of gas coverage of freshly created surfaces varies from zero as in vacuum to unity as in air during this transition which takes place in the range $2 \cdot 10^{-6} < da/dN < 10^{-5}$ mm/cycle.

The conditions determining the access of the active species (H_2O molecules in this case) depends upon many factors, such as the geometry of the crack, the test frequency, the R ratio, the trapping of water molecules by oxide deposits, and the partial pressures of water vapour molecules.

The chemisorption phenomenon describes the formation of hydrogen by the dissociation of the adsorbed water molecules (19). In such a case a hydrogen embrittlement mechanism can be brought into action (20)–(22). The concentration of hydrogen in the process zone can be sufficient only if the adsorption effect is saturated. The attainment of this critical concentration is time dependent (time for hydrogen diffusion by dislocation dragging and time for water vapour transport at the crack tip) and can be favoured by the localized deformation in a single slip system. In the 7075-T651 alloy such a condition seems to prevail only for crack growth rates of the order of $5 \cdot 10^{-6}$ mm/cycle, particularly in N_2.

This hydrogen assisted crack propagation can be associated with a mechanism of crack propagation controlled by the crack tip opening displacement range ΔCTOD (23) described by a relation of the type

$$\frac{da}{dN} = B(\Delta K_{\text{eff}})^2/\mu\sigma^2 \tag{2}$$

The flow stress in this case is characteristic of the embrittled material at the crack tip. Under such conditions the crack growth resistance is much smaller than that in vacuum, resulting in a very low threshold ($\Delta K_{\text{th}} = 2.4$ MPa$\sqrt{m}$, $\Delta K_{\text{eff}\cdot\text{th}} \sim 0.5$ MPa$\sqrt{m}$ in nitrogen).

In air a different situation exists; as in the presence of O_2 molecules at a high partial pressure and at low R values, the crack surfaces are oxidized by a fretting action (24)–(26). In the 7075-T651 alloy the oxide thickness is smaller than the ΔCTOD value (26) which precludes the 'wedging effect' observed in steels (24). At the same time, the presence of this oxidized layer can constitute a water-vapour trap, thus rendering difficult the access of the active species at the crack tip. This effect can explain the existence of a higher threshold in air than in N_2 at $R = 0.1$. The fact that the threshold values at an R value of 0.5 is the same in similar aluminium alloys adds weight to this hypothesis (5).

The short crack behaviour presented in Figs 5, 6, and 9 can be analysed on the basis of these considerations. In vacuum, the crack growth mechanism appears to be the same as the one described for a long crack and is rationalized in terms of ΔK_{eff}. The crack profile presented in Fig. 7 is also consistent with a single slip mechanism in the explored rate range. The fundamental difference between short and long crack data consists in the influence of crack length on closure. Closure for long cracks on this alloy tested in vacuum is essentially related to the very high roughness of the cracked surfaces due to the crystallographic mode of failure. The amount of closure is consequently directly related to the crack wake and the loading history. So qualitatively, a large influence of short crack length must be expected in this alloy.

In air, there exists a more complex situation. Near the threshold crack growth has been shown to be strongly dependent upon the factors which determined the conditions of access of active species (i.e., water vapour) to the crack tip, as indicated above. The respective effects of these various factors which influence $\Delta K_{\text{eff}\cdot\text{th}}$ and the near threshold crack growth are very difficult to analyse. The poor agreement between short and long crack data in Fig. 9 could be related to some difference in the effects of these factors. In addition, for closure measurements, in the low ΔK range, the experimental scatter in air is large (about 30 per cent at K_{op}). Further experiments are necessary to clarify such behaviour. However, the general trend is consistent with the behaviour of a long crack with the observation of an effective threshold higher than the one in nitrogen, which suggests an oxidation effect.

As suggested by Beevers (27) the threshold range can be analysed in two components

$$\Delta K_{\text{th}} = \Delta K_i + \Delta K_c \tag{3}$$

where $\Delta K_i = \Delta K_{\text{eff}\cdot\text{th}}$ is the intrinsic component and ΔK_c the closure contribution.

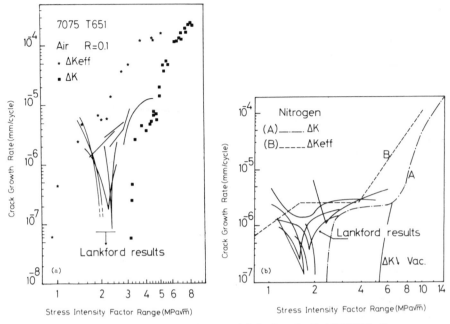

Fig 11   Small cracks compared to long crack behaviour for the 7075 T651 alloy.
(a) in air
(b) in nitrogen (2 to 3 ppm H₂O) (8)

On the basis of the present results and the previous observations on long cracks, this analysis of ΔK_{th} could be extended to a short bidimensional crack with ΔK_c again being dependent on the same factors as a long crack (plasticity, roughness, oxidation, microstructure, slip mechanisms) but now depending also on the crack length. Then the lower bound for short crack thresholds should be ΔK_i. That would mean that no crack growth can be observed in any case for ΔK lower than ΔK_i so long as the effective value of ΔK can be accurately determined.

Another limitation of these concepts is the extent and the shape of the crack considered. The present study was carried out on short through-thickness cracks; for example, Lankford studied semi-elliptic surface cracks, which were typically three-dimensional cracks and were smaller by an order of magnitude. As discussed by different authors the application of LEFM to calculate ΔK in such cases appears to be questionable (**28**). However, it is of some interest to compare the long crack data expressed in terms of ΔK and ΔK_{eff} to the Lankford data obtained on the same alloy in air and in purified nitrogen (2 ppm H₂O) (Fig. 11). It can be seen that, globally, the short crack data fall between the da/dN vs ΔK and da/dN vs ΔK_{eff} curves. This consideration appears to be consistent with a dominant role of closure for this kind of crack. In addition the comparison of the data in air (Fig. 11(a)) and in nitrogen (Fig. 11(b)) illustrates the existence of a large environmental effect in both environmental conditions and explains the absence of environmental influence observed by this author, except for a slightly higher influence in nitrogen, with the crack growing at lower ΔK values than in air.

Conclusions

(1) The growth of physically short bidimensional cracks can be rationalized with that for long cracks in terms of ΔK_{eff}

(2) In the low growth rate range and near the threshold condition, two typical da/dN vs ΔK_{eff} relations can be defined corresponding to pure active or pure inert environments which can be simulated, respectively, at near threshold conditions by a purified inert gas containing water vapour traces (about a few ppm) and high vacuum.

(3) Two characteristic effective stress intensity ranges at the threshold correspond to these two critical environmental conditions.

(4) In ambient air, and specifically at low R ratio, an intermediate growth behaviour is observed due to a more or less pronounced inhibition of the water vapour effect by such closure induced phenomena as crack surface oxidation.

(5) Considering that the threshold range can be analysed in two components, the lower threshold ranges generally observed for short cracks correspond to a decrease in the closure component with crack length, while the intrinsic component is sensitive only to environment.

References

(1) GANGLOFF, R. P. (1985) Crack size effects on the chemical driving force for aqueous corrosion fatigue, *Met. Trans*, **16A**, 963.

(2) SURESH, S. and RITCHIE, R. O. (1984) Near threshold fatigue crack propagation: a perspective on the role of crack closure, *Fatigue crack growth threshold concepts* (Edited by Davidson, D. and Suresh, S.), (Metallurgical Society of AIME), pp. 227–261.

(3) GERDES, C., GYSLER, A., and LUTJERING, G. (1984) Propagation of small surface cracks in Ti-alloys, *Fatigue crack growth threshold concepts* (Edited by Davidson, D. and Suresh, S.), (Metallurgical Society of AIME), pp. 465–478.

(4) ZEGHLOUL, A. and PETIT, J. (1985) Environmental sensitivity of small crack growth in 7075 aluminium alloy, *Fatigue Fracture Engng Mater. Structures*, **8**, 341–348.

(5) PETIT, J. (1984) Some aspects of near-threshold crack growth: microstructural and environmental effects, *Fatigue crack growth threshold concepts* (Edited by Davidson, D. and Suresh, S.), (Metallurgical Society of AIME), pp. 3–24.

(6) BREAT, J. L., MUDRY, F., and PINEAU, A. (1983) Short crack propagation and closure effects in A508 steel, *Fatigue Engng Mater. Structures*, **6**, 349–358.

(7) TANAKA, K. and NAKAI, Y. (1983) Propagation and non-propagation of short fatigue cracks at a sharp notch, *Fatigue Engng Mater. Structures*, **6**, 315–327.

(8) LANKFORD, J. (1983) The effect of environment on the growth of small fatigue cracks, *Fatigue Engng Mater. Structures*, **6**, 15–31.

(9) PETIT, J. and ZEGHLOUL, A. (1982) Gaseous environmental effect on threshold level in high strength aluminium alloys, *Fatigue thresholds* (Edited by Bäcklund, J., Blom, A. F., and Beevers, C. J.), (EMAS, Warley, UK), pp. 563–579.

(10) MINNAKAWA, K., NEWMAN, J. C., and McEVILY, A. J. (1983) A critical study of the crack closure on near-threshold fatigue crack growth, *Fatigue Engng Mater. Structures*, **6**, 359–365.

(11) OHTA, A. and SASAKI (1975) a method for determining the stress intensity threshold level for fatigue crack propagation, *Int. J. Fracture*, **17**, 1040–1050.

(12) JAMES, M. N. and KNOTT, J. F. (1985) Critical aspects of the characterization of crack tip closure by compliance techniques, *Mat. Sci. Engng*, **72**, L1–L4.

(13) JAMES, M. R. and MORRIS, W. L. (1983) Effect of fracture surface roughness on growth of short fatigue cracks. *Met. Trans*, **14A**, 153–155.

(14) McCARVER, J. F. and RITCHIE, R. O. (1982) Fatigue crack propagation thresholds for long and short cracks in 95 nickel-base superalloy, *Mat. Sci. Engng*, **55**, 63–67.

(15) DAVIDSON, D. L. and LANKFORD, J. (1984) Fatigue crack growth mechanisms for Ti-6Al-4V (RA) in vacuum and humid air, *Met. Trans*, **15A**, 1931–1940.

(16) VINCENT, J. N. (1986) *Comportement en fatigue d'un alliage à base de nickel à gros grains: mécanismes de déformation et d'endommagement liés à cristallographie*, Doc. ès. Sci. thesis, Paris XI.

(17) WEERTMAN, J. (1973) Theory of fatigue crack growth based on a BCS theory work hardening, *Int. J. fracture Mech.*, **9**, 125–137.

(18) WESTWOOD, A. R. C. and AHEARN, J. S. (1984) Adsorption sensitive flow and fracture of solids, *Physical chemistry of the solid state: applications to metals and compounds* (Edited by Lacombe, J.), (Elsevier, Amsterdam), vol. 32, pp. 65–85.

(19) BOWLES, C. Q. (1978) The role of environment frequency and wave shape during fatigue crack growth in aluminium alloys, Report L.R. 270 Delft University of Technology, The Netherlands.

(20) CHAKRAPANI, D. G. and PUSH, E. N. (1976) Hydrogen embrittlement in a Mg–Al alloy, *Met. Trans*, **7A**, 173–178.

(21) JACKO, R. J. and DUQUETTE, D. J. (1977) Hydrogen embrittlement of a cyclically deformed high strength Al alloy, *Met. Trans*, **8A**, 1821–1827.

(22) ALBRECHT, J., McTIERNAN, B. J., BERNSTEIN, I. M., and THOMPSON, A. W. (1977) Hydrogen embrittlement in a high-strength aluminium alloy, *Scripta Met.*, **11**, 893–897.

(23) LANKFORD, J. and DAVIDSON, D. L. (1983) Fatigue crack micromechanisms in ingot and powder metallurgy 7XXX aluminium alloys in air and vacuum, *Acta. Met.*, **31**, 1273–1284.

(24) SURESH, S., ZAMISKI, G. F., and RITCHIE, R. O. (1981) Oxide induced crack closure: an explanation for near-threshold corrosion fatigue crack growth behavior, *Met. Trans*, **12A**, 1435–1443.

(25) VASUDEVAN, A. K. and SURESH, S. (1982) Influence of corrosion deposits on near-threshold fatigue crack growth behavior in 2XXX and 7XXX series aluminium alloys, *Met. Trans*, **13A**, 2271–2280.

(26) RENAUD, P., VIOLAN, P., PETIT, J., and FERTON, D. (1982) Microstructural influence on fatigue crack growth near threshold in 7075 Al alloy, *Scripta Met.*, **16**, 1311–1316.

(27) WALKER, N. and BEEVERS, C. J. (1979) A fatigue crack closure mechanism in titanium. *Fatigue Engng Mater. Structures*, **1**, 135–148.

(28) MILLER, K. J. (1982) The short crack problem, *Fatigue Engng Mater. Structures*, **5**, 223–232.

P. E. V. de Miranda* and R. Pascual†

Initial Stages of Hydrogen Induced Short Fatigue Crack Propagation

REFERENCE   de Miranda, P. E. V. and Pascual, R., **Initial Stages of Hydrogen Induced Short Fatigue Crack Propagation**, *The Behaviour of Short Fatigue Cracks*, EGF Pub. 1 (Edited by K. J. Miller and E. R. de los Rios) 1986, Mechanical Engineering Publications, London, pp. 179–190.

ABSTRACT   The initiation of fatigue cracks was studied in an austenitic stainless steel type AISI 304L which was cathodically hydrogenated, outgassed, and then tested. Tests were conducted with R ratio equal to -1 at 30 Hz and stopped at different periods of the fatigue life to permit a survey of the gauge surface of the samples with a scanning electron microscope.

Contrary to the non-hydrogenated fatigued material, slip marks were not evident on the surface of the hydrogenated and outgassed samples. From the beginning of the fatigue life the gauge surface exhibited a 'peeling off' of very thin layers in several regions which was always associated with pre-existing hydrogen-induced cracks.

However, the main fatigue crack was found to initiate sub-superficially probably in the region close to the interface between the hydrogen hardened outermost layer and the inside ductile material. Since the hydrogen-induced surface cracks do not grow themselves, they are not likely to be used as a tool to model the fatigue behaviour of short cracks in austenitic stainless steels.

Introduction

The introduction of hydrogen into the structure of austenitic steels may harm their mechanical properties if they are not outgassed at a high enough temperature, or if mechanical work is applied while hydrogen is still in the material. There is a broad variety of situations in which austenitic stainless steels are used in hydrogen rich environments, especially in the chemical and nuclear industries. These steels are particularly affected by the presence of hydrogen if outgassing takes place at around room temperature (**1**). The low diffusion coefficient of hydrogen in austenite (as compared to the mobility it has in ferrite) leads to an accumulation of a high content of hydrogen in a very superficial layer of the material. As a result, high localized levels of strain are developed (as was shown by the use of X-ray techniques (**1**)–(**6**)) and the following surface effects might be observed, depending on the chemical composition of the steel and the consequent level of austenite instability: (i) during hydrogenation austenite may destabilize and transform partially into an h.c.p. ε martensitic phase; (ii) upon outgassing another martensitic transformation may take place (giving rise to an α' phase) as well as the appearance of very small superficial cracks. Both nucleate and grow with time after hydrogenation, i.e., with ageing time, and, in fact, the b.c.c. α' martensitic phase will occur

* COPPE-EE-Universidade Federal do Rio de Janerio, C.P. 68505 – CEP 21945 Rio de Janeiro, RJ – Brazil.

† Instituto Militar de Engenharia – Praça Gal. Tibúrcio, 80, CEP: 22290 Rio de Janeiro, RJ – Brazil.

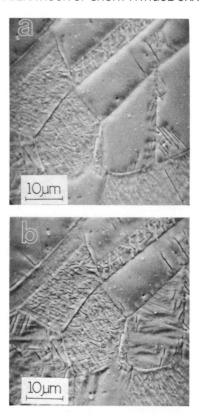

Fig 1 Delayed features associated with the room temperature outgassing of a hydrogenated austenitic stainless steel. Bulged grains, such as in (a), and the delayed appearance of cracks and a martensitic phase, as in (b). Micrographs were made (a) 30 minutes, and (b) 20 hours after hydrogenation. Unetched (7)

preferentially close to the delayed cracks due to the high concentration of deformation at these regions (7). Figure 1 shows an example of the delayed appearance of cracks and phase transformation for an AISI type 304L stainless steel degassed at room temperature. The same area was photographed 30 minutes (Fig. 1(a)) and 20 hours (Fig. 1(b)) after hydrogenation. The kinetics of phase and crack appearance varies from grain to grain. It is also interesting to take into account the fact that the hydrogenated samples show an increase in surface hardening (8).

The effect of hydrogen contained in the material and that of the surface effects induced by hydrogen in austenite have been studied in uniaxial tension (9)–(13). It was found that hydrogen may indeed cause a severe decrease in the ductility of austenite, but the effect is more pronounced in very thin samples due to the very low diffusion coefficient of hydrogen in austenite. That is, the mechanical properties in tension depend on the geometry of the hydrogenated sample, since higher values of surface to volume ratio increases the suscep-

Table 1   Number of cracks (N) per square millimetre as a function of the current density (J) used
for a two-hour cathodic hydrogenation of 304 stainless steel at room temperature (14)

J (Amps/m^2)	50	250	500	1000	2000	4000	10 000
N (mm^{-2} × 10^{-3})	85.2	138.9	177.8	216.7	259.3	307.4	305.6

tibility to hydrogen embrittlement (**12**)(**14**). Tensile testing of outgassed
samples, which contain surface cracks and martensitic phases, showed that the
material recovers most of its ductility (**12**)(**15**). This indicated both the presence
of hydrogen in the structure to cause the embrittlement of the austenite, and
that the hydrogen-induced surface effects in this material are of secondary
importance when considering the mechanical properties in uniaxial tension.
However, the question of how these hydrogen induced surface effects (specially
the delayed cracks) would influence the fatigue properties of austenitic steels,
which are influenced by surface conditions, should be addressed.

The first step towards answering this question involved the morphological
and quantitative characterization of these surface cracks (**2**)(**6**)(**7**)(**14**)(**16**)–
(**19**), which showed that the cracks are very small (having average length
smaller than one grain diameter) and very shallow due to the small depth of
penetration of hydrogen into the austenite. They appear in a crystallographic
fashion, with different orientations in different grains, and are numerous, as is
exemplified by the results shown in Table 1. As a summary of the results
obtained to date, it is widely accepted that: (a) the depth of the hydrogen-
induced cracks in austenitic stainless steel and of the hardened layer, measured
by metallographic sectioning, was found to be of the order of 6 to 15 μm
(**14**)(**17**), 20 μm (**5**) or even 50 μm (**2**) depending on the current density, time,
and temperature of charging, as well as on the alloy's chemical composition and
microstructure; (b) a compressive stress state was found to occur on the surface
of the sample during charging, which, coupled with tensile stress during
outgassing, lead to phase transformations, plastic deformation, and cracking
(**4**)–(**7**)(**16**); (c) the number of cracks differs widely from grain to grain
(**2**)(**5**)(**6**)–(**8**)(**16**), but may reach average densities of thousands of cracks per
square millimetre (Table 1 and references (**14**) and (**18**)).

The second step was to conduct fatigue tests of hydrogen charged and
outgassed samples. These showed (**8**) that the fatigue life of type AISI 304
stainless steel was reduced about 40 per cent by a hydrogenation–outgassing
cycle, due to the introducton of surface cracks. The effect was more significant
for high cycle fatigue conditions. It was also observed that the fatigue fracture
surface presented steps, whereas the non-hydrogenated specimens had a fairly
flat overall fracture surface. This was rationalized as a function of the simultaneous growth of different hydrogen-induced surface cracks during fatiguing. It
was then realized that these facts would be better understood by studying the
initiation phenomenon of the fatigue cracks on the hydrogenated and outgassed
samples. This is one of the objectives of the present paper. Another objective

is to analyse the possibility of using hydrogen-induced surface cracks, which can be introduced in a controlled way, to model the fatigue behaviour of short cracks in an austenitic stainless steel.

Experimental procedures

A hot rolled sheet of an AISI type 304L stainless steel with a thickness of about 4.0 mm was used in this work, having the following chemical composition (wt%): C 0.024, Cr 18.25, Ni 9.50, Mn 1.37, Si 0.51, P 0.037, S 0.006, Mo 0.060, N 0.0379, Fe balance. Flat hour-glass type fatigue specimens with a 20 mm minimum gauge width and a 7 mm gauge radius of curvature were machined parallel to the rolling direction and then heat treated for 30 minutes at 1100°C and quenched in water. This yielded a mean grain size of 74.4 μm (Fig. 2(a)) and a hardness of 148 Vickers. All microhardness measurements were taken with a load of 50 g. The static mechanical properties of this material were: yield stress: 208.90 MPa, tensile strength: 678.80 MPa, elongation: 79.8 per cent.

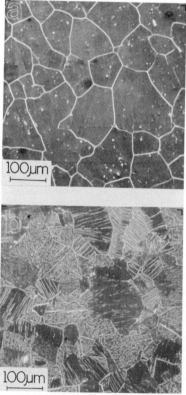

Fig 2   Initial microstructure of the material. (a) Non-hydrogenated. Electrolytically etched with 50 per cent H_2O + 50 per cent HNO_3, J = 4402 amps/m^2, t = 90 s at room temperature (b) Hydrogenated and outgassed. Unetched

Hydrogenation was performed cathodically at room temperature, using an electrolyte of H_2SO_4 1 N containing 100 mg/l of As_2O_3, for 4 hours and with a current density of 1000 A/m^2. After hydrogenation the samples were aged at room temperature and atmospheric pressure for one week before testing, which gave rise to surface cracks and to martensitic phases (Fig. 2(b)) as most of the hydrogen introduced by the electrolitic charging was outgassed. The hardness of the hydrogenated and outgassed sample was 192 Vickers.

Fatigue tests were conducted at room temperature, in a servo-hydraulic machine in tension–compression, at a frequency of 30 Hz and R ratio of -1.

The initiation and growth of fatigue cracks was studied in constant stress range tests, which were interrupted at several points during the fatigue life in order to survey the lateral surfaces of the gauge section with a scanning electron microscope.

Results

All the results selected to describe the fatigue behaviour in this paper

Fig 3   Detail of the gauge surface of a hydrogenated–outgassed sample fatigued to 15 per cent of
its fatigue life. (b) shows an enlargement of the area indicated in (a). Unetched

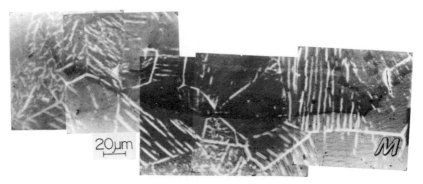

Fig 4 Early indication of the fatigue sub-surface crack appearing on the surface of the hydrogenated and outgassed sample at 49 per cent of its fatigue life. Unetched

correspond to a constant maximum stress of 195 MPa. For this condition the fatigue life of the non-hydrogenated specimens was found to be 317 000 cycles, while that of the hydrogenated–outgassed samples was 101 000 cycles, which corresponds to about 32 per cent of the former.

The first effect that could be observed on the surface of the hydrogenated and outgassed fatigued samples was the 'peeling off' of very thin layers in several small regions throughout the gauge section. This is depicted in Fig. 3(a) and (b) and gives the impression of being due to the fracture of very fragile parts of the outermost surface and associated with the pre-existent hydrogen-induced microcracks. These fractured regions appear whiter in the micrographs because they stand up above the plane of the surface. They show up immediately after a few fatigue cycles and remain without much change throughout the fatigue life. Some of these fractured zones, however, eventually fall off the surface. The hydrogen induced microcracks were not observed to grow in length throughout the fatigue life.

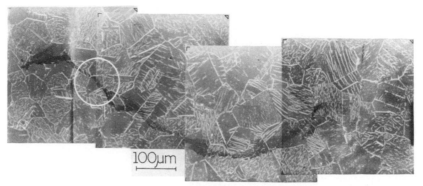

Fig 5 Growth of the feature associated with the sub-surface fatigue crack for the hydrogenated and outgassed specimen at 63 per cent of its fatigue life. The area shown in Fig. 4 is included. Unetched

Fig 6   Fatigue crack on the gauge lateral surface at the termination of fatigue life of a hydrogenated and outgassed sample. This micrograph corresponds to the area indicated by the white circle in Fig. 5. Unetched

In all the specimens observed, the main fatigue crack developed in the way represented in Figs 4, 5, and 6, which correspond to the same region of the surface of a particular specimen. Some dark markings, such as the one shown in Fig. 4, were found to appear on the gauge surface of the specimens and grow in length with increasing number of fatigue cycles. One of the marks was always found to be associated with the main fatigue crack responsible for fracture (Fig. 6). In brief, the main crack developed sub-surface and was visible at first as a dark marking on the surface. Only by the end of the fatigue life did the main fatigue crack open to show a crack on the surface. The sub-surface crack depicted in Fig. 4 was photographed after the sample was cycled 49 per cent of its fatigue life. It is interesting to note that it spread transgranularly, crossing without distinction grains with and without hydrogen-induced surface cracks, and did not follow the direction of these cracks, so that it cannot be thought to be formed mainly as a consequence of the in-depth growth of hydrogen-induced cracks. The same sub-surface crack of Fig. 4 has grown much longer after cycling the sample about 63 per cent of the fatigue life as shown in the micrograph of Fig. 5. In both Fig. 4 and Fig. 5 the white markings (white lines) represent the sites where there are intergranular and transgranular hydrogen-induced surface cracks. In a few regions of Fig. 4 (which was taken with greater magnification) the martensitic phase (marked M) can be noted. No relationship between this phase and the sub-surface cracks was found. Finally, Fig. 6 shows the crack at the end of the fatigue life in the same region circled in Fig. 5.

The fracture surface in the region appearing on the upper left side of Fig. 5 is shown in Fig. 7. It is interesting to observe that that hump in the sub-surface crack appearing on the upper left hand side of Fig. 5 (to the left of the white circle) corresponds exactly to the step found on the fracture surface shown in Fig. 7. A detail of the top part of this step, closer to the surface of the sample (on the left hand side of Fig. 7) is presented in Fig. 8. This micrograph shows

Fig 7   Detail of a step on the fracture surface of a fatigued hydrogenated–outgassed sample. The
step corresponds to the hump on the upper left hand side in Fig. 5

the presence of secondary cracks, one of which appears to start from what may be a hydrogen-induced surface crack (marked C in Fig. 8). S.E.M. observations included careful examination of the lateral and of the fracture surfaces of the hydrogenated–outgassed samples. Despite this, no clear indication of initiation sites for the main fatigue crack was found, probably due to the fact that the fracture surfaces are very irregular.

It was not an objective of this paper to give a detailed analysis of the initiation and propagation features of the fatigue crack in the non-hydrogenated material. However, for the sake of comparison with the hydrogenated–outgassed samples, Fig. 9 shows the features associated with the propagation of a fatigue crack in a non-hydrogenated specimen tested under the same conditions of the former. A great number of slip marks can be observed.

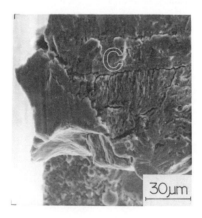

Fig 8   Magnification of a region of Fig. 7 at the top of the step on the fracture morphology, near
to the lateral surface of a fatigued hydrogenated–outgassed sample. A fatigue crack is marked at C

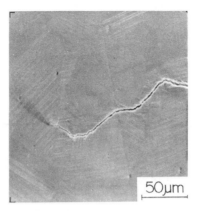

Fig 9 A fatigue crack on the lateral gauge surface of a non-hydrogenated sample. The micrograph shows many slip marks due to accumulated fatigue strain. Unetched

Discussion

The initiation (20)–(22) and propagation (23) of fatigue cracks in austenitic stainless steels and other f.c.c. materials have been studied recently. It was found that deformation markings appear early in the fatigue life, together with a certain level of surface roughness, and are followed by the nucleation of superficial microcracks. These microcracks grow in number and length and coalesce to form the main fatigue crack. The same behaviour is shown in Fig. 9, relative to fatigued, non-hydrogenated samples.

The fatigue behaviour of the hydrogenated and outgassed specimens was found to be quite different. To begin with, for the conditions used in the present work, no slip markings were evident on the surface, even at fracture, so that there is no evidence of plastic deformation on the surface layer. Also, no nucleation of fatigue microcracks on the surface and/or growth of the pre-existing hydrogen-induced cracks was observed throughout the fatigue life.

Surprisingly, the fatigue cracks were found to initiate sub-surface and not as a result of the growth and coalescence of the pre-existing hydrogen-induced cracks, as was expected. Several factors may influence the initiation of the fatigue cracks, such as the existence of a hardened surface layer, the presence of hydrogen-induced surface cracks, and martensitic phases, as well as the role played by the 'peeled off' areas. The hydrogen-induced hardened surface layer may have an irregular depth due to the fact that the penetration of hydrogen into the material varies greatly according to the local microstructure. The mean depth can be estimated for the present hydrogenation condition as follows. The diffusion coefficient (D) of hydrogen in austenite at 296K can be taken to have the value of 1.38×10^{-16} m^2/s (24). The solution of Fick's law for the diffusion in a semi-infinite solid is

$$\frac{C}{C_o} = 1 - \text{erf}\left\{\frac{x}{2\sqrt{(Dt)}}\right\} \tag{1}$$

where C_o is the concentration of hydrogen in the surface, C is the concentration at a depth x, and t is the time of hydrogenation (equal to 14 400 s for the present case). Assuming that the hydrogen-affected region of the material extends to a depth in which the hydrogen concentration is 0.1 per cent of that found in the surface (i.e., $C/C_o = 0.001$), and applying the present conditions in Equation (1), one obtains an average depth of the order of 6 μm which is of the same order of magnitude of the experimental determinations (2)(5)(14)(17) mentioned before. This reiterates the opinion that the effects induced by hydrogen are indeed concentrated in a very thin hardened surface layer. The hydrogenated–outgassed specimens behave, then, as a 'composite material', having a hard and thin surface layer, which contains microcracks, surrounding a ductile core. This layer cannot accommodate, plastically, the strains imposed during the fatigue test and the layer fractures locally along the pre-existing hydrogen-induced cracks. However, this effect is not found to be associated with the initiation or growth of the main fatigue crack. This 'peeling off' effect produced a reduction in the effective cross-section of the sample, but this is too small to explain the observed reduction of the fatigue life. This situation is similar to what is observed in a shot-peened material, but with the difference that in the latter the presence of a hardened surface layer usually improves the fatigue life. In both cases the fatigue crack initiates sub-surface. The hydrogen-induced hardened layer differs from the shot-peened one in at least two aspects. One is the presence of delayed surface cracks and the other is the probable irregularity of the depth profile of this layer, due to microstructural inhomogeneities.

Differences in the microstructure of the steel (such as grain and twin boundaries, second phase particles, slip bands, and localized constitutional segregation) makes hydrogen penetration heterogeneous and gives rise, upon outgassing, to a hardened surface layer with an irregular depth profile, as mentioned earlier. This, in turn, could facilitate the nucleation of fatigue sub-surface cracks in several sites close to the interface between the hardened layer and the ductile material inside. For that reason the initiation of the fatigue cracks would be faster in hydrogenated–outgassed samples than in the non-hydrogenated ones. This may have led to the observed decrease in the fatigue life.

On the other hand, even if the pre-existing hydrogen cracks are found not to grow during the fatigue life, they could eventually be responsible for the nucleation of secondary fatigue cracks, which may, in turn, grow and coalesce to form the main crack. In effect, the crack marked C in Fig. 8 seems to have been initiated in a hydrogen-induced surface crack. This effect may also explain the observed reduction in fatigue life. Despite the fact that the initiation sites for the main fatigue crack were not clearly defined, two possibilities are raised to account for this initiation: (1) the irregular depth profile of the hardened layer, and (2) the stress concentration at the tip of the hydrogen-induced surface cracks. Eventually, they may contribute separately or simultaneously to the decrease in the fatigue life. The existence of multiple secondary fatigue cracks alters the path of the main fatigue crack, which may become wavy, as the one shown in Figs 4 and 5. That is probably why there are features such as the hump on the left hand side of Fig. 5, which gives rise to the steps on the fracture morphology. Finally, due to the complexity of the initiation features in this material, it would not be appropriate to use hydrogen-induced surface cracks as a tool to model the fatigue behaviour of short cracks in austenitic stainless steels.

Conclusions

(1)   The short fatigue cracks in the hydrogenated–outgassed material were found to initiate sub-surface.
(2)   The fact that the hydrogen-induced surface cracks do not grow themselves eliminates the possibility of using them to model the fatigue behaviour of short cracks in austenitic stainless steels.

Acknowledgements

The authors acknowledge the financial support, which made this research possible, through FINEP, CNPq (Grant No. 402289/84-MM) and Ministério do Exército. Thanks are also expressed to Oswaldo Pires Filho for technical support.

References

(1)   NARITA, N., ALTSTETTER, C. J., and BIRNBAUM, H. K. (1982) Hydrogen-related phase transformations in austenitic stainless steels, *Metal. Trans*, **13A**, 1355–1365.
(2)   HOLZWORTH, M. L. and LOUTHAN Jr., M. R. (1968) Hydrogen-induced phase transformations in type 304-L stainless steels, *Corrosion–Nace*, **24**, 110–124.
(3)   MATHIAS, H., KATZ, Y., and NADIV, S. (1978) Hydrogenation effects in austenitic steels with different stability characteristics, *Metal Sci.*, **00**, 129–137.
(4)   TOUGE, M., MIKI, T., and IKEYA, M. (1983) Effects of X-ray irradiation on hydrogen-induced phase transformations in stainless steel, *Met. Trans*, **14A**, 151–152.
(5)   BRICOUT, J. P. (1984) *Contribution a l'etude de la fragilisation par l'hidrogene des aciers inoxydables austenitiques instables*, Ph.D thesis, Université de Valenciennes et du Hainaut Cambresis, France.
(6)   MIRANDA, P. E. V., SAAVEDRA, A., and PASCUAL, R. (1986) Metallographic characterization of hydrogen-induced surface phenomena in an austenitic stainless steel, *Microstructural science*, **13** (Edited by Shiels, S. A., Bagnall, C., Witkowski, R. E., and Vander Voort, G. F.), (IMS, ASM), pp. 349–359.
(7)   DE MIRANDA, P. E. V. (1984) Fenomenologia da fratura retardada e das transformações de fases na austenita hidrogenada, *Proceedings of the VI CBECIMAT*, Rio de Janeiro, pp. 62–67.
(8)   PIERANTONI, P. S., MIRANDA, P. E. V., and PASCUAL, R. (1985) Effect of surface cracks induced by hydrogen on the fatigue properties of AISI 304 stainless steel, *Proceedings of the 7th ICSMA*, Montreal, pp. 1213–1218.

(9)   HÄNNINEN, H. E. and HAKKARAINEN, T. J. (1980) Influence of metallurgical factors on hydrogen-induced brittle fracture in austenitic stainless steels, *Advances in fracture research* (Edited by Francois, D. *et al.*), (Pergamon Press, Oxford), pp. 1881–1888.

(10) HABASHI, M. and GALLAND, J. (1982) Considérations sur la fragilisation par l'hydrogène des aciers inoxidables austénitiques, *Mem. Etudes Sci. Rev. Met.*, pp. 311–323.

(11) ROSENTHAL, Y., MARK-MARKOWITCH, M., STERN, A., and ELIEZER, D. (1984) Tensile flow and fracture behaviour of austenitic stainless steels after thermal aging in a hydrogen atmosphere, *Mat. Sci. Engng*, **67**, 91–107.

(12) LELÉ, M. V. and DE MIRANDA, P. E. V. (1984) Fragilização e Encruamento do Aço Inoxidável AISI 304 Hidrogenado, *Metalurgia ABM*, **40**, 673–678.

(13) HUWART, P., HABASHI, M., FIDELLE, J. P., GARNIER, P., and GALLAND, J. (1985) Proprietés mechaniques d'aciers inoxydables austenitiques stable (ZXNCTD 26-15) et instable (Z 2 CN 18-10). Role des traitements thermiques et de l'hydrogene cathodique, *Proceedings of the 7th ICSMA*, Montreal, pp. 1099–1104.

(14) SILVA, T. C. V., PASCUAL, R., and DE MIRANDA, P. E. V. (1984) Hydrogen induced surface effects on the mechanical properties of type 304 stainless steel, *Fracture prevention in energy and transport systems* (Edited by Le May, I., and Monteiro, S. N.), (EMAS), pp. 511–520.

(15) WHITEMAN, M. B. and TROIANO, A. R. (1965) Hydrogen embrittlement of austenitic stainless steel, *Corrosion*, **21**, 53–56.

(16) WASIELEWSKI, R. C. and LOUTHAN Jr, M. R. (July 1983) Hydrogen embrittlement of type 316 stainless steel, *Proceedings of the 16th Annual IMS Meeting*, Calgary, Canada.

(17) EVANGELISTA, G. E. and DE MIRANDA, P. E. V. (1984) Efeitos superficiais provocados pelo hidrogênio no aço AISI 304, *Metalurgia ABM*, **40**, 501–506.

(18) SILVA, T. C. V. and DE MIRANDA, P. E. V. (1985) Caracterização das microtrincas induzidas pelo hidrogênio no aço inoxidável austenítico, *Proceedings of the 40th Annual ABM Meeting*, pp. 79–94.

(19) TÄHTINEN, S., KIVILAHTI, J., and HÄNNINEN, H. (1985) Crystallography of hydrogen-induced surface cracking in a spherical austenitic stainless steel crystal, *Scripta Met.*, **19**, 967–972.

(20) CHANG, N. S. and HAWORTH, W. L. (1985) Fatigue crack initiation and early growth in an austenitic stainless steel, *Proceedings of the 7th ICSMA*, Montreal, pp. 1225–1230.

(21) SIGLER, D., MONTPETIT, M. C., and HAWORTH, W. L. (1983) Metallography of fatigue crack initiation in an overaged high-strength aluminium alloy. *Met. Trans*, **14A**, 931–938.

(22) BASINSKI, Z. S. and BASINSKI, S. J. (1985) Low amplitude fatigue of copper single crystals-II. Surface observations, *Acta Met.*, **33**, 1307–1317.

(23) SCHUSTER, G. and ALTSTETTER, C. (1983) Fatigue of stainless steel in hydrogen, *Met. Trans*, **14A**, 2085–2090.

(24) LOUTHAN, M. R. and DERRICK, R. G. (1975) Hydrogen transport in austenitic stainless steel, *Corrosion Sci.*, **15**, 565–577.

MICROSTRUCTURAL
CONSIDERATIONS

W. J. Baxter*

The Growth of Persistent Slip Bands During Fatigue

REFERENCE Baxter, W. J., **The Growth of Persistent Slip Bands During Fatigue**, *The Behaviour of Short Fatigue Cracks*, EGF Pub. 1 (Edited by K. J. Miller and E. R. de los Rios) 1986, Mechanical Engineering Publications, London, pp. 193–202.

ABSTRACT One of the early manifestations of fatigue is the appearance of persistent slip bands, some of which eventually become sites for fatigue cracks. This paper describes some of the first systematic quantitative observations of the early stages of growth of individual persistent slip bands (psbs) in polycrystalline materials. Specimens of 6061-T6 aluminium were fatigued in a specially constructed photoelectron microscope. The psbs are shown to consist of a periodic linear array of semicircular extrusions ~1 μm in diameter. Initially a single extrusion appears at a site within a grain. The psb elongates across the grain by the sequential addition of further extrusions. The psbs elongate rapidly at first and then more slowly, following a parabolic dependence on the number of fatigue cycles. A model is proposed based upon a periodic array of 'cells' in the psb, similar to the dislocation structures which have been observed in other materials. The model agrees with the experimental observations, provided that the cyclic strain amplitude in the psb is at least ten times that in the matrix of the grain.

Introduction

The earliest visible manifestation of metal fatigue is the development of surface deformation in the form of so-called persistent slip bands (psbs) within some of the grains of the metal (1)(2). These linear features are rendered visible by their profile, which in cross section consists of extrusions and intrusions. This profile can be very pronounced and is indicative of very severe localized deformation. Since this process eventually results in the formation of a fatigue crack, either along the grain boundary at the tip of a psb, or along the psb itself (3), it clearly plays an important role. Therefore, psbs have been studied very extensively by a variety of techniques. The following findings pertain to the present investigation.

(1)   The cyclic strain within a psb is much greater than that in the surrounding matrix of the grain (4). This holds true not only for alloys but also for pure single crystals, where the creation of psbs correlates with a plateau in the cyclic stress–strain curve (5).

(2)   The microstructure within a fully formed psb is quite different from that in the adjacent matrix. For example, transmission electron microscopy of precipitate-strengthened aluminium alloys has shown that the cyclic motion of dislocations effectively destroys small precipitates within the psb, so that it consists of a thin (~0.1 μm) planar layer of single-phase

* Physics Department, General Motors Research Laboratories, Warren, Michigan 48090-9055, USA.

material (6)(7). As another example, electron microscopy of pure single crystals shows that there is a marked difference in the dislocation structure between psbs and the matrix. Within the psb there are walls of dislocations arranged in patterns known as ladders and cells (8)(10) while in the adjacent matrix dislocations are more randomly dispersed. The presence of similar but less developed dislocation structures has also been reported in psbs in polycrystalline aluminium alloys (6)(11).

(3)   The above knowledge has been obtained primarily from post mort_m examination of well developed psbs. There is much less information on the manner in which the psb forms. In particular there are hardly any measurements of the growth kinetics, which is in marked contrast to the extensive literature on the growth of fatigue cracks (12)(13).

This paper describes measurements of the elongation of psbs in polycrystalline 6061-T6 aluminium by means of a photoelectron microscope equipped with a fatigue stage (14). This technique is sensitive enough to detect the early stages of psb formation and is non-destructive so that it provides sequential information during a fatigue test. It will be shown that psbs appear on the surface as an approximately periodic linear array of small ($\sim$1 μm) extrusions. Initially a single extrusion forms. Then the psb elongates, with the addition of more extrusions, at a rate which varies inversely with its length. This behaviour is explained in terms of a concentration of the cyclic strain within the psb.

Experimental

Small specimens of 6061-T6 aluminium (1% Mg, 0.6% Si, 0.3% Cu, 0.2% Cr) were machined from sheet material 1.5 mm thick, with pancake shaped grains of average diameter $\sim$30 μm. On some of the specimens the surface to be examined was mechanically polished, to aid subsequent examination by conventional microscopy. The majority were studied with the as-received surface texture. All the specimens were degreased with acetone and cleaned by immersion in chromic acid at 75°C for five minutes. After rinsing with water and alcohol, they were anodized in a 3 per cent solution of tartaric acid at a potential of 10 volts to form a surface oxide film 14 mm thick.

The geometry of the specimens for the photoemission electron microscope (PEM) is shown in Fig. 1. These specimens were fatigued by reverse bending *in vacuo* (10^{-3} Pa) in the specimen chamber of the microscope, to produce surface maximum cyclic strains of $\pm 3.0 \times 10^{-3}$. In the photoemission electron microscope, the specimen is illuminated by ultraviolet radiation and electrons emitted from the specimen form a magnified image of the surface on a fluorescent screen. As the specimen is fatigued, the emergence of psbs ruptures the anodic oxide film exposing the fresh metal surfaces of the extrusions. This extruded material emits much more intensely, creating the white spots or lines in the photoelectron micrographs shown in the next section. This phenomenon is also known as exoelectron emission (14).

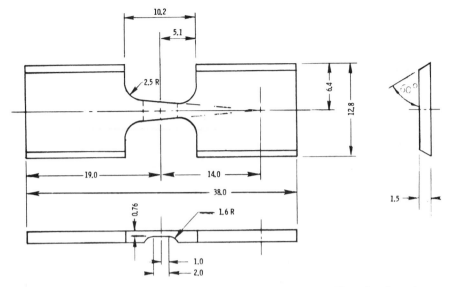

Fig 1   Geometry of specimens used in the photoelectron mciroscope (dimensions in mm)

Results

The photoelectron micrographs in Fig. 2 were obtained at intervals during the fatigue cycling of an unpolished specimen. The three long straight lines of emission in these micrographs correspond to fiducial scratches, which identify the location on the surface of the specimen and provide a calibration of magnification. This sequence shows the development of many sources of intense emission, where the 14 nm oxide film has been ruptured by emerging psbs. A distinctive feature is that each psb appears initially as a small spot of exoelectron emission and subsequently elongates with continued fatigue cycling. This behaviour is clearly illustrated by the three psbs labeled A, B, and C. The positions of these psbs were measured with respect to the fiducial scratches, so that their point of initiation could be identified in subsequent micrographs. These initiation points are marked by the tips of the arrows in Fig. 2(f). In each case the psb elongated in both directions, showing that it originated somewhere in the interior of a grain.

The growth of these three psbs is summarized quantitatively in Fig. 3, which shows that the length of each psb increased as the square root of the number of fatigue cycles. The psb at A grew the most slowly despite having been the first to emerge, while psb C apparently experienced some resistance to growth at the very beginning.

Essentially identical results to the above were also obtained from polished specimens and have been reported elsewhere (15). Subsequent examination of both polished and unpolished specimens in a scanning electron microscope

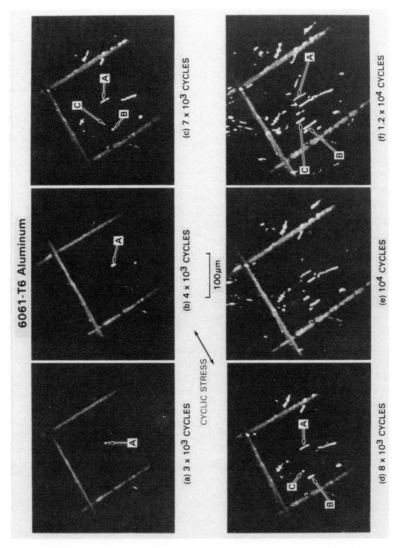

**Fig 2   Photoelectron micrographs showing development of exoelectron emission from psbs pro-
duced by fatigue of an unpolished specimen**

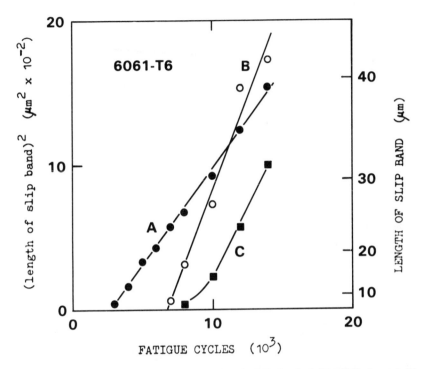

Fig 3   Effect of fatigue cycles on the square of the length of slip bands A, B, and C, shown in Fig. 2

(SEM) revealed that the psb extrusions were nearly identical in appearance. However, the polished specimens were of advantage in that the extrusions could be seen more clearly in the SEM. An example is provided by the scanning electron micrograph in Fig. 4, which shows two psbs consisting of quite regular arrays of individual extrusions. These extrusions are seen more clearly at higher magnification in Fig. 5, each one measuring ~1 μm parallel to the psb. This periodicity was evident on a large number, but not all, of the psbs; it tends to be masked in the later stages of growth when the extrusions become more pronounced. (This process has just started on an extrusion at the lower right in Fig. 5.) On the other hand, in the early stages of growth when a psb is still scarcely visible in the SEM, it often appears as a pair of spots in the PEM (i.e., there are already two extrusions). Thus, this perodicity seems to be an inherent feature of the psbs, and elongation occurs by the sequential addition of individual extrusions.

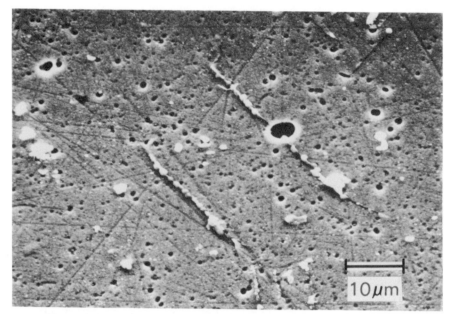

Fig 4   Scanning electron micrograph showing the array of extrusions on two psbs

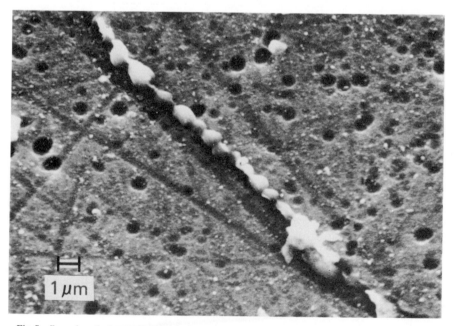

Fig 5   Scanning electron micrograph showing an enlarged view of the lower psb in Fig. 4

Discussion

Let us now consider the above findings in more detail.

1. *Extrusions*

The periodic nature of the extrusions (Fig. 4) is not considered to be unique to this alloy, since an identical periodicity has been reported in iron (**16**). This periodicity shows that within the psb there must be an array of barriers to dislocation motion with a spacing of ~1–2 μm. Unfortunately, transmission electron micrographs of the dislocation structure in psbs in 6061-T6 aluminium are not available at the present time, but a dislocation cell structure has been observed in iron (**16**) and other aluminium alloys (**6**)(**11**) and the cells are typically ~1–2 μm across. In addition, the extrusions are not unlike those observed by Laufer and Roberts (**10**) on single crystals of copper, and which they correlated directly with the sub-grain or dislocation cell structure of the psbs. Thus, there is circumstantial evidence to suppose that such a 1–2 μm cell structure is also present in the psbs in 6061-T6 aluminium, and that the periodic extrusions mark the presence of these cells.

2 *Elongation process*

The initial formation of a single extrusion, followed by the systematic addition of other extrusions, shows that the 'cell' structure is developing either in concert with, or shortly before, the appearance of the individual 'single cell' extrusions. Thus, the location of the initiatory 'cell' and extrusion is probably determined by either an inherent weak spot or stress concentration within the grain, where deformation occurs preferentially. At that site we can expect dispersal of the strengthening precipitates and a build up of dislocation density. It can then be visualized that the dislocation density in that region may increase to the point of instability, so that it is energetically favourable for the dislocations to rearrange into a cell structure, by a mechanism such as that proposed by Kuhlmann-Wilsdorf and Nine (**17**). Once the initiatory cell has formed, the sudden ease of dislocation motion within the soft interior of the cell can now result in the rapid growth of the first extrusion. By the time this initiatory cell has produced a small extrusion, say ~100 nm high, ~300 dislocations will have arrived at the side walls of the cell. This is far too many to be accommodated in a pile up, so the associated stresses will promote the emission of dislocations from the cell walls into the adjacent matrix material of the grain. This flux of dislocations will again produce localized changes in the microstructure of the matrix, so that the process will be repeated. In this way the psb will grow into a two dimensional array of cells with an associated array of surface extrusions.

3 Rate of elongation

The length (l) of a psb increases with the number of fatigue cycles (N) in accordance with the relation

$$L^2 \sim (N - N_o) \tag{1}$$

where N_o is the number of cycles required to produce the first extrusion. Thus, the rate of elongation is of the form

$$\frac{dl}{dN} \sim l^{-1} \tag{2}$$

Thus, the rate at which the new cell and associated extrusion are generated at the tip of a slip band decreases as the slip band elongates.

To account for this behaviour we will start with a simple two phase model (psb and matrix) of the type used previously to describe the cyclic stress strain curve (**18**). According to this model the cyclic strain is not uniformly distributed throughout a grain because the interior of the psb is known to be softer than the matrix. However, the cyclic strain in the psb (ε_c) is assumed to be uniform, i.e., the cells are all the same and the uniform deformation occurs in the soft interior of each of the cells. In addition, we will assume, based upon the physical process of psb elongation outlined above, that

$$\frac{dl}{dN} = \beta \varepsilon_c \tag{3}$$

where β is a constant.

Consider a grain of diameter D in the surface of the specimen which has developed a psb of length l and thickness t. In a constant displacement fatigue test, as in these experiments, the total surface cyclic strain (ε_T) accommodated by the grain is constant and is usually perpendicular to the slip band (see, for example, slip bands in Fig. 2).

Applying a simple rule of mixtures

$$\varepsilon_T = f_c \varepsilon_c + (1 - f_c)\varepsilon_g \tag{4}$$

where ε_g is the cyclic strain in the matrix of the grain, and

$$f_c = tl/D^2 \tag{5}$$

is the areal fraction of the grain occupied by the psb. Combining equations (4) and (5) yields the following expression for the cyclic strain within the psb

$$\varepsilon_c = \frac{D^2}{tl} (\varepsilon_T - \varepsilon_g) + \varepsilon_g \tag{6}$$

Substituting in equation (3)

$$\frac{dl}{dN} = \frac{\beta D^2}{tl} (\varepsilon_T - \varepsilon_g) + \beta \varepsilon_g \tag{7}$$

The first term in this equation is of the same form as the empirical equation (1). Indeed if the second term can be ignored, integration yields

$$l^2 = \frac{2\beta D^2}{t}(\varepsilon_T - \varepsilon_g)(N - N_o) \tag{8}$$

This equation is consistent with the results plotted in Fig. 3. Thus, it may be concluded that the first term in equation (6) is dominant. This term is simply that portion of the cyclic strain within the psb which is in excess of the cyclic strain experienced by the rest of the grain (ε_g). In other words, it is the process of strain localization within the psb that controls the elongation of the psb.

For typical values of $D = 50$ μm and $t = 0.1$ μm, the geometrical strain concentration factor D^2/tl in equation (6) is $\sim 10^4$ for the initial extrusion, and decreases only to 500 by the time the psb extends across the grain (i.e., when $l = D$). Thus, $\varepsilon_c \gg \varepsilon_g$ even for small values of $(\varepsilon_T - \varepsilon_g)$; i.e., the strain in the psb is very large and controls the elongation process but contributes very little to the total strain ε_T.

Finally, from Fig. 3 we see that the parabolic growth relationship holds for $l \leqslant 40$ μm. Thus, even at the end of these observations $\varepsilon_c \sim 10\varepsilon_g$ or $\varepsilon_c = \pm 3 \times 10^{-2}$. When only the first cell and extrusion were formed, $l \sim 2$ μm and $\varepsilon_c \sim \pm 6 \times 10^{-1}$.

Conclusions

The following picture emerges from these measurements and modelling of the growth of persistent slip bands in 6061-T6 aluminium.

(1)   The psb appears as a linear array of surface extrusions with a periodicity of ~ 1 to 2 μm.
(2)   The psb elongates by the addition of further extrusions, the total length increasing as the square root of the number of fatigue cycles.
(3)   Each extrusion is thought to be associated with a characteristic cell in the microstructure of the psb.
(4)   The cyclic strains (ε_c) within the psb are very large, encompassing a range of 10–100 times greater than the applied macroscopic strain.
(5)   The value of ε_c decreases as the psb elongates.
(6)   The cyclic strain in the leading cell at the tip of a psb ejects a high density of dislocations into the adjacent material, which in some manner creates a new cell and associated extrusion.

Acknowledgements

The author is grateful to T. R. McKinney for the photoelectron micrographs in Fig. 2, and for the scanning electron micrographs in Figs 4 and 5.

References

(1) See for example, the review by LAIRD, C. and DUQUETTE, D. J. (1971) Mechanisms of fatigue crack nucleation, *Corrosion fatigue: chemistry, mechanics, and microstructure, NACE-2*, pp. 88–117.

(2) FORSYTH, P. J. E. (1969) *The physical basis of metal fatigue* (Blackie and Šon, London).

(3) MUGHRABI, H., WANG, R., DIFFERT, K., and ESSMANN, U. (1983) Fatigue crack initiation by cyclic slip irreversibilities in high-cycle fatigue, *Fatigue mechanisms: Advances in quantitative measurement of physical damage, ASTM STP 811*, p. 5.

(4) LEE, J. and LAIRD, C. (1983) Strain localization during fatigue of precipitation hardened aluminium alloys, *Phil. Mag. A.*, **47**, 579–597.

(5) MUGHRABI, H. (1979) Plateaus in the cyclic stress–strain curves of single and poly-crystalline metals, *Scripta Met.*, **13**, 479.

(6) STUBBINGTON, C. A. (1964) Some observations on microstructural damage produced by reversed glide in an Aluminum 7.5% zinc 2.5% magnesium alloy, *Acta Met.*, **12**, 931–939.

(7) VOGEL, W., WILHELM, M., and GEROLD, V. (1982) Persistent slip bands in fatigued peak aged Al-Zn-Mg single crystals I and II, *Acta Met.*, **30**, 21, 31.

(8) See the review by LUKAS, P. and KLESNIL, M. (1971) Fatigue damage and resultant dislocation substructures, *Corrosion fatigue: chemistry, mechanics, and microstructure, NACE -2*, pp. 118–132.

(9) WANG, R., MUGHRABI, H., McGOVERN, S., and RAPP, M. (1984) Fatigue of copper single crystals in vacuum and in air. I, *Mater. Sci. Engng*, **65**, 219–233.

(10) LAUFER, E. E. and ROBERTS, W. N. (1966) Dislocations and persistent slip bands in fatigued copper, *Phil. Mag.*, **14**, 65–78.

(11) LYNCH, S. P. (1979) Mechanisms of fatigue and environmentally assisted fatigue, in *Fatigue mechanisms, ASTM STP 675*, pp. 174–213.

(12) SURESH, S. and RITCHIE, R. O. (1984) Propagation of short fatigue cracks, *Int. Met. Rev.*, **29**, 445.

(13) These proceedings.

(14) BAXTER, W. J. (1983) Exoelectron emission, *Treatise on materials science and technology 19B* (Academic Press, New York), p. 1.

(15) BAXTER, W. J. (1984) What are the kinetics of slip band extrusion? *Fundamental questions and critical experiments on fatigue, ASTM STP*, in press.

(16) MUGHRABI, H., ACKERMANN, F., and HERZ, K. (1979) Persistent slip bands in fatigued fcc and bcc metals, *Fatigue mechanisms, ASTM STP 675*, pp. 69–105.

(17) KUHLMANN-WILSDORF, D. and NINE, H. D. (1967) Striations on copper single crystals subjected to torsional fatigue II, *J. Appl. Phys.*, **38**, 1683–1693.

(18) PEDERSEN, O. B., RASMUSSEN, K. V., and WINTER, A. T. (1982) The cyclic stress strain curve of polycrystals, *Acta Met.*, **30**, 57.

P. Mulvihill * *and C. J. Beevers* *

The Initiation and Growth of Intergranularly Initiated Short Fatigue Cracks in an Aluminium 4.5 per cent Copper Alloy

REFERENCE Mulvihill, P. and Beevers, C. J., **The Initiation and Growth of Intergranularly Initiated Short Fatigue Cracks in an Aluminium 4.5 per cent Copper Alloy**, *The Behaviour of Short Fatigue Cracks*, EGF Pub. 1 (Edited by K. J. Miller and E. R. de los Rios) 1986, Mechanical Engineering Publications, London, pp. **203–213**.

ABSTRACT The initiation and growth of short fatigue cracks with surface length in the range 60–1000 μm in a peak aged Al 4.5 wt% Cu alloy with a grain size of 200–400 μm has been studied. Testing was carried out on small plane specimens in three point bend at stress levels below that for gross yield. Observations were made using cellulose acetate replication techniques and a miniature bending rig for use in the scanning electron microscope (SEM).

In all cases crack initiation occurred at grain boundaries. Transmission electron microscope (TEM) examination of the boundaries revealed the presence of the equilibrium θ precipitates and an accompanying precipitate-free zone.

Surface growth of grain boundary cracks was observed to be retarded at inclusions, grain boundary triple points, and positions of transition from intergranular to transgranular growth.

Introduction

In the studies of short fatigue crack propagation in aluminium alloys most of the experimental work has been done on either machined-down long pre-cracks or on contained cracks initiated at intermetallic particles either by cracking of the particle itself or cracking of the particle–matrix interface (**1**)(**2**). However, these particle initiation mechanisms are not the only ones known to occur in aluminium alloys. The other mechanisms are either associated with persistent slip bands or grain boundaries.

Initiation of fatigue cracks at grain boundaries in single phase materials can take place even though the boundaries themselves are not inherently weak. This occurs by the so called 'geometric rumpling' mechanism and is due to severe deformation within the grains causing what are effectively micro-notches along the boundaries and subsequent crack initiation (**3**).

In multi-component materials the situation is complicated by the occurrence of segregation. This paper examines the initiation and growth of cracks in a precipitation hardening system which has a number of special segregation problems associated with it.

After initiation the growth of short cracks has been found to be affected by grain boundaries in a number of materials and at second phase boundaries, particularly in steels (**4**). This work looks at another aspect of short crack growth where the growth mechanism changes from intergranular to trans-granular.

* Department of Metallurgy and Materials, University of Birmingham.

Material and test methods

The Al, 4.5% Cu alloy had a yield stress of 258 MPa, a tensile strength of 340 MPa, and an elongation of 26 per cent.

The material was made from high purity base metals in the laboratory. Solution treatment was carried out at the relatively low temperatures of 500°C to ensure that any furnace temperature fluctuations did not result in localized grain boundary melting. Between solution treatment and ageing no additional deformation was given. An ageing curve was then determined for the material at 160°C. The alloy was aged for 16 hours to obtain peak hardness.

Fatigue testing was carried out using a small three point bend rig, using plain rectangular specimens of dimensions 6 × 15 × 32 mm, on an Amsler vibrophore at a frequency of approximately 100 Hz. The maximum tensile stress at the surface of the specimen is given by the equation

$$\sigma_{max} = \frac{3PL}{2Bw^2} \tag{1}$$

where

P = applied load
L = length between upper loading points (24 mm)
B = width of specimens (15 mm)
w = thickness of specimen (6 mm)

All tests were carried out at surface stress levels below the tensile yield stress at an R ratio of 0.15. Specimens were polished to a 6 μm finish then electro-polished in a 3 per cent perchloric acid, 32 per cent butoxyethanol, 65 per cent methanol solution at -50°C with an open circuit voltage of 60 V for five minutes in an effort to remove any residual stresses resulting from machining.

Crack detection and growth measurements were made using cellulose acetate replicas which were examined at ×400 magnification using a light microscope with a vernier eyepiece. Using this method crack length changes down to 1 μm could be detected, although at least 5 μm growth or 10^6 cycles were generally allowed between measuring increments.

For further detailed examination replicas were sputtered with gold and then plated with copper from a copper sulphate solution. The metal was then stripped from the plastic to give a positive replica of the specimen surface which could be examined in the scanning electron microscope. In addition to this method an *in situ* three point bend rig was made to enable specimens to be deflected in the electron microscope. The loading dimensions of the *in situ* rig were the same as those of the rig on the fatigue testing machine and the loadings were calibrated directly using a strain gauge on each specimen.

In view of the initiation mechanism which was observed a number of TEM foils were thinned using twin electro-jet polishing at -30°C with a closed circuit voltage of 14 V in a 30 per cent nitric acid in methanol solution. The foils were then examined in a Philips EM 400 scanning transmission electron microscope.

Results and discussion

It was observed in more than 30 specimens that crack initiation invariably took place at grain boundaries. The initiation event, that is, the number of cycles after which a crack could be unmistakably observed at lengths of around 50 μm, occurred after approximately 10^5 cycles at a maximum stress of 80 per cent of the tensile yield stress.

The result of the TEM examination of the material is shown in Fig. 1. The micrograph shows the area around a grain boundary. It can be seen that along the boundary there were a number of elongated θ precipitates surrounded by a light area, the precipitate-free zone, with the remainder of the material being made up of the θ'' and θ' precipitates associated with peak hardness in the matrix.

The reasons for the occurrence of precipitate-free zones are now well established and are essentially bound up with the fact that the boundaries are low strain and surface energy sites for precipitation and sinks for vacancies (5).

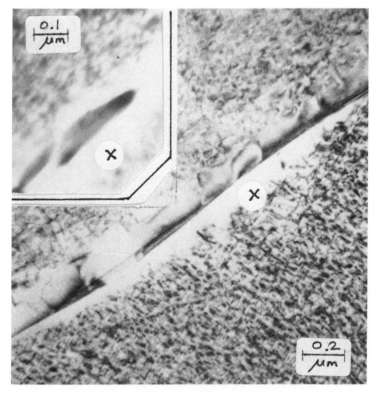

Fig 1   Transmission electron microscope micrographs of the experimental material showing θ precipitates (x) lying in a grain boundary precipitate free zone

In later experiments on an alloy of 0.5 per cent lower copper content, fatigue testing in the solution-treated and quenched condition produced extensive persistent slip band formation with no grain boundary cracking. When subsequently peak aged and re-tested, initiation took place at large intermetallic intragranular inclusions in the more usual way. This result would tend to confirm the conclusion that in the peak aged alloy the precipitate-free zones rather than any stray embrittling impurities cause crack initiation since the presence of any impurities on the boundaries would be just as likely in the solution-treated as in the aged condition. Fatigue testing in the solution-treated condition produced plastic strains within the grains creating additional vacancies and favourable sites for intragranular precipitation thereby making precipitation on the boundaries and the formation of associated precipitate-free zones less likely.

The crack propagation data has been presented in terms of both log c, where c is the half surface crack length from tip to tip, and log ΔK, where, for a semi-circular surface crack, $K = 0.8847\ \sigma\sqrt{(\pi c)}$ (6). Neither parameter is entirely satisfactory for a number of reasons (7) but both are useful in comparing cracks of a similar length in the same material.

Figure 2(a) shows growth rate data for a crack growing in the initiating grain boundary. It was found that, when uninterrupted by microstructural features, the crack growth rate increased with increasing crack length, as would be expected.

Figure 2(b) shows data for a crack which grew along the initiating boundary, section A, until it encountered a triple point at B. At this point the crack growth was retarded until continued growth took place in a transgranular manner at C after a 6 per cent increase in stress range. This type of retardation at triple points was observed a number of times. The reason for this behviour is unclear, but it is perhaps useful to regard the material as being made up of what are essentially two 'regions': the first being a soft grain boundary region of a dilute aluminium copper solid solution, and the second being the harder matrix region of the precipitation-hardened material. Hence, for crack growth at the same rate, the crack length or ΔK required would be higher in the matrix region than in the grain boundary. This explanation is similar to that put forward by de Los Rios et al. (4) for growth of cracks along prior austenite grain boundaries in a steel.

Alternatively the problem may be seen as similar to the grain boundary retardations reported by other workers (8) for entirely transgranular cracks. This explanation requires that a crack builds up plasticity and reorientates itself for growth in the next grain.

Although growth mechanism changes took place at triple points in an approximately equal number of cases the growth mechanism remainded unchanged at the first triple point. This continued intergranular growth could also result in growth rate changes which were due to two factors. Since the crack lengths are measured normal to the stress axis any increased deviation from this

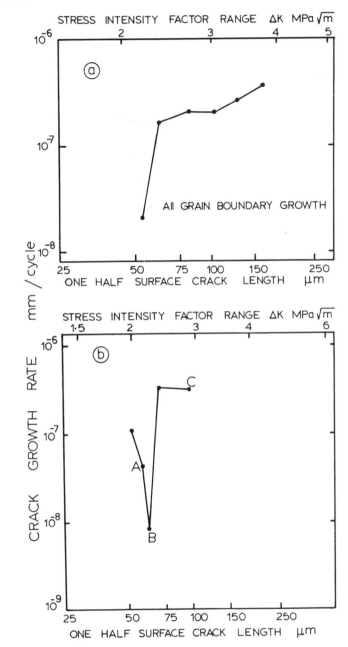

Fig 2 Fatigue crack propagation data.
(a) An uninterrupted crack growing in the initiating boundary
(b) A grain boundary crack (A) retarded at a triple point (B). Transgranular growth took place at
    (c) after a 6 per cent increase in stress

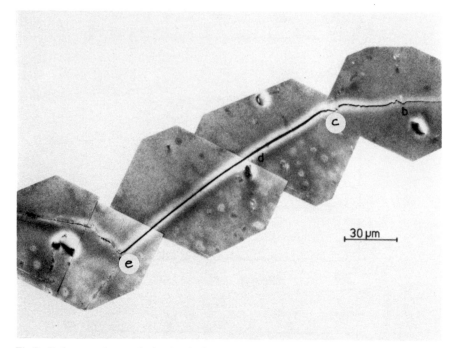

Fig 3 **Fatigue crack growth showing both transgranular and intergranular growth from triple points (c) and (e), respectively. The photograph was taken at maximum load in the *in situ* rig and the tensile stress axis is vertical**

direction results in an apparent decrease in growth rate. (Conversely, any decreased deviation will result in an apparent increase in growth rate). Hence, for individual cracks, apparent propagation rates may change markedly. Again, associated with deviation of the crack growth direction from the stress axis normal is the mechanism of crack deflection described by Suresh (**9**), where the change in crack plane results in an overall reduction in the stress intensity range and a real growth-rate reduction.

An example of a crack showing both types of triple point behaviour is shown in Fig. 3.

Figure 4 shows a number of examples of short crack behaviour. The data in the centre is for a crack which was retarded (A) at an inclusion on the initiating boundary, a replica of which is shown in Fig. 5. The surface length, $2c$, of this crack was 95 μm and the inclusion was 9.5 μm in diameter. It has been observed that only inclusions of a size greater than 5 μm had any effect on propagation rates, and the maximum length of crack for which any significant interactions with microstructure took place was 500 μm.

The retardation of grain boundary cracks at surface inclusions may be due to prevention of slip at the crack tip. Since any inclusion will only occupy a small fraction of the crack front – in the case described about 3 per cent if it is assumed

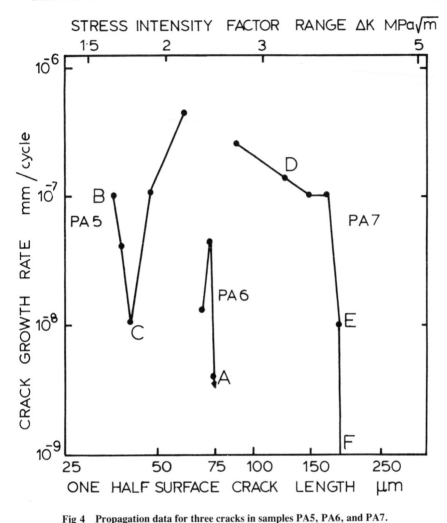

Fig 4   Propagation data for three cracks in samples PA5, PA6, and PA7.
A   one crack tip at an inclusion the other tip at a grain boundary triple point
B and D   grain boundary growth
C   crack branching
E   one crack tip growing transgranularly and the other tip growing intergranularly
F   transgranular crack tip arrested near grain boundary, the other tip arrested in
           the boundary

to be spherical – then it could be argued that the important slip for growth takes place close to the surface of the specimen. However, it may also be possible that any inclusion of a reasonable size on a crack front of this length would cause an arrest or retardation. If this is so it would tend to indicate that at short crack lengths the difference between growth and non-growth is only marginal.

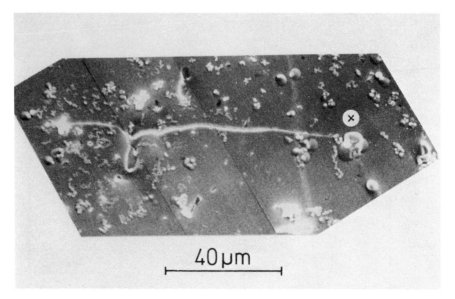

Fig 5   SEM micrograph of a replicated crack retarded at an inclusion (x). The tensile stress axis is vertical

The data on the left in Fig. 4 is for a crack which changed growth mechanism from intergranular (B) to transgranular (C) remote from a triple point. Often with such growth mode changes was a period of zero surface growth, possibly to allow plasticity to be set up in the grain neighbouring the boundary best orientated for slip. The initial transgranular growth was then observed to be by a Mode II opening in a crystallographic manner which then changed to a Mode I opening for Stage II-type transgranular growth. An example of such growth is shown in Fig. 6.

The data on the right in Fig. 4 is for a crack in which plasticity was observed, by interference contrast, to have been built up in a grain neighbouring the boundary; however, a change in growth mechanism did not take place and no crack length increase was observed in over 10^7 cycles.

The greatest disadvantage of any surface observations of contained cracks is that no knowledge can be gained of any sub-surface events; hence, the reason why any grain boundary crack should change growth mode remote from a triple point at any particular point must remain an unknown. It seems likely, however, that some sub-surface disruption of growth, possibly at an inclusion or a triple-point, arrests a crack for a sufficient time for a mature plastic zone to be set up and transgranular growth to occur.

An example, of a crack which had been marked with ink at maximum load and then fatigued to fracture in an effort to look at sub-surface events, is shown in Fig. 7. It can be seen that the crack had a somewhat eccentric shape and this

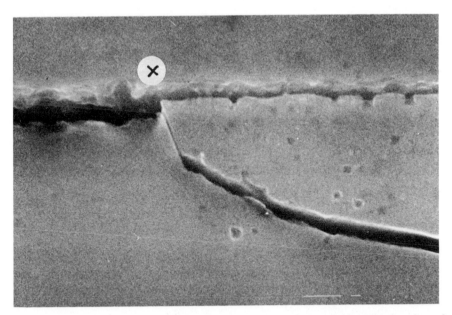

Fig 6   SEM micrograph (at maximum load) of the area in which the growth mechanism changed at X from intergranular to transgranular showing Mode II displacement. The tensile stress is vertical

has been associated with the change in growth mechanism from intergranular to transgranular at a triple point. The intergranular facets were smooth while those in the transgranular growth region were stepped. The non-uniform crack shape also illustrates some of the difficulties of taking propagation rate measurements from surface cracks. One such difficulty is the fact that comparing growth rates using 1/2 surface length c or ΔK relies on cracks having a known shape. In addition to this it would appear from the micrograph that even after a crack tip has passed a triple point on the surface the sub-surface influence of the triple point continues and disrupts the crack shape, thereby changing the surface growth rate.

Conclusions

(1)   Crack initiation in this peak aged Al 4.5% Cu alloy took place at grain boundaries. TEM examination revealed the presence of preciptate free zones at the boundaries.

(2)   Growth rates of the cracks were primarily affected by grain boundary triple points. Approximately 50 per cent of such interactions were associated with growth mechanism changes from intergranular to transgranular and 50 per cent were associated with crack deflection when the growth mechanism remained unchanged.

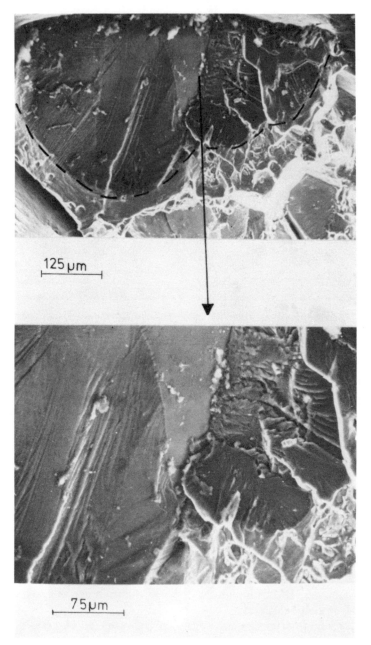

Fig 7   SEM micrographs of the non-uniform crack front associated with an intergranular to transgranular growth mode transition at a triple point

(3)   Intergranular growth was also inhibited by inclusions lying on the grain boundaries. Only inclusions greater than 5 μm in diameter had any observable effect on cracks less than 500 μm in length.

(4)   Growth mechanism changes remote from triple points were also significant. This type of transition was accompanied by a period of incubation followed by crystallographic transgranular growth with a Mode II opening.

References

(1)   MORRIS, W. L. (1978) The effect of intermetallics composition and microstructure on fatigue crack initiation in A12219 T851, *Met. Trans*, **9A**, 1345–1348.

(2)   PEARSON, S. (1975) Initiation of fatigue cracks in commercial aluminium alloys and subsequent propagation of very short cracks, *Engng Fracture Mech.*, **7**, 235–247.

(3)   LAIRD, C. and SMITH, G. C. (1963) Initial stages of damage in high stress fatigue in some pure metals, *Phil. Mag.*, **8**, 1945–1963.

(4)   DE LOS RIOS, E. R., TANG, Z., and MILLER, K. J. (1984) Short crack fatigue behaviour in a medium carbon steel, *Fatigue Engng Mater. Structures*, **7**, 97–108.

(5)   SMALLMAN, R. E. *Modern physical metallurgy*, 3rd Edition (Butterworths, London), pp. 425–431.

(6)   PICKARD, A. C. Private Communication.

(7)   MILLER, K. J. (1982) The short crack problem, *Fatigue Engng Mater. Structures*, **5**, 223–232.

(8)   LANKFORD, J. (1982) The growth of small fatigue cracks in 7075-T6 aluminium, *Fatigue Engng Mater. Structures*, **5**, 233–248.

(9)   SURESH, S. (1983) Crack deflection: implications for the growth of long and short fatigue cracks, *Met. Trans*, **14A**, 2375–2385.

A. Plumtree* and S. Schäfer*

Initiation and Short Crack Behaviour in Aluminium Alloy Castings

REFERENCE Plumtree, A. and Schäfer, S., **Initiation and Short Crack Behaviour in Aluminium Alloy Castings**, *The Behaviour of Short Fatigue Cracks*, EGF Pub. 1(Edited by K. J. Miller and E. R. de los Rios) 1986, Mechanical Engineering Publications, London, pp. 215–227.

ABSTRACT The initiation and early growth of fatigue cracks in strain cycled squeeze-formed aluminium–7% silicon alloy castings have been investigated using single-stage replication techniques. Cracks initiated at the silicon particles in the interdendritic region, generally at triple points. Cracking of the silicon particles took place together with debonding of the aluminium–silicon interface at high cyclic strains, whereas at the low strain ranges only the latter took place. In all cases, fatigue cracks propagated through the interdendritic regions.

The critical length of a short fatigue crack was directly related to the dendrite spacing; cracks of about twice this length and less displayed initially fast growth rates. At high cyclic strains they continued to grow at a constant rate. At low cyclic strains the cracks experienced deceleration until they passed through the first triple points in the interdendritic regions after initiation. Subsequently they accelerated to blend with the crack growth rate for long cracks. This crack growth behaviour was explained in terms of larger plastic strains existing at the tips of short cracks when compared to long cracks.

Introduction

The behaviour of short fatigue cracks has been described as 'anomalous' because the propagation rate does not correspond to that seen for long cracks. For the latter, the log. crack propagation rate is a linear function of log. crack length. In contrast, short fatigue cracks demonstrate different types of propagation behaviour, namely, acceleration, deceleration to crack arrest, or deceleration followed by acceleration. Long cracks do not propagate at levels below the threshold stress intensity factor range (ΔK_{th}), whereas it is known that short cracks grow below this level (1)(2). In fact, it is debatable whether linear elastic fracture mechanics (LEFM) is applicable to short cracks.

One attractive model for the prediction of propagation rates of short fatigue cracks based on crack tip strains has been propounded by Lankford (2) and developed by de los Rios *et al*. (3). This model allows for arrest, deceleration followed by subsequent acceleration, and continuous acceleration of short cracks. It includes the effect of grain boundary retardation of short fatigue cracks and is based on a larger plastic strain range at the tips of short cracks when compared with long cracks. The size of the crack-tip plastic zone is of the order of half the crack length (4). As the small crack grows and approaches a grain boundary or another barrier, the crack propagation rate will be reduced if crack-tip slip is restricted by this obstacle. The continued growth of the small crack requires the propagation of crack-tip slip into neighbouring grains.

* Department of Mechanical Engineering, University of Waterloo, Waterloo, Ontario, Canada.

The present work was undertaken to investigate the short fatigue crack growth behaviour in a squeeze-formed aluminium alloy and to examine the results in terms of the crack-tip slip model.

Experimental procedure

A commercial squeeze-formed aluminium–7% silicon casting alloy was used for the testing programme. The alloy conformed to British specification L99, which is similar to the Aluminum Association A356 designation. The material was heat-treated to the T-6 condition resulting in a 0.2 per cent offset yield stress of 256 MPa and ultimate tensile strength of 342 MPa.

Completely reversed strain controlled uniaxial cyclic testing ($R = -1$) was carried out at room temperature on smooth cylindrical specimens of 5 mm diameter with a gauge length of 14 mm. A sinusoidal waveform was used at frequencies between 0.1 Hz and 20 Hz. All tests were conducted on an MTS servo-controlled closed-loop electrohydraulic testing machine.

Fatigue crack propagation rates were determined using replication techniques. This also permitted a critical evaluation of crack initiation sites and crack propagation paths to be made. The replicas were examined using a JEOL JSM-840 scanning electron microscope (SEM).

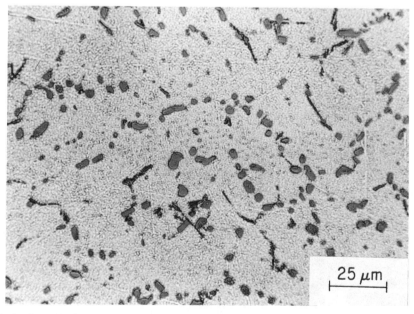

Fig 1   General microstructure of the squeeze-formed aluminium–7% silicon alloy. Etched with 5% HF in water

Results

The results of the cyclic strain-life tests allowed the relationship between total true strain range ($\Delta\varepsilon_T$) and number of reversals to failure ($2N_F$) to be expressed as follows

$$\Delta\varepsilon_T = 0.057(2N_f)^{-0.40} + 0.018(2N_f)^{-0.12} \tag{1}$$

The first term on the right-hand side of equation (1) is the plastic strain component and the second, the elastic component of total strain range.

During each test the stress response was monitored continually. For the lowest strain range employed ($\Delta\varepsilon_T = 0.36$ per cent) the stress range at half life (1×10^6 cycles) was 250 MPa.

The microstructures consisted of primary aluminium-rich dendrites surrounded by an interdendritic eutectic consisting of silicon particles (dark grey) and the aluminium-rich α-phase, as shown in Fig. 1. The dendrite spacing ranged between 25 μm and 45 μm. On strain cycling, crack initiation was associated with the silicon particles, and in particular those present at triple points of the interdendritic network. Figure 2 shows a silicon particle intact before fatigue testing and which cracked after 100 cycles at a total strain range of 1.0 per cent. Besides fracture of the silicon particles, debonding occurred at the high strain ranges, as also seen in Fig. 2. At the low strain ranges cracks

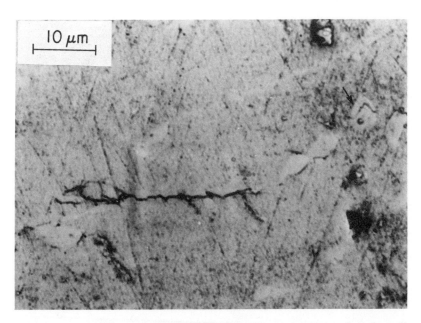

Fig 2 Fracture of silicon particle (indicated by arrow) as consequence of strain cycling at $\Delta\varepsilon_T = 1$ per cent after 80% life with stress response of 270 MPa. Also note fracture and debonding of other silicon particles

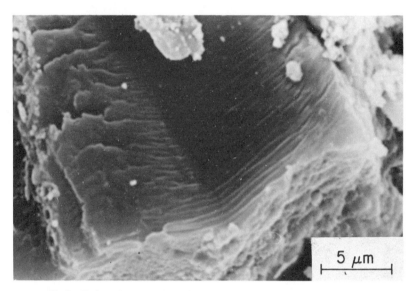

Fig 3   Fatigue fracture of eutectic aluminium in interdendritic region

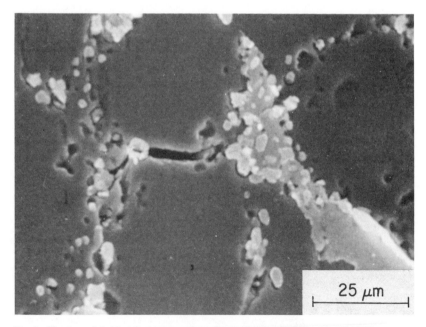

Fig 4   Short crack held at first triple points after initiation. Etched with 5% HF in water

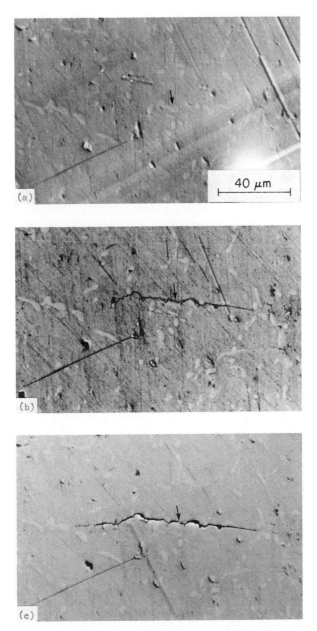

Fig 5   Surface crack development. Crack originated at triple point indicated by arrow
(a) 50 per cent life   (b) 65 per cent life   (c) 80 per cent life

formed by interfacial decohesion between the silicon particles and the matrix. Once initiated, the cracks progressed by fatigue failure through the eutectic α-aluminium in those parts of the interdendritic areas where silicon particles were not present. Figure 3 shows transgranular fracture which has occurred in the eutectic aluminium within this interdendritic region.

In general, cracks were found to be macroscopically transverse to the stress axis, yet microscopically they meandered a considerable amount. The cracks followed the eutectic areas rather than cutting directly through the primary aluminium-rich dendrites and this is illustrated in Fig. 4. A small interdendritic crack is seen which has grown until it reached the first triple points and then stopped.

The evolution of the catastrophic surface crack from initiation to final fracture is given in Fig. 5. These replica photographs show the crack at various stages in the fatigue life. The light coloured areas are silicon particles. No crack is visible in Fig. 5(a), which represents the surface of the specimen after 10^6 cycles. After 1.3×10^6 cycles a crack initiated at an interdendritic triple point and grew to a length of less than twice the triple point spacing, as seen in Fig. 5(b). It is important to note that the right side of the crack is held at the first triple point, after initiation, in the interdendritic network. After a further 0.3×10^6 cycles the crack grew to a length of two triple point spacings by progressing to the nearest triple point on the left side (Fig. 5(c)). No further growth had obviously taken place past the triple point on the right side. This specimen finally fractured after 2×10^6 cycles, indicating a rapid crack growth rate during the remaining 0.4×10^6 cycles.

The surface crack propagation rate (dl/dN) can be approximated by using the secant method (5)

$$dl/dN = \Delta l/\Delta N \qquad (2)$$

A log–log plot of dl/dN vs l, given in Fig. 6 shows the representative growth behaviour of fatigue cracks for low and high strains. The short cracks displayed anomalous behaviour. For low strains, a decrease in dl/dN was observed as the cracks approached the first triple points in the interdendritic network after initiation. In the majority of cases, since the cracks originated at interdendritic triple points, the minimum crack growth rate corresponded to two interdendritic facets. Once these triple points were overcome, the cracks accelerated and eventually their surface crack propagation rate became the same as that for long cracks. On the other hand, at high strains, the short cracks grew faster than those at the low strains yet they displayed a relatively constant or slightly increasing growth rate until they blended with the long crack growth behaviour.

In order to compare the present work on aluminium alloys with that of other workers, such as Lankford (2) who use stress intensity range factors (ΔK), it was necessary to convert the surface crack lengths (l) to crack depths (a) and express the results in terms of ΔK. Pearson (6) measured the shape of a fatigue crack initiated at a second phase particle from a plane surface in aluminium

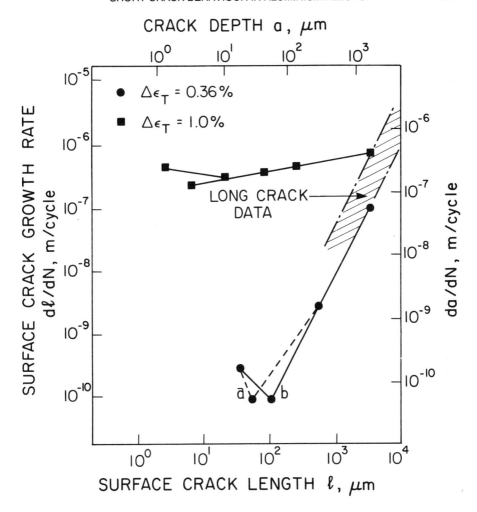

Fig 6 Variation of crack rate with crack length. Curve (a) represents cracks which have initiated between interdendritic triple points, whereas curve (b) represents those which have started at triple points

alloy and found that the crack, which had a 400 μm surface length, was approximately semi-circular in shape. Dowling (7) has reported that for axially loaded smooth specimens of approximately the same size as those used in this investigation, the crack depth was 'approximately equal to half the surface length' for all cracks examined. These cracks were between 250 μm and 1800 μm in surface length. Thus, it is taken that $a = l/2$ and this relationship is used for the plots involving crack depths in Figs 6 and 7.

By applying both the edge correction factor (8) and the approximation for an

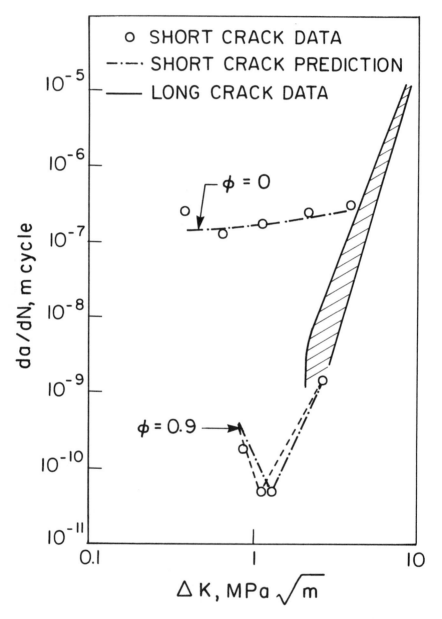

Fig 7   Application of the crack-tip strain model to the representative crack growth curves

embedded circular crack (9), and assuming that there are no crack closure effects so that the crack is open when the nominal applied load is tensile and closed when the applied load is compressive, the stress intensity factor range may be expressed by:

$$\Delta K = 0.713\sigma_t \sqrt{(\pi a)} \tag{3}$$

where σ_t is the applied tensile stress and da/dN is the mean crack depth corresponding to a. It should be noted that for the fully-reversed tests performed here, $\sigma_t = \Delta\sigma/2$. Thus, the crack propagation rate can be related to the fracture mechanics quantity ΔK at the midpoint of the crack growth interval.

The tensile stress σ_t may be determined from the hysteresis loops plotted during cyclic testing. The results of the calculations of da/dN and ΔK are plotted in Fig. 7 which includes a linear least squares regression for all of the long crack data in the form of the Paris law ($da/dN = C(\Delta K)^m$). The equation of the line is given by

$$da/dN = 4.7 \times 10^{-11}\Delta K^{5.8} \text{ (m/cycle)} \tag{4}$$

The long crack data tend towards a threshold stress intensity factor range near 2 MPa$\sqrt{\text{m}}$. This is similar to the threshold of 2 MPa$\sqrt{\text{m}}$ reported by Lankford for 7075-T651 (2).

Also included in Fig. 7 are the short crack results, which are emphasized by broken lines. The short crack results indicate two different types of behaviour depending upon strain range. For those specimens cycled at high strains, fatigue crack growth occurs at stress intensities below ΔK_{th}, and the cracks accelerate slowly and merge into long crack results. At the low strains, the cracks decelerate until they reach the first barriers. This is equivalent to one or two interdendritic facets depending on the position of the crack initiation site. Once these barriers are overcome, the cracks accelerate, as shown by the broken lines connecting the data points. This type of behaviour is predicted by the crack-tip plastic strain model for fatigue crack growth of short cracks.

Discussion

A number of fracture processes have been identified. These include decohesion between silicon particles and the aluminium-rich α-phase, cleavage of silicon particles, and striated growth of the aluminium-rich phase. Cracks grew preferentially through the eutectic rather than through primary aluminium dendrites. Failure of silicon particles took place when the cyclic stress amplitude was about 270 MPa, representing the stress response of the specimens cycled at high strains of $\Delta\varepsilon_T = 1.0$ per cent. Debonding between the aluminium-rich phase and the silicon particles occurred at both low and high strain ranges.

In general, the mechanism of crack propagation and hence crack growth rate depends on the micro-morphology of the material. In high strength aluminium alloys at low crack growth rates, the crack path usually follows the interface between inclusions and the matrix. At higher growth rates, inclusions may fracture ahead of the crack front (due to the larger plastic zone size) and cause ductile tearing of the aluminium matrix surrounding the inclusion (10). Similar effects can be found in other aluminium casting alloys. Fracture has been observed to occur by interfacial fracture between the aluminium rich phase and silicon particles in both hypereutectic (11) and hypoeutectic (12) aluminium–silicon alloys.

De los Rios et al. (3) have developed a shear model involving crack-tip plastic strains to describe the initial crack growth stages, taking into account the location, path and obstacles which must be surmounted. In this manner the initial high crack growth rate of small cracks and subsequent deceleration as they approached crystallographic barriers was represented for a fully annealed 0.4% C steel cycled in torsion. The crack growth rate (da/dN) could be expressed as follows

$$da/dN = f(2\pi aD)^{1/2}\{1 - \phi(D - X)/D\}^3\tau/\mu \qquad (5)$$

where f is a constant for the material, τ is half the stress range in torsion, μ the shear modulus, D is the distance from barrier to barrier, X is the total distance from the crack tips to the nearest barrier, and ϕ is a function of the relative crystallographic orientation between neighbouring grains or barriers through which the crack must pass. For instance, when ϕ is formulated in terms of the resolved shear stress (τ) along slip bands in neighbouring grains A and B, then $\phi = 1 - (\tau_B/\tau_A)$, assuming that the crack develops in grain A and propagates into grain B. The maximum value for ϕ is unity and represents the most unfavourable case for crack advance, resulting in arrest. The minimum value for ϕ is zero and is tantamount to no barrier to the crack being present since the orientation in each grain would be the same. Using this model, it has been shown that small cracks are either arrested ($\phi = 1$) or temporarily halted ($0 < \phi < 1$) depending upon stress level (3).

In the present work a similar effect was seen. At the high strains no decrease in growth rate of the small cracks was observed because the corresponding stress level was sufficiently high to overcome the barriers to crack advance, yet at low strains significant deceleration in the small cracks was observed as they approached these barriers. Unfortunately the model of de los Rios et al. (3) cannot be applied directly because the constant f has not been determined independently for aluminium alloys. However it is possible to apply a similarly based model which has been verified for these alloys. This is the model of Lankford (2). The concepts of a larger plastic strain range $\Delta\varepsilon_p$ associated with the tip of a small crack, grain boundary blockage of crack tip slip, and the same low cycle fatigue crack tip failure process established for large cracks, have been combined (13) into a crack growth expression for small cracks, i.e.

$$\frac{da}{dN} = \frac{\Delta aC}{\Delta\varepsilon_p(0)} \Delta K^{n'}[1 - \phi\{(D - X)/D\}^m] \qquad (6)$$

where C and n' are experimentally measured constants describing the ΔK dependence of the strain at microcrack tips far from grain boundaries; m also is an experimentally established constant related to the rate at which the strain decreases as a boundary or barrier is approached, $\Delta\varepsilon_p(0)$ is the accumulated plastic strain at the crack tip, and Δa is an increment of crack advance. The other terms have been defined previously.

The available experimental data for $m(=2)$ and $n'(=0.2)$ have been applied to the present work. Since no measured values for Δa and $\Delta\varepsilon_p(0)$ were available, a constant C_1 has been introduced such that

$$C_1 = \frac{\Delta a}{\Delta\varepsilon_p(0)} \, C \tag{7}$$

Hence equation (6) becomes

$$\frac{da}{dN} = C_1 \Delta K^{0.2}[1 - \phi\{(D - X)/D\}^2] \tag{8}$$

The values of ϕ can be deduced from the shape of the log da/dN vs log ΔK plot and these are included in Fig. 7. For the high strains $\phi = 0$, with the consequence that there was no impediment to crack growth since the associated stress response was sufficient to continue crack-tip slip past the triple points. On the other hand for the low strains $\phi = 0.9$, indicating a strong barrier to short crack growth. In this case, the distance between barriers, D, was either one or two interdendritic spacings, respectively, depending whether the crack initiated between interdendritic nodes (as in Fig. 4) or at an interdendritic node (as in Fig. 5). Hence equation (8) appears to be applicable to the present results. For specific use of equation (8), however, C_1 takes on different numerical values according to the strain range, i.e., $C_1 = 1.7 \times 10^{-7}$ for $\Delta\varepsilon_T = 1.0$ per cent and $C_1 = 5.8 \times 10^{-10}$ for $\Delta\varepsilon_T = 0.36$ per cent. This is to be expected since both Δa and $\Delta\varepsilon_p(0)$ depend upon strain range. In particular, it indicates that the increment of crack advance, Δa, is greater for the higher strain range.

Figures 6 and 7 show that the minimum growth rate for short cracks at the low strain range occurs when the crack depths, a, are approximately 25 μm and 45 μm. Above the larger critical value, LEFM analysis may be considered. The limiting case would result in an equivalent crack depth, a_0, which may be calculated for the present set of experimental conditions by rearranging equation (3) in the following manner

$$a_0 = \left(\frac{\Delta K_{th}}{0.713\Delta\sigma_0}\right)^2 \frac{1}{\pi} \tag{9}$$

where ΔK_{th} is the threshold and $\Delta\sigma_0$ is the endurance limit ($= 125$ MPa for 2×10^6 cycles). From Fig. 7, if ΔK_{th} is taken as 2 MPa$\sqrt{m}$ then a_0 is 80 μm. If ΔK_{th} is taken as 1.1 MPa$\sqrt{m}$, then a_0 is 48 μm, which is approximately the equivalent of two interdentritic spacings. This value has strong metallurgical implications since ΔK_{th} and $\Delta\sigma_0$ may be influenced by thermo-mechanical treatment, as well as by surface preparation and testing conditions. However when the crack depth is less than a_o, short crack growth behaviour should be expected, whereas when $a > a_o$ short crack growth would not be expected. These effects were observed in the present investigation.

Conclusions

(1) Fatigue cracks in a squeeze-formed aluminium–silicon alloy casting have been observed to initiate at silicon particles either by debonding at the interface with the aluminium matrix or by cracking of the silicon particles. The former is associated with low cyclic strain ranges and long lives.

(2) Crack growth occurred preferentially in the interdendritic regions rather than through the dendrites.

(3) Short crack growth behaviour may be expressed using a model based on crack-tip plasticity where the barriers are triple points in the interdendritic regions. These barriers are significant at low strain ranges. In this case, as the short cracks approach the first triple point after initiation, the growth rate is drastically reduced and on overcoming these barriers the cracks accelerate rapidly to eventually comply with the long crack behaviour.

Acknowledgements

The authors would like to thank GKN Technology Ltd, Wolverhampton, UK, for providing the squeeze-formed aluminium alloy. One of the authors, S. Schäfer, wishes to acknowledge the Natural Sciences and Engineering Research Council of Canada (NSERC) for a Scholarship. This work was financed through grants from GKN Technology Ltd, and NSERC.

References

(1) TAYLOR, D. and KNOTT, J. F. (1981) Fatigue crack propagation behaviour of short cracks; the effect of microstructure, *Fatigue Engng Mater. Structures*, **4**, 147–155.

(2) LANKFORD, J. (1983) Material aspects of crack tip yielding and subcritical crack growth in engineering materials, *Mechanical Behaviour of Materials – IV* (Edited by J. Carlsson and N. G. Ohlson), (Pergamon Press, Oxford), Vol. 1, p. 3.

(3) DE LOS RIOS, E. R., MOHAMED, H. J., and MILLER, K. J. (1985) A micromechanics analysis for short fatigue crack growth, *Fatigue Fracture Engng Mater. Structures*, **8**, 49–63.

(4) LANKFORD, J. (1982) The growth of small cracks fatigue cracks in 7075-T6 aluminum, *Fatigue Engng Mater. Structures*, **5**, 233–248.

(5) *Annual Book of ASTM Standards* (1983) Standard Test Method for Constant-Load-Amplitude Fatigue Crack Growth Rates above 10^{-8} m/cycle, ASTM E647–83, 03.01 (American Society for Testing and Materials).

(6) PEARSON, S. (1975) Initiation of fatigue cracks in commercial aluminum alloys and the subsequent propagation of very short cracks, *Engng Fracture Mech.*, **7**, 235–247.

(7) DOWLING, N. E. (1977) Crack growth during low-cycle fatigue of smooth axial specimens, *Cyclic stress–strain and plastic deformation aspects of fatigue crack growth*, STP 637 (American Society for Testing and Materials), p. 97.

(8) BROEK, D. (1983) *Elementary engineering fracture mechanics*, Third revised edition (Martinus Nijhoff, The Hague), p. 77.

(9) BROEK, D. (1983) *Op. cit.*, p. 81.

(10)   BROEK, D. (1969) The effect of intermetallic particles on fatigue crack propagation in aluminum alloys, *Fracture 1969* (Edited by P. L. Pratt) (Chapman and Hall, London), p. 734.

(11)   CULVER, L. E., RADON, J. C., and BALTHAZAR, J. C. (1984) The influence of silicon on the cyclic crack growth in a cast aluminum alloy, *Life assessment of dynamically loaded materials and structures* (Edited by L. Faria), (National Laboratory of Engineering and Industrial Technology, Lisbon), Vol. 1, p. 495.

(12)   OGILVY, I. M. and ROBINSON, P. M. (1975) The fracture of aluminum die casting alloys – the role of morphology and casting defects, 8th SDCE International Die Casting Exposition and Congress, Paper No. G-T75-014 (Society of Die Casting Engineers).

(13)   CHAN, K. S. and LANKFORD, J. (1983) A crack tip strain model for the growth of small fatigue cracks, *Scripta. Met.*, **17**, 529–534.

*C. Howland**

The Growth of Fatigue Cracks in a Nickel Base Single Crystal

REFERENCE   Howland, C., **The Growth of Fatigue Cracks in a Nickel Base Single Crystal**, *The Behaviour of Short Fatigue Cracks*, EGF Pub. 1 (Edited by K. J. Miller and E. R. de los Rios) 1986, Mechanical Engineering Publications, London, pp. 229–239.

ABSTRACT   The growth of fatigue cracks in a precipitation strengthened single crystal superalloy has been investigated using a specimen design with a unique crystallographic orientation. Testing was limited to room temperature, with peak and mean stress being the primary variables.

In addition to providing essential data for component lifing purposes, the work has provided valuable information which can be used in modelling short fatigue crack growth in fcc materials which also occurs on specific slip planes. In particular it examines whether the rate of fatigue crack growth across a single grain can be characterized by conventional linear elastic fracture mechanics.

Introduction

Significant improvements in the creep strength of conventionally cast nickel based superalloys can be made by directional solidification. Further improvements in strength can be made by suitably modifying the alloy chemistry, and casting as the single crystal form. Components, such as turbine blades and nozzle guide vanes, designed in such high strength, high temperature materials, however, are not just loaded under conditions of static creep, and may often be subjected to significant fatigue loads during service.

The work described in this paper, is part of an ongoing programme aimed at gaining a fundamental understanding of fatigue crack growth phenomena in fcc (face centred cubic) single and poly crystalline materials, which are currently under consideration for several aero-engine applications. Load ratio and peak stress effects were explored as the primary variables.

In view of the crystallographic nature of fatigue cracking in fcc superalloy single crystals (**1**), it is pertinent to examine previous work on faceted fatigue crack growth in single and poly-crystals. In general, two classes of materials might be expected to support extensive crystallographic fatigue crack growth, namely (i) single phase materials of low stacking fault energy, and (ii) two phase materials with coherent precipitates.

The term 'persistent slip band' was first introduced by Thompson *et al.* (**2**) in studies of copper single-crystals oriented for single slip. In general, such slip bands and their co-planar fatigue cracks are limited in length. Two phase, precipitation strengthened materials support more extensive cracking, and it is

* Rolls-Royce Limited, PO Box 31, Derby DE2 8BJ, UK.

in this area that the majority of research has taken place. Extensive faceted fatigue crack growth occurs as a consequence of either the dissolution of sheared precipitates or by the destruction of precipitate order on the operating slip planes. Materials which have been investigated include nickel based superalloys in equiaxed, columnar grained, and single crystal forms (1)(3)–(10), aluminium–zinc–magnesium single crystals (11)–(13), copper–cobalt (14), copper (15), and nickel–chrome austenitic stainless steel (16). Fatigue crack growth has been observed predominantly on {111} in fcc materials, however, growth on {100} has also been reported by Sadananda and Shahinian (4)(5) in Udimet 700 in air and vacuum at room temperature and 850°C, and by Crompton and Martin (10) at 600°C. Vincent and Remy (7) showed the fracture plane to be temperature dependent, and King (8) indicated changes in growth mechanism at room temperature with applied stress intensity. King concluded that under conditions of predominantly planar slip (low temperature, restricted plasticity), octahedral slip on {111} was favoured and that more uniform deformation, as might be promoted by high temperature or extensive crack-tip plasticity, resulted in cube slip on {100}. Previous work (1) on the single crystal alloy under study here reported only {111} planar slip and fatigue crack growth.

The single crystal provides a means of examining crack growth within a grain, something very difficult experimentally in commercial polycrystalline materials. Furthermore crack closure effects which are known to affect long crack growth rates at low stress intensities were expected to be minimal due to the single, flat fracture path. This may be particularly informative in examining the difference between long and short crack behaviour which several authors have attributed solely to differences in closure.

The phenomenon of fatigue crack closure in relation to short cracks has been recently reviewed (17)(18). It is evident in both reviews that closure and threshold determinations for short cracks, and in particular, crystallographic short cracks are very limited.

Hicks and Brown (19) in a study which compared the long and short crack growth behaviour of Astroloy and SRR99 suggested that crack propagation rates in single crystals may be the limiting maximum crack growth rate which may be observed for any particularly crystal system.

Experimental details

The nickel based superalloy used in this research had the following bulk chemical composition (per cent weight): 8.5 Cr, 5 Co, 5.5 Al, 2.2 Ti, 9.5 W, 2.8 Ta, 0.015 C and the balance Ni.

Single crystal bars of 12 mm diameter were cast, with the $\langle 001 \rangle$ crystal growth direction aligned along the axis of the bars. Heat treatment consisted of a 4 hour – 1300°C solution treatment, followed by a 16 hour – 870°C ageing treatment, which produced a microstructure of fine cuboidal γ', of size approximately 0.5 μm on edge, in a matrix of γ, as shown in Fig. 1. The volume fraction of γ' was

Fig 1   Alloy microstructure

80 per cent. The 0.2 per cent yield strength and ultimate tensile strength, measured in the ⟨001⟩ direction were 1000 and 1100 MPa, respectively.

A Laué back reflection technique was used to determine the precise crystallographic orientation of the single crystal bars, thus enabling specimens of specific orientation to be machined from them.

Fatigue crack growth rates were monitored in specimens of the corner notched form, shown in Fig. 2, using a standard replication technique (20). The test pieces had been machined such that the four planar faces were of {001} form. The notch was ground to a depth of 0.25 mm. Prior to test the faces of the test pieces were mechanically polished to a 6 μm finish to faciliate crack length measurement.

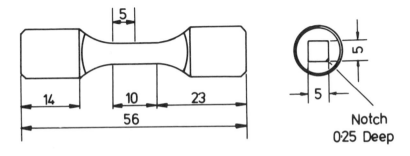

Fig 2   Specimen geometry (dimensions in mm)

Table 1   Test details

Test number	R	Max. stress (MPa)
1	0.1	500
2	0.3	500
3	0.5	500
4	0.8	500
5	0.8	750
6	0.1	250

Fatigue tests were performed using a closed loop servo-hydraulic load frame, at a frequency of 1 Hz. All tests were conducted in air at 20°C. Fatigue cracks were initiated at a load ratio of 0.1, i.e., until the first appearance of a fatigue crack, with subsequent crack propagation at the load ratio of interest. Cracks were monitored at intervals of at least 50 μm of crack extension. A constant peak stress was utilized for each test, during both the initiation and crack propagation phases, with the exception of test numbers 5 and 6 where a change in peak stress was necessary to produce growth rates over the required range of ΔK. A total of four load ratios (P min/P max) was investigated. Full test details are given in Table 1.

Crack growth rates were determined by dividing successive surface increments of crack extension by the number of cycles taken for each increment. The stress intensity calibration was taken from Pickard (**21**), and was a three dimensional linear elastic finite element analysis which modelled the observed fracture geometries.

Results

The propagation rates, calculated from the raw data, are presented as a function of the applied stress intensity range, and the maximum stress intensity, in Fig. 3. The crack growth rates presented are those measured along the direction of propagation (rather than those resolved normal to the applied stress axis) and the mixed mode stress intensity is similarly that at the crack tip in the direction of crack growth on the specimen surface, i.e., the ($\langle 110 \rangle (111)$) system. Full details for the calculation of K are given in reference (**21**).

The growth rates measured at load ratios of 0.1 and 0.3 are equivalent when plotted against K_{max}. Data collected from tests performed at a load ratio of 0.5 are marginally slower, whilst data from a test performed at a load ratio of 0.8 is slower still. If the data is plotted against the alternating stress intensity the trend is reversed, with the R of 0.8 data being the fastest with data from tests conducted at load ratios of 0.5, 0.3 and 0.1 being progressively slower.

The observed crack growth mechanism was in all cases crystallographic, occurring in the wake of intense shear bands on {111} planes (Fig. 4). Typical fracture surfaces are shown in Fig. 5, the views being parallel to the stress axis and normal to the plane of fracture in Fig. 5(a) and (b) respectively. The

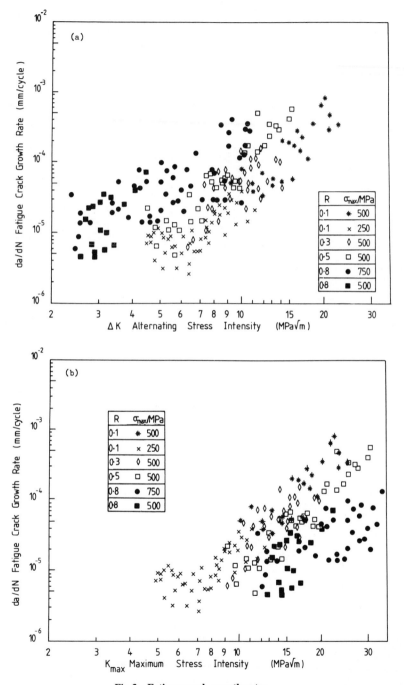

Fig 3   Fatigue crack growth rates
(a) as a function of alternating stress intensity   (b) as a function of maximum stress intensity

Fig 4   Deformation band ahead of the crack tip

fracture surfaces, were planar when observed macroscopically, but displayed a small degree of surface roughness when viewed by scanning electron microscopy. A similar study of high and low load ratio test pieces revealed no major difference in fracture morphology (Fig. 6).

Discussion

The propagation rate of fatigue cracks along co-planar slip bands has been monitored at four different load ratios. All have resulted in crack growth on {111} crystallographic planes. Fatigue crack growth rates have been plotted against a crack tip stress intensity factor which as been utilized previously (1), and which demonstrated a good correlation between the growth rates of similar crystallographic cracks growing in three separate specimen geometries, and at various stress levels between 500 and 1000 MPa.

The results show a sensitivity to mean stress which is similar to that reported for 'long' cracks in polycrystalline materials at low stress intensities. Comparison of this data with other work (19)(22) reveals the crack growth rates measured at $R = 0.1$ to be only a factor of two to three times faster than short crack growth rates measured in Astroloy, a Nickel base superalloy having a 50 μm grain size, and three to four times faster than those in ultra-fine grained Astroloy of 10 μm grain size, suggesting that within a single alloy system such as γ–γ' superalloys the growth rates measured within a single crystal represent

Fig 5   Fatigue fracture surface at $R = 0.1$ and maximum stress 500 MPa
(a) view parallel to applied stress axis
(b) view normal to the fracture surface

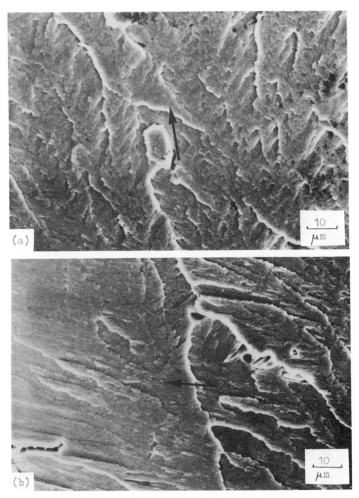

Fig 6   Fracture surface details for a maximum stress of 500 MPa; the arrow shows the direction of crack propagation
(a) $R = 0.8$
(b) $R = 0.1$

the maximum growth rate, with growth rates being reduced as the number of grain boundaries is increased. It should be pointed out, however, that there are fundamental differences in the analysis methods employed. In the case of the single crystal, growth rates and stress intensities have been suitably modified so that they are truely representative of the observed fracture path, whereas polycrystalline materials are usually treated in a manner such that standard Mode I stress intensity solutions are used, even where significant crack path tortuosity exists, and growth rates are derived from resolved crack lengths. Both of these latter factors are sources of error in the analysis of short crack growth phenomena.

One reason frequently used to account for the anomalously fast propagation rate of short cracks at a given level of ΔK is the breakdown of continuum mechanics when the crack is contained within the relevant microstructural dimension. In such cases the crack tip plastic zone is likely to be distorted away from the classical shape in favour of a shape which reflects the nature of the available deformation. The deformation zone shown in Fig. 4, and those discussed more fully by Vogel *et al.* (**11**)(**12**) for Al–Zn–Mg single crystals are clearly planar in nature, with material only a few microns either side of the deformation band being totally undisturbed. Previous work on the growth of cracks in nickel base single crystals (**1**)(**3**)–(**10**) suggested that the influence of crack length in single crystal material is very small because of the uniformity of yield strength throughout the test piece which is unaffected by weak grain orientations and grain boundaries. Hence, plasticity, although not of classical shape, will be confined to a zone which is well defined by K. It will only be at very short crack lengths, which appoach the scale of crack tip plasticity that differences in crack growth rate might be seen. This is borne out by the demonstrated independence of growth rate, at constant R and K, of maximum stress (Fig. 3).

A number of theories have been put forward to explain the effect of mean stress on crack propagation rates, the most widely accepted being that of closure. Current models of crack closure rely upon either plastic wake effects, roughness or oxidation of the fracture surfaces. By far the most commonly observed of these mechanisms, particularly at room temperature, is roughness induced closure (**23**). Contact of opposing surface asperities would inevitably result in frettage damage of the fracture faces, especially in cases where significant in plane shear is occurring. However, scanning electron microscopy of the high (0.8) and low (0.1) load ratio fatigue test pieces showed no such damage. Fine fracture surface detail from the two tests are shown for comparison in Fig. 6. The absence of frettage damage on the low load ratio tested specimen, although not conclusive proof of the absence of closure, does suggest that closure effects are not responsible for the results shown in Fig. 3. Although the growth mechanism involves a large shear component there are insufficient surface asperities to influence the behaviour of the crack tip; also one would have expected to see a better agreement with the $R = 0.5$ and $R = 0.8$ crack growth rates (Fig. 3) which, in view of the high minimum loads, should be free from closure.

The only differences in fracture morphology between the high and low load ratio tests which support a closure argument come from macroscopic observations. The $R = 0.1$ test piece shows markings (Fig. 5(b)), possibly caused by frettage along directions corresponding to $\langle 112 \rangle$ on the fracture surface, which are absent on the $R = 0.8$ test piece shown in Fig. 7.

The precise mechanism by which the accelerated growth rates at high load ratios are generated, such as crack opening response, or dislocation/particle interactions is not clear. Further work is required to examine these effects in detail, and quantify crack opening behaviour.

Fig 7   Normal view of the fracture surface for the $R = 0.8$ test at 500 MPa maximum stress

Conclusions

(1) Fatigue crack growth at room temperature was observed to be exclusively crystallographic, occurring on {111} planes.
(2) Presentation of the data in terms of the alternating stress intensity, ΔK, revealed a simple progression with data at $R = 0.8$ being fastest, and $R = 0.1$ being slowest.
(3) Although the observed trends in the data suggest the operation of a simple closure mechanism, the inherent smoothness of the fracture surfaces tends not to support this concept.

Acknowledgements

Thanks are due to Dr M. A. Hicks for invaluable discussions and help with the manuscript, to Dr A. C. Pickard for the finite element stress intensity solutions, and to D. Scott and M. Elliot for assistance with the test programme.

References

(1) HOWLAND, C. and BROWN, C. W. (1984) The effect of orientation on fatigue crack growth in a nickel based single crystal superalloy, *Proc. 2nd Int. Conf. on Fatigue and Fatigue Thresholds* (Edited by C. J. Beevers) (EMAS Publications), p. 1349.

(2) THOMPSON, N., WADSWORTH, N. J., and LOUAT, N. (1956) The origin of fatigue fracture in copper, *Phil. Mag.*, **1**, 113–118.

(3) LEVERANT, G. R. and GELL, M. (1968) The influence of temperature and cyclic frequency on the fatigue fracture of cube oriented nickel base superalloy single crystals, *Met. Trans,* **6A**, 367–371.

(4) GELL, M. and LEVERANT, G. R. (1968) The fatigue of the nickel base suparalloy Mar M200 in single crystal and columnar grained forms at room temperature, Trans. *Met. Soc. AIME*, **242**, 1869–1879.

(5) SADANANDA, K. and SHAHINIAN, P. (1979) A fracture mechanics approach to high temp fatigue crack growth in Udimet 700, *Engng Fracture Mech.*, **11**, 73–86.

(6) SADANANDA, K. and SHAHINIAN, P. (1981) Analysis of crystallographic high temp fatigue crack growth in a nickel base alloy, *Met. Trans*, **12A**, 343–351.

(7) VINCENT, J. N. and REMY, L. (1981) The temperature dependence of the pseudo cleavage mechanism in the threshold regime of a superalloy, *Proc. 1st Int. Symp. on Fatigue Thresholds* (Edited by J. Backlund, A. Blom, and C. J. Beevers) (EMAS publications) p. 441.

(8) KING, J. E. (1981) Crystallographic fatigue crack growth in Nimonic AP1, *Fatigue Engng Mater. Structures*, **4**, 311–319.

(9) GELL, M. and LEVERANT, G. R. (1969) The effect of temperature on fatigue fracture in a directionally solidified nickel base superalloy, *2nd Int. Conf. on Fracture*, Brighton, England, p. 565.

(10) CROMPTON, J. S. and MARTIN, J. W. (1982) Fatigue crack propagation at elevated temperature in MarM002 single crystals, *High temperature materials for gas turbines* (Edited by D. Brunet *et al.*) (Reidel Publishing Company), p. 611.

(11) VOGEL, W., WILHELM, M. and GEROLD, V. (1982) Persistent slip bands in fatigued peak aged Al–Zn–Mg single crystals – I, *Acta met.*, **30**, 21–30.

(12) VOGEL, W., WILHELM, M., and GEROLD, V. (1982) Persistent slip bands in fatigued peak aged Al–Zn–Mg single crystals – II, *Acta Met.*, **30**, 31–35.

(13) AFFELDT, E. and GEROLD, V. (1984) Micromechanism controlling the fatigue crack propagation in Al–Zn–Mg single crystals, *Proc. ICF 6* (Edited by S. R. Valluri *et al.*) (Pergamon Press, Oxford), p. 1587.

(14) MEYER, R., GEROLD, V., and WILHELM, M. (1977) Stage I fatigue crack propagation in age-hardened Cu–Co single crystals, *Acta Met.*, **25**, 1187–1190.

(15) MUGHRABI, H. and WANG, R. (1980) Cyclic strain localization and fatigue crack initiation in persistent slip bands in face centred cubic metals and single phase alloys, *Proc. 1st Int. Symp. on Defects*, Tuczno, Poland, p. 15.

(16) RIEUX, P., DRIVER, J., and RIEU, J. (1979) Fatigue crack propagation in austenitic and ferritic stainless steel single crystals, *Acta Met.*, **27**, 145–153.

(17) LEIS, B. N., KANNINEN, M. F., HOPPER, A. T., AHMAD, J., and BROEK, D. (1983) A critical review of the short crack problem in fatigue, AFWAL TR-83-4019.

(18) SURESH, S. and RITCHIE, R. O. (1983) The propagation of short fatigue cracks, University of California, Berkeley, Report UCB/RP/83/10014.

(19) HICKS, M. A. and BROWN, C. W. A comparison of short crack growth behaviour in engineering alloys, *Proc. 2nd Int. Conf. on Fatigue and Fatigue Thresholds* (Edited by C. J. Beevers) (EMAS Publications), pp. 1337–1347.

(20) BROWN, C. W. and SMITH, G. C. (1982) A two stage plastic replication technique for monitoring crack initiation and early fatigue crack growth, *Advances in Crack Length Measurement* (Edited by C. J. Beevers) (EMAS Publications), p. 41.

(21) PICKARD, A. C. (1986) The application of three dimensional finite element methods to fracture mechanics and fatigue life prediction, in press.

(22) BROWN, C. W., KING, J. E., and HICKS, M. A. (1984) Effects of microstructure on long and short crack growth in nickel base superalloys, *Metal Sci.*, **18**, 374–380.

(23) WALKER, N. and BEEVERS, C. J. (1979) A fatigue crack closure mechanism in titanium, *Fatigue Engng Mater. Structures*, **1**, 135–148.

J. M. Kendall,* M. N. James,† and J. F. Knott*

The Behaviour of Physically Short Fatigue Cracks in Steels

REFERENCE Kendall, J. M., James, M. N., and Knott, J. F., **The Behaviour of Physically Short Fatigue Cracks in Steels**, *The Behaviour of Short Fatigue Cracks*, EGF Pub. 1 (Edited by K. J. Miller and E. R. de los Rios) 1986, Mechanical Engineering Publications, London, pp. 241–258.

ABSTRACT The behaviour of engineering or physically short fatigue cracks in two structural steels has been examined, using both through-thickness and thumbnail short crack geometries. The results indicate that much of the 'anomalous behaviour' of physically short cracks may be due to differences in the amount of closure experienced by long and short cracks. The prediction of fatigue life using a fracture mechanics approach is discussed and it is concluded that conservative life predictions may be based on a closure-free value of the threshold stress intensity range.

Introduction

The application of fracture mechanics to fatigue is based upon the assumption that the fatigue crack growth rate, da/dN, is a function of stress intensity range, $\Delta K = K_{max} - K_{min}$, and mean stress, often expressed as the load ratio, $R = K_{min}/K_{max}$. This approach provides a powerful method of calculating fatigue lives by integration of the da/dN versus ΔK relationship. For the integration to be performed, it is essential that the relationship holds for the whole range of crack lengths spanned by the limits of integration, which are usually taken as the initial defect size and the final crack length to cause failure. In these terms, the most general definition of the 'short crack problem' in fatigue is the situation when the functional relationships established between da/dN, ΔK, and R for long cracks do not hold.

Particular interest has been shown in the low growth rate or near-threshold region in which anomalies appear to be most in evidence. In this region the linear relationship between $\log da/dN$ and $\log \Delta K$, which is observed at higher growth rates, is no longer applicable, and the value of da/dN at successively lower values of ΔK tends asymptotically to the threshold stress intensity range, ΔK_{th}, below which long crack growth does not occur. It is still possible to calculate fatigue lives in the near-threshold region by integration if the da/dN versus ΔK long crack curve is fitted by a number of straight lines and each line is integrated separately (1). The behaviour of short cracks in the near-threshold

* Department of Metallurgy and Materials Science, University of Cambridge, Pembroke Street, Cambridge CB2 3QZ.
† Now at Department of Metallurgy and Materials Engineering, University of the Witwatersrand, 1 Jan Smuts Avenue, Johannesburg 2001, South Africa.

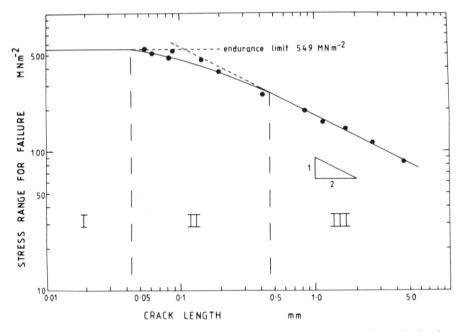

Fig 1   The 'Kitagawa plot' of applied stress range to cause failure against crack length, showing data of Kitagawa and Takahashi (8)

region is of importance because it has been observed in a large number of studies (see, for example, (1)–(7)) that fatigue crack growth can occur at elastically calculated values of ΔK that are lower than the long crack threshold ΔK_{th}.

The object of this paper is to explore the significance of physically short cracks in steels, with respect to engineering applications and the determination of the fatigue lives of structures and components. The experimental work presented has been carried out using two structural steels and it is of interest first of all to set the 'short crack problem' in the context of the engineering applications of these materials. Two general types of application which illustrate the S–N approach and the LEFM approach to fatigue design will be considered: machine components and engineering structures. One of the most convenient representations of short crack data is the 'Kitagawa plot' of applied stress range to cause failure, $\Delta \sigma_o$, against crack length, a, using logarithmic scales (8), as shown in Fig. 1. This shows clearly the regions of behaviour described by the S–N approach and the long crack LEFM approach to fatigue.

Machine components

In general, the moving components of machines are required to withstand a large number of fatigue cycles at relatively low applied stresses. For instance, an automobile engine that has travelled 100 000 miles at an average speed of 60 miles per hour and at 4000 revolutions per minute will have experienced 4×10^8 cycles. The fatigue design procedure is conventionally based on the endurance limit of the S–N curve, with appropriate correction factors for regions of stress concentration. Account may also be taken of the effect of surface hardening treatments on fatigue lifetimes.

The S–N curve is influenced by the size and distribution of defects in the material. A cast or wrought component typically contains an evenly distributed multiplicity of small defects of 1–100 μm in size. This is well below the limits of non-destructive testing (NDT) techniques, but a measure of defect control ('process control') can be achieved by controlling the size of defects produced during processing routes such as casting or hot working. Material containing a consistent population of small defects determined by process control fails when $\Delta\sigma_0$ exceeds the endurance limit (region T in Fig. 1).

In these applications fracture mechanics at present is conceptual and not quantitative, and the fatigue life of the component is determined directly from the S–N data. The 'short crack problem', in terms of design, is to determine whether the S–N curve for 'process-controlled' material can be non-conservative. There is, of course, scientific interest in examining the extent to which fracture mechanics can be extended to multiplicities of extremely small defects and situations in which components are designed to operate at applied stress ranges that exceed the endurance limit, when the number of cycles to failure may be less than 10^6 cycles (**9**)(**10**). The behaviour of 'microstructurally' short cracks originating at these small defects does not then follow the behaviour illustrated in region I in Fig. 1. The present paper, however, is concerned with 'physically' short cracks which are significantly larger than the microstructural unit, as considered in the following section.

Engineering structures

Many steel structures are fabricated by welding processes or by the use of rivets, and it is frequently found that defects of 0.1–2 mm in size are located in the region of these joints. Since static design stresses are typically of the order of two thirds of the yield stress, it should be possible to calculate the fatigue life by integration using a fracture mechanics approach. The loading spectrum in service will include a spread of applied stress ranges. If ΔK_{th} is appropriate, the integration can be carried out only for those stress ranges that produce ΔK values exceeding ΔK_{th}.

It can be seen from Fig. 1 that material initially containing a long crack would fail when ΔK exceeds the threshold ΔK_{th} (region III). The 'anomalous' behaviour of cracks which are physically short but several times larger than the microstructural unit is represented by region II. The significance of the position of the experimental data in this region is that both the S–N prediction and the LEFM prediction of $\Delta\sigma_0$ are non-conservative.

For long cracks, ΔK_{th} can be measured experimentally and related to the applied stress range at threshold. For short through-thickness cracks the value of ΔK is given by

$$\Delta K = Y \Delta\sigma(\pi a)^{1/2} \tag{1}$$

where Y is a constant for a given testpiece geometry, e.g. $Y = 1.12$ for an edge crack in a semi-infinite body. For short thumbnail cracks the value of ΔK is given by

$$\Delta K = Q \Delta\sigma(\pi a)^{1/2} \tag{2}$$

where Q refers to the ellipticity of the thumbnail crack. Therefore, by taking logarithms an expression of the following general form can be obtained

$$\log \Delta\sigma_0 = \log \Delta K_{th} - \tfrac{1}{2} \log a + \text{constant} \tag{3}$$

In a number of critical applications, fatigue design and integrity in service are dependent on the sensitivity and reliability of the NDT technique that is employed, and on the ability of periodic NDT inspections to reveal fatigue crack growth. A general estimate of the limit of NDT for the majority of engineering structures is detection of a defect of 1–2 mm in size. Defects larger than the NDT limit can be eliminated by 'NDT control', but it must be assumed that defects up to that size exist in the structure, such as pre-existing weld defects which are physically small but large with respect to the microstructure generally.

It should be noted that the positions of the lines in Fig. 1 are altered by the magnitude of the mean tensile stress superimposed on the stress range applied during the fatigue cycle. It has long been recognized that the endurance limit is reduced by increasing the mean stress. Similarly, a reduction in the measured value of ΔK_{th} for long cracks is observed as the mean stress (i.e., load ratio, R) is increased. At high R values (R > approximately 0.6) there is minimal effect of closure on the near-threshold crack growth rate. An 'intrinsic material threshold', ΔK_c, may be defined as the measured ΔK_{th} at high load ratio, which is equal to the effective stress intensity range at threshold for all load ratios. From equation (3) it can be seen that a reduction in ΔK_{th} would lower the long crack LEFM line in Fig. 1 by reducing the log ΔK_{th} term on the right hand side of the equation. This discussion will be developed later.

Types of short crack

Short cracks in steel structures and components are not uniform in their morphology, origin or stress state, and these factors may influence subsequent growth in fatigue. The common features from which short fatigue cracks may grow in steels can be broadly divided as follows.

(a) *Casting and weld defects.* Cast components frequently contain shrinkage cavities that are formed on solidification and which may be associated with second phase particles. Weld defects such as lack-of-fusion defects are also formed on solidification of the weld metal. The residual stress fields associated with such defects may range from virtually zero (for a large, slowly-cooled casting in which sufficient diffusion can occur during cooling to relieve stresses induced by thermal contraction or phase changes) to yield point magnitude (for a rapidly-cooled weld in which negligible stress relief occurs during cooling). Studies of fatigue crack growth from natural casting defects in nickel–aluminium bronze (1) revealed no effect on propagation rate which could be attributed to residual stresses.

(b) *Hydrogen cracks.* Hydrogen cracks may form as a result of hydrogen embrittlement of steels and they may be associated with welds where 'hard spots' develop (11). The arrest of such cracks is associated with relieving plastic flow at the crack tips. The residual plasticity associated with short hydrogen cracks in iron – 3 per cent silicon single crystals has been revealed by etching (12) and found to be concentrated at the ends of the crack. However, the residual stresses are likely to be removed completely by a stress-relief heat treatment, e.g., a few hours at temperatures above 600°C.

(c) *Inclusions.* Second phase particles, including sulphides, oxides, inter-metallic compounds and slag particles, frequently act as sites for fatigue crack initiation. During steel production the differing coefficients of thermal expansion in the inclusion and matrix give rise to differential strains at the interface between them. On cooling, these strains will result in residual tesselated stresses around the inclusions, which may be tensile or compressive (9)(13).

(d) *Quench cracks and surface damage.* A variety of surface hardening treatments may be applied to steel structures and components to extend their fatigue lives. Thermal, chemico-thermal, or mechanical treatments may be employed to provide both surface hardening and favourable compressive residual stresses. However, although the processes are designed to improve fatigue properties, they may be accompanied by the production of defects such as quench cracks or roughened surface profiles. Fatigue cracks originating from these sites will experience the effect of the residual stress fields resulting from the hardening treatment employed (14). Surface damage in the form of machining marks may also be produced by processes such as turning or grinding, where the depth of cut controls not only the size of a defect, but also the local stress field associated with it.

It is clear that the characteristics of the site at which a fatigue crack initiates can influence its early growth. Before any attempt is made to apply experimentally determined short crack data to engineering situations the types of short cracks involved must be taken into account. The methods of production of short cracks for experimental work are generally concerned with two main geometries: through-thickness cracks and thumbnail (semi-elliptical) cracks.

Short through-thickness cracks may be prepared by growing a long crack and machining away the top surface to leave a crack that is short in the depth dimension (15)–(19): short thumbnail cracks may be initiated at small surface notches (8) or produced by growing a crack from a ridge that is subsequently removed by machining (20)(21). However, the most common method of obtaining short surface cracks that are approximately semi-elliptical in shape is by initiation at inclusions using smooth bar or hour-glass testpieces (1)–(6)(22). These testpieces may be subjected to stresses that exceed the yield stress at the surface, and so the short pre-crack grows under conditions of bulk plasticity. Even when the macroscopic applied stress is below the yield stress, the individual grains within which fatigue cracks initiate are locally yielded, and their early growth takes place within the local plastic field (4).

After preparation of short pre-cracks by these methods a stress-relief heat treatment may be applied to anneal out the plastically deformed material around the cracks. The aim is to leave a short pre-crack that is simply a geometric discontinuity. However, the effectiveness of common stress-relief treatments in achieving this is not certain, and a residual stress field may still exist around the short crack, particularly if the applied stresses have been high.

To summarize, if a fracture mechanics approach is to be used to describe the behaviour of physically short cracks in fatigue, it is essential to consider both the geometry of the cracks and the residual stress fields around them, whether produced during fabrication, surface-hardening, or testpiece preparation. The aim of the present paper is to attempt to do this using two well-characterized structural steels to examine the extent of the short crack problem in engineering structures. The work on the low strength steel is presented first.

Experimental work on a low strength steel

The material was a low carbon steel with a yield strength of 280 MNm^{-2} and an ultimate tensile strength of 420 MNm^{-2}, similar to the weldable structural steels used for the construction of railway bridges and railway vehicle bogies. The composition is shown in Table 1. The cast ingot was rolled and forged to 20 × 40 mm cross-section bars from which the testpieces were machined. The material had a ferrite–pearlite microstructure, as shown in Fig. 2(a).

Experimental procedure

Long and short crack data were obtained using a 60 kN servo-hydraulic testing machine operating at a frequency of 40 Hz, and the load ratio, R, was 0.5 in all tests. The crack length was measured using the direct current potential drop technique (23). Closure measurements were carried out using a back face strain gauge system incorporating an offset elastic displacement circuit (24).

Long crack data were obtained using single edge notched bend specimens with a width of 25 mm, thickness of 15 mm, and notch depth of 5 mm, tested in

Table 1 Composition of materials

Steel type	C	Mn	Si	Ni	Cr	Mo
Low strength steel	0.15	0.57	0.15	0.03	<0.02	<0.02
High strength steel	<0.2	—	—	2.5	1.5	0.5

four point bend loading. The loads were reduced in 5 per cent decrements and the crack was allowed to grow a distance of at least four times the size of the reversed plastic zone at each load range. The load-shedding procedure was continued until no further crack growth could be detected in 10^6 cycles and then the loads were incremented in 5 per cent steps.

The short through-thickness cracks were initiated at central slots in testpieces with a width of 25 mm and thickness of 15 mm. The pre-crack was grown using a load-shedding sequence with 10 per cent decrements to achieve near-threshold growth at a pre-crack length of approximately 5 mm. Then 2.5 mm was removed from the side faces of the testpiece by machining, and approximately 5 mm removed from the top, to leave a straight-fronted short through-thickness pre-crack of between 0.1 and 1.0 mm in length in a testpiece with a width of 20 mm and thickness of 10 mm. The specimens were stress-relieved at 650°C for one hour in a vacuum furnace prior to testing.

The short pre-cracks were subjected to fatigue loading in four point bend

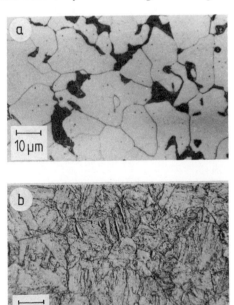

Fig 2 Microstructures: (a) low strength steel, (b) high strength steel Q1N

loading at initial ΔK values in the range 2–12 MNm$^{-3/2}$. If no crack growth was detected in 10^6 cycles the loads were increased by 5 per cent until crack growth occurred.

Results

The long crack data are shown in Fig. 3. When the applied ΔK was greater than approximately 7 MNm$^{-3/2}$, the opening stress intensity, K_{op}, measured by the back face strain gauge on reloading, was lower than the minimum stress intensity during the fatigue cycle, K_{min}, so that there was no effect of closure on the measured growth rate. When the applied ΔK was less than 7 MNm$^{-3/2}$, K_{op}

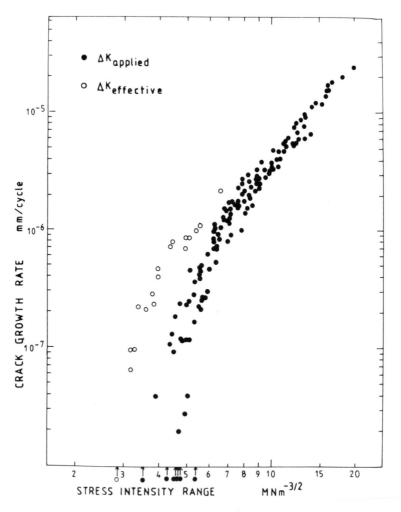

Fig 3   Long crack results in the near-threshold region for low strength steel. $R = 0.5$

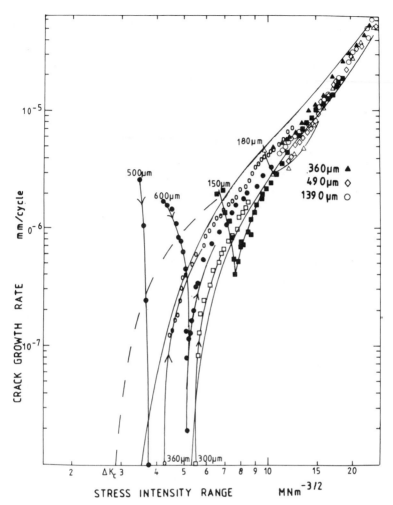

Fig 4 Short crack results compared with long crack results for low strength steel. Dashed line shows ΔK_{eff} data for long cracks. The initial lengths of short cracks are given in microns

was greater than K_{min} and the measured growth rate was reduced by the effect of closure. The threshold at $R = 0.5$ was equal to approximately 4.5 MNm$^{-3/2}$. The effective stress intensity range at which crack growth could not be detected, as determined by closure measurements, was approximately 3.0 MNm$^{-3/2}$.

The short crack data are compared with the long crack data in Fig. 4. Considering first the short cracks tested at lower initial ΔK values, it can be seen that some cracks grew at applied ΔK values below the measured long crack ΔK_{th} of 4.5 MNm$^{-3/2}$, sometimes at an initially decreasing growth rate. Other short cracks simply followed the long crack curve after commencing growth at ΔK values close to the long crack ΔK_{th}.

In order to apply the required stress intensity range to the short cracks that were tested at initial ΔK values greater than 7 MNm$^{-3/2}$, it was necessary to apply such high loads that the specimen was yielded over the whole width, although the change in applied moment-arm as a result of specimen deformation was negligible. Nonetheless, the growth rates for the short cracks (0.1–1.0 mm long) coincided with the long crack data. That is, within the sensitivity of the apparatus (which could detect 30 μm of crack growth), no short crack growth was observed for cracks of these sizes when the value of the initial applied ΔK was in the range in which long cracks were not affected by closure at a load ratio of 0.5.

Experimental work on a high strength steel

The material studied in a second set of experiments was a weldable alloy steel, Q1N, with a yield strength of 653 MNm^{-2} and an ultimate tensile strength of 743 MNm^{-2}. The composition is shown in Table 1. The material was water quenched and tempered at 640°C to give a tempered martensitic microstructure, as shown in Fig. 2(b).

Experimental procedure

Threshold ΔK_{th} values were measured for long cracks and for short cracks of both through-thickness and thumbnail geometries. The experimental procedures have been described previously (**15**)(**24**). Briefly, the long crack data were obtained using single edge notched bend specimens tested in four point bend loading and the short through-thickness crack data were obtained by growing a long pre-crack and removing the top of the specimen by machining. The short surface cracks were initiated at inclusions on unnotched testpieces, under loading conditions in which the surface of the testpiece was yielded. A stress-relief heat treatment was carried out before testing at load ratios between 0.2 and 0.7. Closure measurements were made during the long crack tests.

Results

The measured ΔK_{th} values are shown in Fig. 5 (**15**), plotted against the crack length on a logarithmic scale. It was found that the short through-thickness cracks of length 0.27–1.0 mm had ΔK_{th} values lower than the measured long crack ΔK_{th} of 4.8 MNm$^{-3/2}$, but higher than the value of the effective stress intensity range at threshold (ΔK_c) of approximately 3.0 MNm$^{-3/2}$. Therefore, the short through-thickness crack data fell in the region of ΔK in which long cracks were affected by closure.

However, for the short surface cracks of 0.1–0.4 mm in length, ΔK_{th} values close to the measured long crack ΔK_{th} of 4.8 MNm$^{-3/2}$ were obtained. This indicates that there is a variation in the ΔK_{th} values measured for short cracks of different geometries prepared by different techniques.

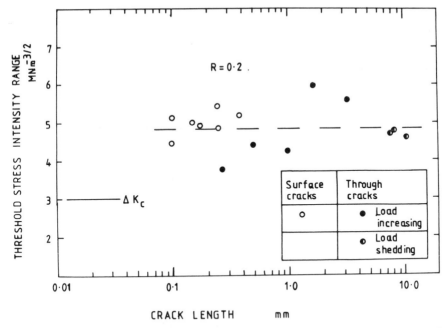

Fig 5   Threshold values of surface thumbnail and through-thickness cracks for high strength
steel. Dashed line shows measured long crack threshold

Discussion

Discussion of results

Plasticity-induced closure arising from forces within the plastic wake is one of the most significant factors affecting the growth of fatigue cracks at low ΔK values. The concept has been used to explain the effect of load ratio on long crack ΔK_{th} values determined by conventional load-shedding threshold tests (25). If K_{op} is greater than K_{min} during the fatigue cycle, the effective stress intensity, ΔK_{eff}, will be lower than the applied ΔK, producing an elevation in the measured value of ΔK_{th}. It has been suggested that during a long crack threshold test the increase in closure as ΔK_{th} is approached is primarily a load-shedding effect (15). At high load ratios, however, and at higher applied ΔK values, there is minimal effect of plasticity-induced closure.

Experiments have been carried out in which part of the plastic wake behind the tip of a long crack is removed by electro-discharge machining (EDM) (15)(26). This process is shown in Fig. 6 for a long crack growing in the near-threshold region. The crack growth rate was approximately 6×10^{-8} mm/cycle for a crack that was 2.43 mm in length after a load-shedding sequence. Part of the wake was removed by EDM to leave 0.5 mm of wake behind the crack tip. The subsequent growth rate, at almost the same applied

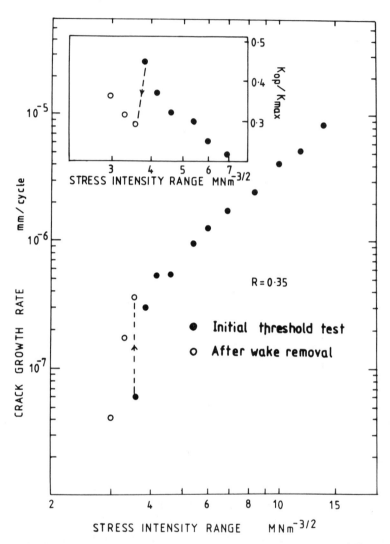

Fig 6   Growth rate and closure data showing the effect of wake removal (15)

ΔK, was found to have increased to approximately 4×10^{-7} mm/cycle, associated with a reduction in K_{op}/K_{max} from 0.45 to 0.29. This demonstrates that, in these experiments, significant effects of plasticity-induced closure were produced by parts of the plastic wake located more than 0.5 mm behind the crack tip. It has been calculated that the distance over which there is contact between the opposing fatigue faces may be up to 1–2 mm during a long crack threshold test (15).

The short through-thickness cracks were prepared by growing long cracks under a load-shedding sequence into the threshold region and then machining away the material behind the crack tip. As for the EDM experiments described above, this method of preparation removed a large part of the plastic wake and it is deduced that the difference in behaviour between long and short cracks, shown in Fig. 4, arises simply from the fact that less plasticity-induced closure is associated with the short crack growth. At higher stress intensities, plasticity-induced closure becomes less significant and the results for long and short cracks lie on a single curve.

The experimental values of ΔK_{th} for short surface cracks in Q1N steel, however, were found to fall within the scatter-band of long crack values. The surface cracks were initiated at inclusions and grown to lengths in the range 0.1–0.4 mm under conditions of gross surface yielding ($R > 0.2$). Unloading then induces a compressive residual stress and if the stress-relief treatment of one hour at 650°C does not completely remove this stress, clamping forces could produce a closure effect during the subsequent determination of ΔK_{th}, such that the measured values were similar to those for long cracks, where plasticity-induced closure was operative. Support for this is given by crack closure measurements made on small thumbnail cracks grown in a region of surface plasticity. These gave K_{op}/K_{max} values identical to those determined for long cracks growing under elastic conditions at a low load ratio (22)(27).

The difference in behaviour observed for the two kinds of short cracks examined in the present work is therefore rationalized in terms of the effects of different residual stress fields on threshold behaviour.

The 'Kitagawa plot'

The 'anomalous behaviour' of physically short cracks is illustrated by region II of Fig. 1, in the graph of log $\Delta\sigma_o$ against log a. Here, experimental values of $\Delta\sigma_o$ are lower than would be predicted by either the LEFM calculations based on ΔK_{th} (which works well for region III) or the endurance limit, for 10^7 or 10^8 cycles as appropriate (region I).

One reason for the anomalous behaviour is that the plasticity associated with crack growth increases rapidly with applied stress level, so that the LEFM analysis becomes increasingly inaccurate. Miller (28) has shown that deviation from region III behaviour first occurs at a crack length such that the applied stress level exceeds one third of the cyclic yield stress. For monotonic loading, the maximum extent of the plastic zone in plane strain, R_{IY}, is given by (29)

$$R_{IY} = 0.16(K/\sigma_y)^2 = 0.16(\sigma_{app}^2/\sigma_y^2)\pi a \qquad (4)$$

where σ_{app} is the applied stress. For $\sigma_{app} = \sigma_y/3$, the ratio of plastic zone size to crack length. $R_{IY}/a = 0.056$. The size of the reversed plastic zone is often obtained by substituting $2\sigma_y$ for σ_y in equation (4), so that the ratio of reversed plastic zone to crack length is approximately 0.014. The size of plastic zone increases initially with σ_{app}^2, but then more rapidly as σ_{app} approaches σ_y.

The results given in Figs 3 and 4 show that for values of $\Delta K > 7 \, \text{MNm}^{-3/2}$ the calculated values of da/dN for initially quite short cracks in structural steel of modest yield strength are identical to those measured on specimens containing long cracks, even though the short crack testpieces have undergone substantial general plastic yielding. It is remarkable that the results agree so well, for a value of ΔK calculated as if the loading were elastic. It is possible that, after the first few cycles, the general plasticity in the specimen as a whole 'shakes down' through cyclic hardening processes, eventually approaching saturation, so that the overall specimen response progressively becomes quasi-elastic, with the elastically calculated ΔK characterizing behaviour in the crack tip region.

If cyclic hardening can rapidly lead to a quasi-elastic response in specimens which have undergone gross yielding, it seems unlikely that the main cause of anomalous short crack behaviour is the increase of plastic zone size with applied stress level. An alternative possibility is that the different amounts of closure associated with short cracks produce relationships between the applied and effective stress intensity ranges which are different from those for long cracks. As described earlier in the paper, different types of residual stress distribution are associated with different sources of short cracks, so that details of behaviour must be assessed on an individual basis. Presumably, however, a lower bound is given by the complete absence of closure, e.g., for solidification cracks, or cracks for which stress-relief annealing has been completely effective.

The essential point is whether or not the line based on ΔK_{th} in Fig. 1 represents the 'no-closure' situation, because it is clear that the distribution of plasticity in the plastic wake can strongly affect the value of ΔK_{th} in a load-shedding test, unless the stress ratio, R, is high. The results of Kitagawa and Takahashi (Fig. 1) give a ΔK_{th} value of 15.5 Kgmm$^{-3/2}$ (4.8 MNm$^{-3/2}$) for $R = 0.04$. This is compatible with their results, recognizing that the cracks of interest were semi-circular and much smaller than testpiece dimensions. The appropriate value of Q in equation (2) is 0.66 so that the relationship between ΔK and $\Delta \sigma_0$ becomes

$$\Delta K = 0.66 \, \Delta \sigma_0 (\pi a)^{1/2} \tag{5}$$

The value of $\Delta K_{th} = 4.8 \, \text{MNm}^{-3/2}$ for $R = 0.04$ is comparable with (although perhaps a little lower than might be expected from) a value of $\Delta K_{th} = 4.8 \, \text{MNm}^{-3/2}$ for $R = 0.2$ obtained by James and Knott for a quenched and tempered steel of similar yield strength (15). In the latter case, measurements of closure indicated that plastic wake closure was highly significant in the $R = 0.2$ results and that the 'closure-free' value of ΔK_{th}, i.e., ΔK_c, was approximately 3.0 MNm$^{-3/2}$. A similar value applies to the lower strength structural steel, for which the experimental data are given in Fig. 3.

Recalculation of the long crack data in Fig. 1 using $\Delta K_c = 3.0 \, \text{MNm}^{-3/2}$ reduces all the values of $\Delta \sigma_0$ by a factor of (3/4.8), i.e., the position of the line in the log:log plot is lowered by a constant amount (see equations (1) and (2)

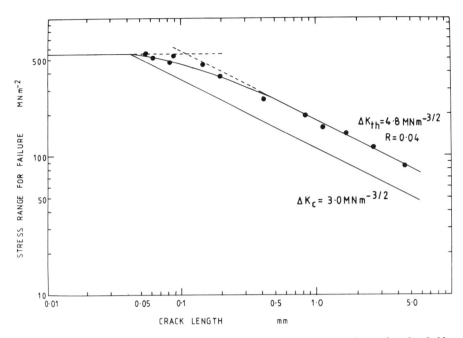

Fig 7   Effect on 'Kitagawa plot' (8) of recalculating long crack line using a closure-free threshold, ΔK_c

and Fig. 7). The effect of this shift is such as to provide an effective lower bound to all the data points. The inference is then that most of the 'anomalous' short crack behaviour observed here is attributed to an artefact of the long crack threshold test, i.e., the effect of closure at low stress ratio.

In terms of engineering design, it would seem that a rather good lower bound is obtained from the endurance limit and the value of stress calculated from the 'effective' threshold, $\Delta K_c = 3.0$ MNm$^{-3/2}$. If a modest safety factor is applied to these bounds, integrity should be assured for the stressing situations envisaged, provided that gross variations from the accepted material properties (defect populations) do not occur. This can be achieved either by attention to quality control in fabrication and NDT or by consistency of materials processing.

The above discussion refers to cracks grown in air. It has been found that ΔK_c for cracks in vacuum is higher than that in air (31) and the position of the lower bound line in Fig. 7 would be altered accordingly. Tests in vacuum simulate the growth of cracks from buried defects.

An example of the use of ΔK_c in 'lifing'

The detailed assessment of the fatigue life of a welded component which is part of a vehicle demands a knowledge of the loading spectrum to which it is subjected during service. In general, this is likely to comprise a large number of low amplitude cycles, each produced by a revolution of the vehicle's wheels on smooth roads or rail-tracks, together with a smaller number of higher amplitudes, associated with rough surfaces, such as pot-holes in roads or joints and welds on rail-tracks. Additionally, a few very high amplitudes may be experienced, due to loading by passengers or freight, or to the occasional onset of resonance. The lifing procedures involved may be appreciated by an illustrative example, using simplified figures taken from a paper by McLester, which shows the loading spectra for locomotive and freight bogies (30).

In approximately 200 miles of travel, the number of wheel revolutions is of the order of 10^5. McLester's figures show that the spectra contain many cycles at stress levels below 50 MNm^{-2} and rather few at higher levels. For simplicity, we take 20 cycles at 100 MNm^{-2} and 10^3 cycles at 65 mNm^{-2}. The assumptions made are that NDT is capable of detecting an edge crack of 1 mm in size and that a proposal has been made that a safe inspection period is every 10^8 cycles (approximately 2×10^5 miles or 1–2 years). A traditional LEFM analysis would wish to ignore all cycles at stresses such that the values of ΔK were below the threshold ΔK_{th}, but this philosophy would be open to question in the light of 'anomalous' short crack growth behaviour, which tends to indicate that cracks of less than 1 mm in length (the NDT limit) could grow at ΔK values less than ΔK_{th}.

If, however, the short crack behaviour is 'anomalous' only because closure effects operate in long crack threshold tests, it would appear that a threshold philosophy could be maintained, provided that the 'closure-free' value of ΔK_{th}, ΔK_c, were employed. Our results suggest that ΔK_c is approximately 3.0 MNm$^{-3/2}$ for structural steel, which, for a 1 mm edge crack, corresponds to an applied stress of approximately 50 MNm^{-2}. We conclude, therefore, that in 10^5 total cycles for the simplified loading spectrum, the only cycles of importance are the 20 at 100 MNm^{-2} and the 10^3 at 65 MNm^{-2}. During the proposed inspection period of 10^8 cycles, the component is therefore subjected to 2×10^4 cycles at 100 MNm^{-2} and 10^6 cycles at 65 MNm^{-2}. Using the expression

$$da/dN = 10^{-11} \Delta K^3 \qquad\qquad (6)$$

where $\Delta K = 1.12 - \Delta\sigma(\pi a)^{1/2}$ for an edge crack, it may be calculated using a simple Miner's Law summation, that a crack, initially just below 1 mm in length (i.e., just undetectabe by NDT) grows to approximately 2.5 mm in 10^8 cycles (ignoring the fact that ΔK_c will be exceeded by stresses in the range 50–30 MNm^{-2} as the crack length increases from 1 to 2.5 mm). Final crack lengths of this order are unlikely to cause failure by plastic collapse or fast fracture in tough, structural steels, and so the proposed inspection period is deemed to be appropriate. Any crack found during inspection would be repaired.

The above calculation is illustrative, rather than precise, but does demonstrate the lifing principles involved. Specifically, we suggest that, if the 'closure-free' value, ΔK_c, is employed, a threshold philosophy may be maintained, uncomplicated by any 'anomalous' short crack behaviour.

Conclusions

In considering the behaviour of physically short cracks it is essential to take into account the residual stress fields surrounding the cracks. In long crack threshold tests at low load ratios, plasticity-induced closure in the wake of the growing crack influences crack growth rates. The results of this study indicate that different amounts of closure in short crack specimens may account for much of the 'anomalous' behaviour of physically short cracks. Predictions of fatigue life using a closure-free value of the threshold may therefore be satisfactory in applications in which reliance is placed on NDT control.

Acknowledgements

The authors wish to thank Professor R. W. K. Honeycombe, FRS, and Professor D. Hull for the provision of research facilities. They are grateful to the SERC and British Rail Research Laboratories, Derby (JMK) and the Ministry of Defence, Procurement Executive (MNJ) for financial support.

References

(1) TAYLOR, D. and KNOTT, J. F. (1982) Growth of fatigue cracks from casting defects in nickel-aluminium bronze, *Metall. Technol.*, 9, 221–228.

(2) PEARSON, S. (1975) Initiation of fatigue cracks in commercial Aluminium alloys and the subsequent propagation of very short cracks, *Engng Fracture Mech.*, 7, 235–247.

(3) TANAKA, K., NAKAI, Y., and YAMASHITA, M. (1981) Fatigue growth threshold of small cracks, *Int. J. Fracture*, 17, 519–533.

(4) LANKFORD, J. (1982) The growth of small fatigue cracks in 7075-T6 aluminium, *Fatigue Engng Mater. Structures*, 5, 233–248.

(5) DE LOS RIOS, E. R., TANG, Z., and MILLER, K. J. (1984) Short crack fatigue behaviour in a medium carbon steel, *Fatigue Engng Mater. Structures*, 7, 97–108.

(6) BROWN, C. W., KING, J. E., and HICKS, M. A. (1984) Effects of microstructure on long and short crack growth in nickel base superalloys, *Metal Sci.*, 18, 374–380.

(7) SURESH, S. and RITCHIE, R. O. (1984) The propagation of short fatigue cracks, *Int. Met. Rev.*, 29, 445–476.

(8) KITAGAWA, H. and TAKAHASHI, S. (1976) Applicability of fracture mechanics to very small cracks or cracks in the early stage, *Proceedings of the 2nd International Conference on the Behaviour of Materials*, Boston, pp. 627–631.

(9) TSUBOTA, M., KING, J. E., and KNOTT, J. F. (1984) Crack propagation and threshold behaviour in Ni-base superalloys and its implications for component life assessment, *First Parsons International Turbine Conference*, Dublin (Institution of Mechanical Engineers, London), pp. 189–195.

(10) COLES, A. (1979) Material Considerations for Gas Turbines, *Proceedings of the Third International Conference on Mechanical Behaviour of Materials*, Cambridge, Vol. 1, pp. 3–11.

(11) THOMPSON, A. W. and BERNSTEIN, I. M. (1977) Selection of structural materials for hydrogen pipelines and storage vessels, *Int. J. Hydrogen Energy*, 2, 163–173.

(12) TETELMAN, A. S. and ROBERTSON, W. D. (1963) Direct observation and analysis of crack propagation in iron-3% silicon single crystals, *Acta Met.*, 11, 415–426.

(13) BROOKSBANK, D. and ANDREWS, K. W. (1972) Stress fields around inclusions and their relation to mechanical properties, *J. Iron Steel Inst.*, 210, 246–255.

(14) CLARK, G. and KNOTT, J. F. (1977) Effects of notches and surface hardening on the early growth of fatigue cracks, *Met. Sci.*, 11, 345–350.

(15) JAMES, M. N. and KNOTT, J. F. (1985) An assessment of crack closure and the extent of the short crack regime in Q1N (HY80) steel, *Fatigue Engng Mater. Structures*, 8, 177–191.

(16) CHAUHAN, P. and ROBERTS, B. W. (1979) Fatigue crack growth behaviour of short cracks in a steam turbine rotor steel – an investigation. *Metall. Mater. Technol.*, **11**, 131–136.

(17) USAMI, S. and SHIDA, S. (1979) Elastic–plastic analysis of the fatigue limit for a material with small flaws, *Fatigue Engng Mater. Structures*, **1**, 471–481.

(18) ROMANIV, O. N., SIMINKOVICH, V. N., and TKACH, A. N. (1981) Near-threshold short fatigue crack growth, *Proceedings of the 1st International Conference on Fatigue Thresholds*, Stockholm, Vol. 2, pp. 799–807.

(19) McCARVER, J. F. and RITCHIE R. O. (1982) Fatigue crack propagation thresholds for long and short cracks in Rene 95 nickel-base superalloy, *Mater. Sci. Engng*, **55**, 63–67.

(20) KING, J. E. and KNOTT, J. F. (1980) The effects of crack length and shape on the fracture toughness of a high strength steel 300M, *J. Mech. Phys Solids*, **28**, 191–200.

(21) BYRNE, J. and DUGGAN, T. V. (1981) Influence of crack geometry and closure on the fatigue threshold condition, *Proceedings of the 1st International Conference on Fatigue Thresholds*, Stockholm, Vol. 2, pp. 759–775.

(22) JAMES, M. N. and SMITH, G. C. (1984) Short crack behaviour in A533B and En8 steels, *Proceedings of 6th International Conference on Fracture* (ICF6), New Delhi, Vol. 3, pp. 2117–2124.

(23) RITCHIE, R. O., GARRETT, G. G., and KNOTT, J. F. (1971) Crack-growth monitoring: optimisation of the electrical potential technique using an analogue method, *Int. J. Fracture Mech.*, **7**, 462–467.

(24) JAMES, M. N. and KNOTT, J. F. (1985) Critical aspects of the characterization of crack tip closure by compliance techniques, *Mater. Sci. Engng*, **72**, L1–L4.

(25) NAKAI, Y., TANAKA, K., and NAKANISHI, T. (1981) The effects of stress ratio and grain size on near-threshold fatigue crack propagation in low-carbon steel, *Engng Fracture Mech.*, **15**, 291–302.

(26) MINAKAWA, K., NEWMAN, J. C., and McEVILY, A. J. (1983) A critical study of the crack closure effect on near-threshold fatigue crack growth, *Fatigue Engng Mater. Structures*, **6**, 359–365.

(27) JAMES, M. N. and SMITH, G. C. (1983) Surface microcrack closure in fatigue: a comparison of compliance and crack sectioning data, *Int. J. Fracture*, **22**, R69–R75.

(28) MILLER, K. J. (1984) Initiation and growth rates of short fatigue cracks, *Eshelby Memorial Symposium*, Sheffield, pp. 477–500.

(29) RICE, J. R. and TRACEY, D. M. (1973) *Numerical and computational methods in structural mechanics* (Edited by Fenves, S. J. *et al.*), (Academic Press, New York).

(30) McLESTER, R. (1977) Fatigue problems in land transport, *Met. Sci.*, **11**, 303–307.

(31) JAMES, M. N. and KNOTT, J. F. (1985) Near-threshold fatigue crack closure and growth in air and vacuum, *Scripta Met.*, **19**, 189–194.

NOTCHES

K. Yamada,* M. G. Kim,* and T. Kunio*

Tolerant Microflaw Sizes and Non-Propagating Crack Behaviour

REFERENCE Yamada, K., Kim, M. G., and Kunio, T., **Tolerant Microflaw Sizes and Non-Propagating Crack Behaviour,** *The Behaviour of Short Fatigue Cracks*, EGF Pub. 1 (Edited by K. J. Miller and E. R. de los Rios) 1986, Mechanical Engineering Publications, London, pp. 261–274.

ABSTRACT The critical length of non-propagating cracks in plain carbon steel specimens with three different carbon contents of 0.36%C, 0.55%C, and 0.84%C are compared with the critical size of artificially induced micropits prepared by electro-discharge-machining in order to discuss the physical meaning of the endurance limit and the evaluation of the tolerant microflaw size at the stress level of the endurance limit.

It is found that there exists a particular size of micropit below which there is no effect on the endurance limit. This critical size of micropit can be regarded as equivalent to a tolerant microflaw which would not reduce the endurance limit.

It is also found that this critical micropit size does not necessarily coincide with the critical length of a non-propagating crack but coincides with the critical length of an annealed pre-crack.

Local residual stress around artificially introduced microflaws are cited as the reason for discrepancies when compared to natural cracks in plain un-notched specimens.

Introduction

It is well known that the endurance limit of plain carbon steel smooth specimens corresponds to the critical stress for the onset of growth of non-propagating cracks(NPCs), but does not correspond to the critical stress for the initiation of such a crack (1)(2). The appearance of a NPC in a fatigued specimen implies the possible existence of a tolerant flaw which would not reduce the original endurance limit.

Murakami and others (3)–(5) have made a comprehensive study of the relationship between the size of a NPC and an artificial micropit size. However, difficulties in finding the critical size of a crack, namely, the maximum size of a NPC in a smooth specimen (2), has perturbed a proper understanding of the relationship between the size of a NPC and the tolerant flaw size.

The authors have recently reported that the critical size of NPCs in plain carbon steel grow up to a length ranging from 200 μm to 300 μm during stage II growth, which is equivalent to several orders of magnitude greater than the matrix ferrite grain size (2)(6), and also that the non-propagation of such cracks can result from the closure associated mainly with localized residual compressive stresses which are produced around the crack tip during fatigue loading (7).

Fatigue loading history also has an influence on the critical size of a NPC. Even a large crack which would not become a NPC under a conventional stress

* Department of Mechanical Engineering, Keio University 3-14-1 Hiyoshi, Kohoku-ku, Yokohama 223, Japan.

amplitude at the endurance limit, can become a NPC under the same stress level if the crack tip is changed from an open state to a closed state by coaxing effects (7)(8). This evidence may suggest that loading history may have an important role on the critical size of a NPC and that the tolerant flaw size is not directly related to the size of a NPC. It may be said therefore that there still remain some uncertainties in characterizing the critical size of a NPC; a requirement in evaluating the level of the endurance limit from a mechanistic viewpoint.

In this paper special emphasis is placed on the nature of NPCs and the size of tolerant flaws when considering the relationship between the critical size of a micropit and a NPC in plain carbon steel specimens subjected to constant stress amplitude.

Materials and experimental methods

The materials used in this experiment are plain carbon steels with three different carbon contents as given in Table 1. This table also shows the heat treatment condition for each material. These materials were machined into solid hourglass shape specimens having a 9 mm diameter and a 20 mm radius of curvature at the gauge length so as to make microscopic observation easy. Details of the specimen are given in Fig. 1. The stress concentration factor of this geometry is about 1.06, so that these specimens may be regarded as smooth test pieces.

Every specimen was vacuum-annealed at 640°C for 1 hour and then electro-polished by approximately 10 μm in depth in order to eliminate residual stresses and strains due to machining. Rotating bending fatigue tests were carried out on these specimens at a frequency of 48 Hz.

Mechanical properties and microstructural parameters of the specimens are given in Table 2. Volume fraction of pearlite and ferrite grain size were measured by the line counting method (9). A micro-Vickers indentor was selected with a 20 g and 50 g mass for the ferrite and pearlite grains, respectively. Micrographs of the surface microstructure of each specimen are given in Fig. 2. The white portion in the micrograph of the 0.84%C specimen represents the ferrite phase which has decarburized due to the high temperature annealing at 1200°C; the volume fraction of this ferrite is 6 per cent of the

Table 1   Chemical composition and heat treatments of materials

Chemical composition wt%	0.84%C-Steel	C:0.84 Si:0.22 Mn:0.45 P:0.014 S:0.027 Cu:0.03 Ni:0.02Cr:0.10
	0.55%C-Steel	C:0.55 Si:0.22 Mn:0.69 P:0.017 S:0.021
	0.36%C-Steel	C:0.36 Si:0.27 Mn:0.53 P:0.014 S:0.011
Heat treatment condition	0.84%C-Steel	1200°C 2 hr annealing
	0.55%C-Steel	810°C 2 hr annealing
	0.36%C-Steel	1000°C 2 hr annealing

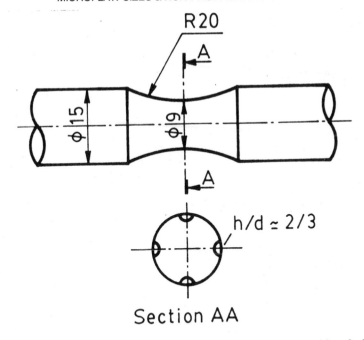

R20

A

$\phi 15$

$\phi 9$

A

$h/d \simeq 2/3$

Section AA

Fig 1   Details of specimen geometry and a schematic illustration of the position of micropits

surface microstructure. Some specimens then had four artificial micropits introduced as shown in Fig. 1, with axial symmetry, on their surface. Micropits may be considered as microflaws in the material and they were inserted by electro-discharge-machining (EDM) and had a dimensional ratio of $h/d \simeq 2/3$ where d and h are the diameter and the depth, respectively. The measurement of crack length was by an optical microscope and was determined in a direction perpendicular to the specimen axis.

The endurance limit of each specimen was determined experimentally as the maximum stress, below which the specimen will not fracture after a repetition of 10^7 stress cycles. Approximately 10 specimens were used to determine each endurance limit.

Table 2   Mechanical properties and microstructural parameters of three materials

Specimen	0.84%C Steel	0.55%C Steel	0.36%C Steel
Volume fraction of pearlite (%)	94	71	45
Ferrite grain size (μm)	—	5	26
Micro-vickers hardness of ferrite	150	132	142
Micro-vickers hardness of pearlite	257	239	247
0.2% proof stress (MPa)	277	314	184
Ultimate tensile strength (MPa)	714	649	512
Endurance limit (MPa)	210	245	194

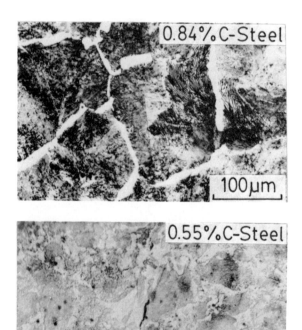

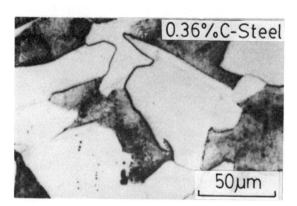

Fig 2   Surface microstructure of specimens

Results and discussion

An estimation of the critical size of a tolerant microflaw

The endurance limits, σ_e, of smooth specimens obtained from rotating bending fatigue tests are 210 MPa, 245 MPa, and 194 MPa for 0.84%C, 0.55%C, and 0.36%C steel specimens, respectively.

Tests indicate that a relatively large NPC, of stage II growth, has grown to a length several orders of magnitude greater than the matrix ferrite grain size. Thus the existence of tolerant microflaws equivalent to the critical length of a NPC may be expected at the endurance limit.

In order to discuss the relationship between the length of a NPC and the size of a tolerant microflaw, the following experiment was carried out. Several 0.84%C steel specimens, with micropits ranging from 100 to 350 μm, were tested at 210 MPa, the original endurance limit of smooth specimens. Results are shown in Fig. 3 by a solid line with open symbols. This figure shows that the largest micropit diameter which does not cause a fatigue fracture at the endurance limit is approximately 230 μm. This particular size can be regarded

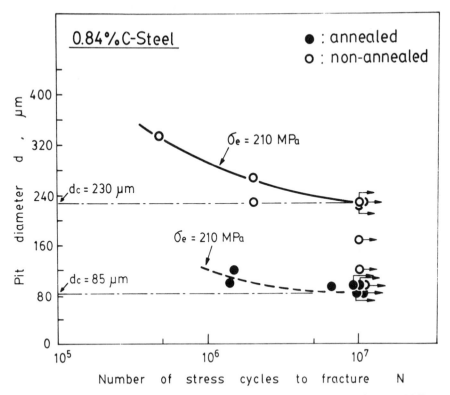

Fig 3   An evaluation of tolerant microflaws at the original endurance limit of $\sigma_e = 210$ MPa

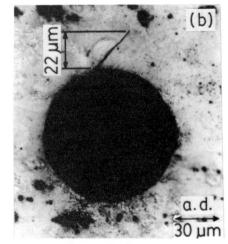

Fig 4   (a) A NPC which initially propagated from the edge of a non-annealed EDM pit at $\sigma = 210$ MPa and $N = 10^7$ cycles

(b) A NPC which initially propagated from the edge of an annealed EDM pit at $\sigma = 210$ MPa and $N = 10^7$ cycles

as the critical size of a tolerant microflaw which does not affect the original level of the endurance limit of this material.

Furthermore, it is found from microscopic observations around the pit that NPCs originated from pits and propagated to a length comparable with the pit size; see, for example, Fig. 4(a). This means that the condition for the onset of fatigue fracture cannot be determined by the critical stress to initiate a crack at the periphery of a pit, but must be determined by the critical stress to cause the growth of the NPC.

It should be noted that if we take 'pit diameter + NPC length' as an effective length of a NPC, the critical length of a NPC can be regarded as approximately 580 μm, which is fairly large compared with the critical length of a NPC on a smooth specimen, which was approximately 340 μm. A reason for the appearance of such a large NPC can be explained by the possible effects of residual compressive stress associated with localized plasticity at the crack tip during cyclic loading and also from the machining by EDM (10). To examine the effects of residual stress on the non-propagating behaviour of a crack, a particular specimen was used with a NPC of 200 μm length which had propagated from the edge of the pit thus giving an effective length of 400 μm.

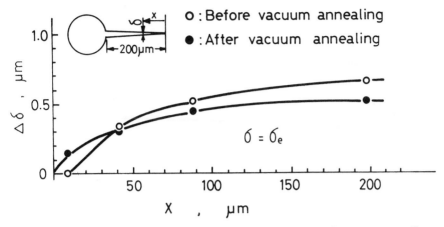

Fig 5   The range of crack opening displacment of a NPC before and after vacuum annealing

A stress relief annealing was performed in vacuum at 640°C for 1 hour, but after measurements of crack opening displacement were made using a method given in a previous study (7). Measurements of crack opening displacement were made again on the same crack after this heat treatment. Results are as shown in Fig. 5.

This figure explains that the crack tip of a NPC which was closed, opens after vacuum annealing under the application of a static bending stress equivalent to the endurance limit. Therefore it can be interpreted that the non-propagation behaviour of a crack nucleated from the pit is caused by closure effects (11)–(14) due mainly to residual compressive stresses in the same manner as ordinary NPCs. However, it is difficult to distinguish which residual stress is dominant, the residual stresses due to localized plasticity or that created in the process of making a pit.

Since an effect of residual stress around the pit cannot be negligible even in the case of an EDM pit(10), the critical size of a pit should be examined under experimental conditions excluding the effects of EDM residual stresses. Therefore, newly prepared specimens were all vacuum-annealed after EDM in order to provide specimens with pits free of residual stress. These specimens were fatigue tested at the stress level of the original endurance limit. Results are shown in Fig. 3. This figure shows the results of fatigue tests on two different types of specimen: the solid line with open symbols is for specimens having conventional EDM pits while the broken line with solid symbols is for specimens having pits free of residual stress. These results show that the critical size of a tolerant micropit decreases from 230 μm to 85 μm when the residual stress due to EDM is removed.

This value of 85 μm should be regarded as the tolerant microflaw size in this material rather than the conventional EDM micropit size of 230 μm.

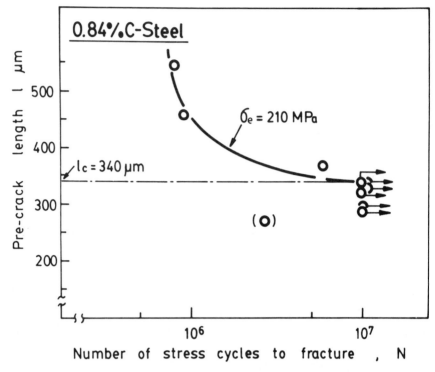

Fig 6 An evaluation of the critical size of a NPC from a fatigue pre-cracked specimen (7)
(○): fracture caused by a coalescence of neighbouring cracks

*Relationship between the critical size of a tolerant microflaw and the critical NPC
length of a smooth specimen*

The critical length of a NPC of the present material was obtained as 340 μm as
shown in Fig. 6, which the authors have already reported in a previous
paper (7). This result indicates that the critical size of that micropit in this
material, 85 μm, does not coincide with the critical NPC length of 340 μm
obtained from pre-cracked specimens having various sizes of natural surface
fatigue cracks in a smooth specimen.

Since a NPC in a plain specimen has resulted from the accumulation of cyclic
loading effects, including crack tip closure, and the EDM micropit is free from
any fatigue loading history, it is perhaps understandable that the results do not
coincide.

As the above critical micropit size of 85 μm generates a NPC of 22 μm, as
shown in Fig. 4(b), the effective crack length amounts to 85 + 22 μm; however,
this is still much less than the critical NPC length of 340 μm.

Figure 6 provides data on un-annealed specimens and so it may be suggested
that the fatigue pre-cracks should be subjected to annealing before fatigue

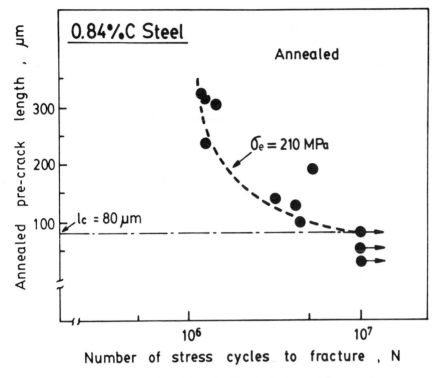

Fig 7   An evaluation of the critical length of an annealed pre-crack

testing. We have previously obtained this evidence by annealing at 640°C for 1 hour, after which the crack is no longer a NPC but a propagating crack at the stress level of the original endurance limit (7).

For the purpose of examining the change in the critical length due to annealing, specimens with a fatigue pre-crack were prepared at a stress level 10 per cent above the endurance limit. A desired length of fatigue pre-cracks can be obtained by stopping the fatigue test at a certain number of cycles prior to a vacuum-anneal at 640°C for 1 hour which makes the fatigue pre-cracks free of residual stress. These specimens were then fatigue tested at the stress level of the original endurance limit. Results are given in Fig. 7. This figure shows that the largest crack length, which will not fracture the specimen after 10^7 cycles of repetition at the stress equivalent to the original endurance limit, can be determined as approximately 80 μm.

The critical size of a micropit which is free from EDM residual stress, agrees well with this critical size from an annealed pre-cracked specimen.

In order to examine the generality of the above coincidence for a wide range of carbon content, two other kinds of specimen were used for similar fatigue experiments as those for the 0.84%C steel specimens. Results are given in Fig. 8.

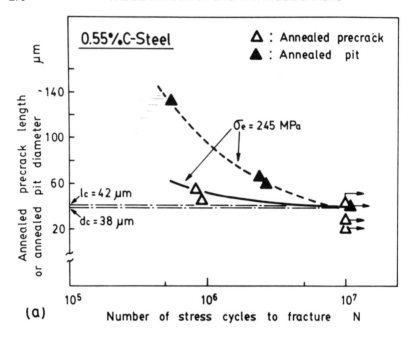

(a)

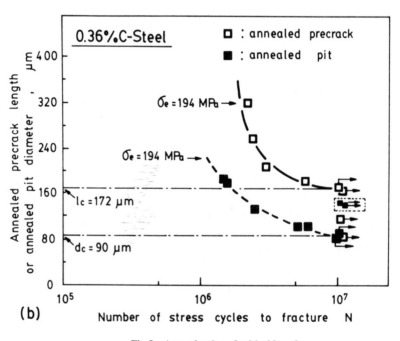

(b)

Fig 8 An evaluation of critical lengths

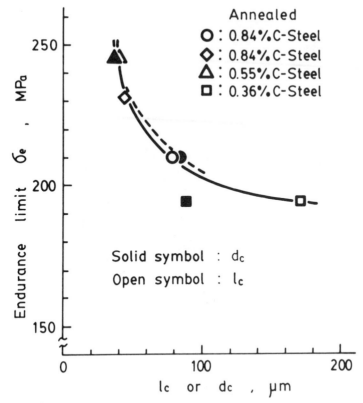

Fig. 9   **Relationships between the endurance limit and defect sizes (micropits and plain specimen non-propagating cracks) in annealed steels**
    ◇: unpublished work by the authors

In the case of the 0.55%C steel, good agreement is obtained; compare 42 μm and 38 μm, respectively. However, in the case of the 0.36%C steel a relatively poor agreement is recognized between 172 μm and 90 μm, respectively. These results, together with that of the 0.84%C specimen, are represented in Fig. 9.

In this figure, a solid line with open marks indicates the relationship between plain specimen data and σ_e while the broken line with solid marks indicates the relationship between the critical micropit size and σ_e. Both relationships show good agreement with each other for the two higher carbon contents, but in the case of the 0.36%C steel there is no agreement.

The disagreement in 0.36%C steel samples could be associated with the decrease in the volume fraction of pearlite, which acts effectively to suppress the crack propagation around the pit. Since the initiation and propagation of a crack is associated with the local volume fraction of pearlite around the pit rather than the average volume fraction, we may expect good agreement between the two parameters in the 0.36%C steel if the local volume fraction

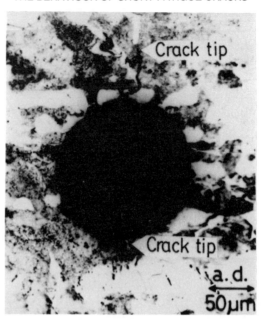

Fig 10   The 144 μm diameter EDM pit which lies within the locally high carbon domain of the 0.36%C steel specimen and the short non-propagating crack which is impeded in its propagation by the surrounding pearlitic structure; $\sigma = 194$ MPa, $N = 10^7$ cycles

around the pit is pre-arranged by choosing a location of the EDM pit within a pearlite colony.

On the basis of this consideration, two specimens were prepared having annealed pits of 140 μm and 144 μm diameters. These pits lie within a locally high carbon domain as shown in Fig. 10. These specimens were then tested at the endurance limit of 194 MPa. These specimens did not fail after 10^7 stress cycles as expected. The two results are shown surrounded by a broken line inset in Fig. 8(b). Thus it is found that the critical micropit size changes from 90 to at least 144 μm when the microstructure surrounding the pit is changed from a mixture of pearlite and ferrite to one of pearlite.

It may be concluded that the critical size of a micropit is closely associated with the local state of mixture of pearlite and ferrite around the pit rather than the average volume fraction of pearlite, i.e., the difference of crack growth resistance between the pearlite and ferrite constituents.

Conclusions

Studies have been performed on the physical meaning of the endurance limit and the tolerant microflaw size of plain carbon steel specimens with three different carbon contents.

The results are summarized as follows.

(1)   There exists a particular size of artificial micropit which would not alter the original level of the endurance limit in plain carbon steel specimens. This particular size of micropit can be regarded as the critical size of a tolerant microflaw which would not reduce the original level of the endurance limit.

(2)   The critical size of an artifically induced micropit does not necessarily coincide with the length of a non-propagating crack due to micro-residual stress effects in the former and different crack closure histories of the latter. Annealing of both types of specimen provides better agreements.

(3)   A local residual stress around the micropit apparently increases the size of the critical tolerant microflaw size.

(4)   The critical size of an artificial micropit in a plain carbon steel is influenced by the local volume fraction of pearlite around the pit rather than the average volume fraction.

Acknowledgement

The authors would like to thank Professor M. Shimizu for his valuable discussion and encouragement.

References

(1)   KUNIO, T., SHIMIZU, M., and YAMADA, K. (1969) Microstructural aspects of the fatigue behaviour of rapid heat-treated steel, *Proceedings of the 2nd International Conference on Fracture* (Chapman & Hall, London), pp. 630–642.

(2)   KUNIO, T. and YAMADA, K. (1979) Microstructural aspects of the threshold condition for the non-propagating fatigue cracks in martensitic and ferritic steel, *ASTM STP 675*, 342–370.

(3)   MURAKAMI, Y., FUKUDA, S., and ENDO, T. (1978) Effect of micro-hole on fatigue strength: (1st Report) Effect of micro-hole of 40 μm–200 μm in diameter on the fatigue strength of annealed 0.13% and 0.46% carbon steel, *Trans. Jap. Soc. mech. Engrs*, **44**, 4003–4013.

(4)   MURAKAMI, Y., KAWANO, H., and ENDO, T. (1979) Effect of micro-hole on fatigue strength: (2nd Report) Effect of micro-hole of 40 μm–200 μm in diameter on the fatigue strength of quenched and tempered 0.46% carbon steel, Loc. cit. **45**, 1479–1486.

(5)   MURAKAMI, Y., TAZUNOKI, Y., and ENDO, T. (1981) Existence of coaxing effect and effect of small artificial holes of 40 μm–200 μm diameter on fatigue strength in 2017S-T4 Al alloy and 7:3 Brass, Loc. cit. **47**, 1293–1300.

(6)   KUNIO, T., SHIMIZU, M. YAMADA, K., and TAMURA, M. (1984) Endurance limit and threshold condition for microcrack in steel, *Proceedings of the 2nd International Conference on Fatigue and Fatigue Thresholds,* Birmingham, UK, pp. 817–826.

(7)   KUNIO, T., YAMADA, K., and KIM, M. G. (1984) Critical behaviour of non-propagating crack in steel, *Proceedings of the International Symposium on Fundamental Questions and Critical Experiments on Fatigue*, ASTM, Dallas, USA.

(8)   KIM, M. G., YAMADA, K., and KUNIO, T. (1985) Effect of cyclic stress history on the threshold condition for microcrack in structural steel, *Trans. Jap. Soc. mech. Engrs*, **51**, 1529–1533.

(9)   De HOFF, R. T. and RHINES, F. N. (1968) *Quantitative Microscopy* (McGraw-Hill, New York), pp. 55–59, 269–270 (Japanese Edition).

(10)   *Technical Handbook of EDM* (1963), Nikkan Kogyo Shimbun Co., Japan, pp. 73–74.

(11)   ELBER, W. (1971) The significance of fatigue crack closure, *ASTM STP 486*, pp. 230–242.

(12) SURESH, S., ZAMISKI, G. F., and RITCHIE, R. O. (1981) Oxide-induced crack closure: An explanation for near-threshold corrosion fatigue crack growth behaviour, *Met. Trans.* **12A**, 1435–1443.
(13) MINAKAWA, K. and McEVILY, A. T. (1981) On crack closure in the near-threshold region, *Scripta Met.*, **15**, 633–636.
(14) KOBAYASHI, H. (1982) Recent research on fatigue crack growth, *Bull. Jap. Inst. Metals*, **21**, 329–338.

Y. Murakami* and M. Endo†

Effects of Hardness and Crack Geometries on ΔK_{th} of Small Cracks Emanating from Small Defects

REFERENCE Murakami, Y. and Endo, M. **Effects of Hardness and Crack Geometries on ΔK_{th} of Small Cracks Emanating from Small Defects**, *The Behaviour of Short Fatigue Cracks*, EGF Pub. 1 (Edited by K. J. Miller and E. R. de los Rios) 1986, Mechanical Engineering Publications, London, pp. 275–293.

ABSTRACT The dependence of ΔK_{th} on crack size and material properties at a stress ratio $R = -1$ was studied on various materials and microstructures. The values of ΔK_{th} of all materials investigated were unified with one geometrical and one material parameter.

The geometrical parameter is the square root of the area determined by projecting defects or cracks onto the plane normal to the maximum tensile stress. The relationship

$$\Delta K_{th} \propto (\sqrt{area})^{1/3}$$

is derived.

The most relevant material parameter to unify data is the Vickers hardness and the relationship

$$\Delta K_{th} \propto (H_v + C)$$

is obtained. The constant C reflects the difference of non-propagation behaviour of small cracks in soft and hard metals.

Combining these equations, experiments show that ΔK_{th} and the fatigue limit σ_w of cracked members are given by

$$\Delta K_{th} = 3.3 \times 10^{-3}(H_v + 120)(\sqrt{area})^{1/3}$$

and

$$\sigma_w = 1.43(H_v + 120)/(\sqrt{area})^{1/6}$$

Here ΔK_{th} is in MPa$\sqrt{m}$, $\sqrt{area}$ in μm, and σ_w in MPa. These equations are applicable to cracks having $\sqrt{area}$ approximately less than 1000 μm.

Notation

area	The area which is occupied by projecting a defect or a crack onto the plane normal to the maximum tensile stress
α, C, C_1, C_2, n	Constants independent of material
C', n'	Constants dependent on material
H_B	The Brinell hardness number (BHN)
H_v	The Vickers hardness number (DPH)
K_I	Stress intensity factor (mode I)
K_{Imax}	The maximum value of the stress intensity factor along the front of a three-dimensional crack (mode I)

* Department of Mechanics and Strength of Solids, Faculty of Engineering, Kyushu University, Higashi-ku, Fukuoka, 812 Japan
† Department of Mechanical Engineering, Fukuoka University, Jonan-ku, Fukuoka, 814-01 Japan

ΔK_{th}	Threshold stress intensity factor range ($R = -1$)
ΔK_{eff}	Effective stress intensity factor range
$\Delta K_{eff,th}$	Threshold effective stress intensity factor range
l	Length of a two-dimensional crack
l_f	Fictitious crack length
l_0	The maximum length of a non-propagating crack observed in unnotched specimens at the fatigue limit
ν	Poisson's ratio
R	Stress ratio
σ_U	Ultimate tensile strength
σ_w	Fatigue limit
σ_{w_0}	Fatigue limit of an unnotched specimen
σ_Y	Yield stress
σ_0	The maximum tensile stress

Introduction

The threshold stress intensity factor range ΔK_{th} is required in order to determine the maximum allowable stress when cracks or defects are detected in machine parts and structures under service loading. However, recent experimental studies (1)–(6) show that the value of ΔK_{th} is dependent on crack size, i.e., the smaller the crack, the smaller the value of ΔK_{th}. On the other hand, the value of ΔK_{eff} characterizes the growth behaviour of not only a large crack but also a small one, and, accordingly, $\Delta K_{eff,th}$ is almost independent of crack size (7)(8). However, in practice, the application of ΔK_{eff} or $\Delta K_{eff,th}$ is inconvenient, because we must measure or assume the opening ratio of a crack in real structures, and this measurement is very difficult. Therefore, it is preferable in practice to determine the value of ΔK_{th} as a function of crack size and geometry and then to estimate the allowable stress from ΔK_{th} rather than to measure ΔK_{eff}.

The correlation between the fatigue limit σ_w and the crack length l was first obtained by Frost (9). But at the time when Frost carried out his fatigue tests and obtained the relationship $\sigma_w^n l = $ constant, the fatigue threshold phenomena and the dependence of ΔK_{th} on crack length were not recognized.

Kitagawa and Takahashi (3) pointed out that ΔK_{th} decreases with decreasing crack size. However, an expression for ΔK_{th} as a function of crack size was not given explicitly. El Haddad et al. (10) added the fictitious crack length l_f to the real crack length l in order to compensate for the decreasing value of ΔK_{th} with decreasing crack length. The value of l_f, considered to be a material constant, was determined from ΔK_{th} for a large crack, and from the fatigue limit of an unnotched specimen, σ_{w_0}. However, the physical meaning of l_f and the rule for determining l_f for various three-dimensional cracks are not clear.

In many previous studies, comparison of ΔK_{th} values for different materials did not consider the dependence of ΔK_{th} on crack size and geometry. Such comparisons are likely to induce erroneous conclusions. One objective of the present paper is to elucidate the dependence of ΔK_{th} on the shape and size of cracks with special emphasis on small cracks. The large amount of available data on rotating bending fatigue in various materials is analysed. A geometrical parameter $\sqrt{area}$, which is defined by the square root of the projection area of defects or cracks onto the plane perpendicular to the maximum tensile stress, is proposed in order to unify the effects of various notches, holes, and cracks. An explicit relationship between ΔK_{th} and $\sqrt{area}$ is confirmed for more than ten materials.

Another objective is to find the most appropriate material parameter which reflects the threshold behaviour. It should be noted that the dependence of ΔK_{th} on material parameters can be made clear only after the most appropriate geometrical parameter is found. Although various material parameters such as yield stress (σ_Y), ultimate tensile stress (σ_U) and hardness (H_v or H_B), may be correlated with ΔK_{th}, the Vickers hardness number H_v was chosen after observing the trend of many data and also for the reasons of simplicity of measurement and availability.

Finally, a simple formula for predicting ΔK_{th} in terms of one material and one geometrical parameter, i.e., H_v and $\sqrt{area}$, is derived.

A geometrical parameter for small defects or cracks

Effects of small defects, cracks, and inclusions on fatigue strength have been investigated by many researchers (11)–(19). However, their effects are so complicated that no method for unifying them quantitatively has been established. The crucial cause for the absence of a unifying method was an incorrect understanding of the threshold condition for such small defects, cracks, and inclusions. Murakami and Endo (20) showed how to overcome this difficulty by interpreting the fatigue limit not as the critical condition for crack initiation but as the condition for the non-propagation of a crack emanating from defects, cracks, and inclusions; previously Miller et al. (21)(22) had applied a similar consideration to notches. For example, the fatigue limit of a structural component containing a small defect must not be interpreted as a notch problem in which the critical condition of crack initiation is questioned, but should be understood as a problem of a crack which emanates from the defect and stops propagating. Only the interpretation of problems in this manner leads us to find the geometrical parameter for defects, cracks, and even sharp notches. It is reasonable to seek the geometrical parameter from the standpoint that the effects of shapes and sizes of cracks on fatigue strength may be correlated with stress intensity factors, especially with the maximum stress intensity factor along the three-dimensional crack front. Previous studies by Murakami et al. (20)(23)–(31) regarding this problem can be summarized as follows.

First, the stress intensity factors, K_I, for elliptical cracks in an infinite body under uniform tension were investigated and the approximate relationship between the maximum value (K_{Imax}) at the tip of the minor axis of an ellipse and the crack area was given as follow (**20**)

$$K_{Imax} \propto (\sqrt{area})^{1/2} \tag{1}$$

Afterwards, three-dimensional stress analyses by the body force method were carried out for surface cracks having various shapes, and the maximum value of the stress intensity factor along their crack front was correlated with the crack areas to give the following equation (**23**)–(**25**) (see Appendix 2):

$$K_{Imax} \cong 0.65\, \sigma_0 \sqrt{(\pi\sqrt{area})} \tag{2}$$

The error in equation (2) may be estimated to be less than 10 per cent. Equation (**2**) implies that the square root of the crack area projected onto the plane perpendicular to the maximum tensile stress should be adopted as the most relevant geometrical parameter for three-dimensional cracks.

When a specimen has a three-dimensional defect other than a planar crack, the fatigue limit is determined by the threshold condition of the crack emanating from the defect (**20**)(**26**)–(**30**). In this case, the initial three-dimensional shape of the defect is not directly correlated with the stress intensity factor. Rather, the planar domain (area) which is occupied by projecting the defect onto the plane perpendicular to the maximum principal stress should be regarded as the equivalent crack and the stress intensity factor should be evaluated from the equivalent crack, see Fig. 1(b). This may be easily understood from a simple two-dimensional example; the stress intensity factor for an elliptical hole having cracks at the ends of the major axis can be approximately estimated from those of an equivalent crack with a total length of (major axis + crack length). Here, it should be noted that the area of crack emanating from a three-dimensional crack occupies only a small portion of the total projected area (**28**)(**31**), see Fig. 1(a), and accordingly the area of an equivalent crack

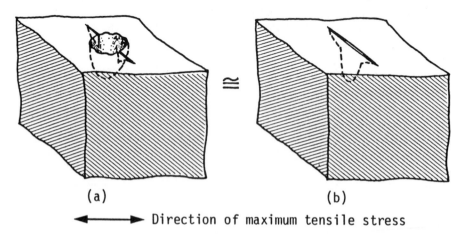

(a)   (b)

◀━━━━▶ Direction of maximum tensile stress

Fig 1   Defect with cracks and its equivalent crack

which should be used for equation (2) may be estimated by the projected area of the initial defect.

From the above discussion, it can be hypothesized that the square root of crack area or the projected area of defects is the relevant geometrical parameter to use when determining quantitatively the fatigue strength of structural components containing various cracks and defects in which no apparent mutual correlations are obvious. With regard to an estimation of $\sqrt{area}$ for irregularly shaped cracks, see Appendix 1.

On the basis of this hypothesis, previous data on rotating bending fatigue were analysed using the parameter $\sqrt{area}$. The artificial defects investigated in this study are very small drilled holes (20)(28)–(30)(32)–(35) with diameters ranging from 40 μm to 500 μm and depths greater than 40 μm, also very small and shallow notches (35)–(46) with depths ranging from 5 μm to 300 μm, and very shallow circumferential cracks (47) with depths greater than 30 μm, and finally a Vickers hardness indentation (35) of 72 μm surface length. The shapes of defects and cracks considered are shown in Fig. 2. The effects of work hardening and residual stress by introducing the drilled holes were examined and found to be small (20). In those tests almost all notched specimens were electropolished after introducing the notches (35)–(46) and the cracked specimens were annealed after introducing the fatigue cracks (47). Accordingly the effect of work hardening would be expected to be negligible.

The relationship between ΔK_{th} and $\sqrt{area}$ is illustrated in Fig. 3. the data in the figure were adopted from various references as well as previous studies by the authors' group. Although this figure may be similar to that of Kitagawa and Takahashi (3), the parameter of abscissa is not crack length but the new parameter $\sqrt{area}$ which can unify the size effect of three-dimensional defects and cracks. Moreover, the adoption of the new parameter $\sqrt{area}$ characterizes the threshold behaviour particularly for the data on very small cracks.

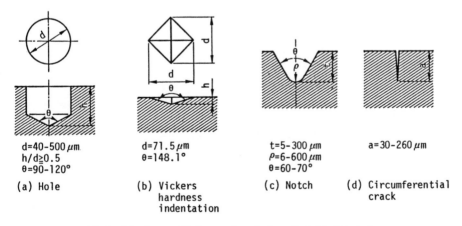

d=40–500 μm	d=71.5 μm	t=5–300 μm	a=30–260 μm
h/d≧0.5	θ=148.1°	ρ=6–600 μm	
θ=90–120°		θ=60–70°	
(a) Hole	(b) Vickers hardness indentation	(c) Notch	(d) Circumferential crack

Fig 2   The shapes of defects and cracks investigated in this study

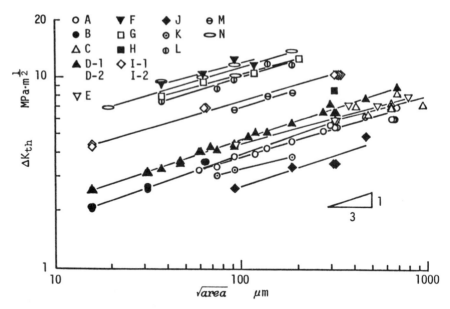

Fig 3   Relationship between ΔK_{th} and $\sqrt{area}$ for various defects and cracks. Letters correspond to the materials given in Table 1

For the region $\sqrt{area} \leqslant 1000\ \mu$m, the relationship between ΔK_{th} and $\sqrt{area}$ is approximately linear and the following equation holds regardless of material

$$\Delta K_{th} \propto (\sqrt{area})^{1/3} \tag{3}$$

Equation (3) characterizes the distinct dependence of ΔK_{th} on the geometrical parameter $\sqrt{area}$. Equation (3) implies also that if we investigate ΔK_{th} of specimens containing a very short two-dimensional crack of length l, we have

$$\Delta K_{th} \propto l^{1/3} \tag{4}$$

If we convert the empirical formula obtained by Frost (**9**), i.e., $\sigma_w^3 l =$ constant, to the same form as equation (4), we have

$$\Delta K_{th} \propto l^{1/6} \tag{5}$$

Equations (4) and (5) contradict each other. This is because the experiments by Frost contain longer or larger cracks than those of the data in Fig. 3. With increasing crack length ΔK_{th} approaches a constant value depending on the specific material. This indicates that the slope of the relationship between ΔK_{th} and $\sqrt{area}$ (or l) on a log–log scale varies from one-third to zero with increasing crack size and, therefore, the slope of one-sixth in equation (5) corresponds to a transient value from one-third to zero. This point was recently discussed in detail by Murakami and Matsuda (**48**).

Vickers hardness as a representative material parameter

Figure 3 presents data on an aluminum alloy and 70/30 brass in addition to various steels. The Vickers hardness H_v of these materials is shown in Table 1. It ranges over from 70 to 720 covering a range of a factor of ten.

From the tendency of ΔK_{th} in Fig. 3, it may be noted that the materials having higher Vickers hardness show higher values of ΔK_{th} (and accordingly higher fatigue strength). However, the tendency cannot be expressed in a simple form such as $\Delta K_{th} \propto H_v$. It has been empirically observed that the fatigue limit of a specimen containing a notch or a defect is not directly proportional to the Vickers hardness. This is presumably because the occurrence of non-propagating cracks may have a different dependency. In other words, a crack is likely to show non-propagating behaviour in soft materials while, on the contrary, it is difficult to find non-propagating cracks at the fatigue limit of hard steels. With increasing hardness non-propagating cracks are experienced only within a narrow range of stress amplitude, and in this case the length of non-propagating cracks is usually very short. Therefore, it may be concluded that ΔK_{th} is not a function of the form $\Delta K_{th} \propto H_v$ or $\Delta K_{th} \propto H_v^{\alpha}$. The difference of threshold behaviour between soft and hard materials may rather be expressed by the following formula

$$\Delta K_{th} \propto (H_v + C) \tag{6}$$

where C is a constant independent of materials. In order to check this prediction the values of $\Delta K_{th}/(\sqrt{area})^{1/3}$ were plotted against H_v from consideration of equation (3) and the validity of equation (6) was confirmed by many data with only a few exceptional data on stainless steels. (See Appendix 3). Now considering equations (3) and (6), the following equation is expected to hold for a wide range of materials

$$\Delta K_{th} = C_1(H_v + C_2)(\sqrt{area})^{1/3} \tag{7}$$

where C_1 and C_2 are constants independent of material.

The constants C_1 and C_2 in equation (7) can be determined by the least square method applied to the data in Fig. 3 and we have

$$\Delta K_{th} = 3.3 \times 10^{-3}(H_v + 120)(\sqrt{area})^{1/3} \tag{8}$$

where the units of ΔK_{th} are MPa$\sqrt{m}$ and that of $\sqrt{area}$ is μm.

Figure 4 shows the comparison of the experimental data in Fig. 3 and the correlation by equation (8). It is pleasing to note that the various data for H_v ranging from 70 to 720 are well represented by equation (8). The reason why equation (8) is not good in predicting ΔK_{th} for two kinds of stainless steels (34) is presumably because non-propagating cracks are unlikely to be observed in stainless steels even at a sharp notch (49)–(51). The existence of non-propagating cracks in the data of (34) was not checked in the present study.

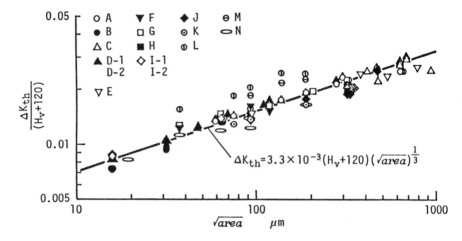

Fig 4   Relationship between $\Delta K_{th}/(H_v + 120)$ and $\sqrt{area}$. Letters correspond to the materials listed in Table 1

Combining equations (8) and (2), the fatigue limit σ_w of a cracked specimen can be expressed as

$$\sigma_w = 1.43(H_v + 120)/(\sqrt{area})^{1/6} \tag{9}$$

where σ_w is the nominal stress defined for the gross diameter and has units of MPa.

Murakami and Endo (20) previously proposed an equation of the form $\sigma_w^{n'}\sqrt{area} = C'$ for predicting the fatigue limit of materials containing defects or cracks. Although this equation is very accurate, the negative aspect of it is that we need fatigue tests for individual materials to determine n' and C' (52). This disadvantage is overcome by equations (8) or (9).

Table 1 shows the comparison of the values predicted by equations (8) and (9) with experimental results. For most materials except two kinds of stainless steels, the error is less than 10 per cent.

It should be noted that Fig. 4 and Table 1 include many results of specimens containing an extremely shallow notch (depths ranging from 5 to 20 μm) or a crack, or very small holes (diameters ranging from 40 to 500 μm), and although it may be said that the theory of the notch effect has been established for medium or deep notches, conventional theories (22)(43)(44)(53)–(57) may not be applicable to extremely shallow notches. From the viewpoint of the present study however, extremely shallow notches may be placed in the category of small cracks, and so the prediction of the fatigue limit can be simply made by equations (8) or (9).

Table 1 Comparison of predicted values by equations (8) and (9) with experimental results

Material	Defect	H_v	$\sqrt{area}$ (μm)	ΔK_{th} ($MPa\sqrt{m}$) Expt	ΔK_{th} ($MPa\sqrt{m}$) Pred.	σ_w (MPa) Expt	σ_w (MPa) Pred.	Error (%)
A: S10C (annealed) (36)	Notch	120	632	6.1	6.8	105	117	11.8
A: S10C (annealed) (37)	Notch	120	632	6.1	6.8	105	117	11.8
		120	632	6.1	6.8	105	117	11.8
A: S10C (annealed) (38)	Notch	120	316	5.5	5.4	134	132	−1.6
		120	316	5.5	5.4	134	132	−1.6
A: S10C (annealed) (20)(28)	Hole	120	74	3.4	3.3	172	168	−2.5
		120	60	3.2	3.1	181	174	−3.9
		120	93	3.8	3.6	172	162	−6.0
		120	136	4.2	4.1	157	152	−3.5
		120	119	4.0	3.9	157	155	−1.3
		120	185	4.6	4.5	147	144	−2.0
		120	272	5.2	5.1	137	135	−1.5
		120	298	5.7	5.3	142	133	−6.4
		120	463	6.3	6.1	128	124	−3.4
		120	681	7.1	7.0	118	116	−1.8
A: S10C (annealed) (32)	Hole	120	632	7.1	6.8	123	117	−4.6
B: S30C (annealed) (39)	Notch	153	16	2.0	2.3	220	247	12.3
		153	16	2.1	2.3	225	247	9.9
		153	16	2.1	2.3	225	247	9.9
		153	32	2.6	2.8	199	220	10.7
		153	32	2.6	2.8	204	220	8.0
		153	32	2.7	2.8	208	220	5.5
		153	63	3.6	3.6	196	196	−0.2
		153	63	3.6	3.6	196	196	−0.2
		153	63	3.6	3.6	196	196	−0.2
		153	316	5.7	6.1	140	150	7.3
		153	316	5.7	6.1	140	150	7.3
		153	316	5.9	6.1	144	150	4.0
C: S35C (annealed) (40)	Notch	160	632	7.0	7.9	120	137	14.0
		160	632	7.1	7.9	122	137	12.4
		160	632	7.3	7.9	126	137	8.4
C: S35C (annealed) (41)	Notch	160	316	5.9	6.3	144	154	6.5
		160	474	6.4	7.2	127	144	13.1
		160	949	7.2	9.1	101	128	26.3
C: S35C (annealed) (27)	Hole	160	409	7.1	6.9	152	147	−3.2
		160	681	8.2	8.1	137	135	−1.3
D-1: S45C (annealed) (42)–(44)	Notch	180	16	2.6	2.5	280	271	−3.1
		180	16	2.5	2.5	275	271	−1.4
		180	16	2.5	2.5	275	271	−1.4
		180	32	3.2	3.1	245	242	−1.4
		180	32	3.2	3.1	250	242	−3.4
		180	32	3.2	3.1	245	242	−1.4
		180	316	6.6	6.7	160	165	2.9
		180	316	6.6	6.7	160	165	2.9
		180	316	6.2	6.7	151	165	8.9

Continued

Table 1   (Continued)

Material	Defect	H_v	$\sqrt{area}$ (μm)	ΔK_{th} ($MPa\sqrt{m}$)		σ_w (MPa)		Error (%)
				Expt	Pred.	Expt	Pred.	
D-2: S45C (annealed) (20)(28)	Hole	170	37	3.3	3.2	235	228	−3.2
		170	46	3.5	3.4	226	219	−3.0
		170	68	4.3	3.9	226	206	−9.1
		170	48	3.7	3.5	230	218	−5.2
		170	74	4.2	4.0	211	203	−3.9
		170	109	4.8	4.6	201	190	−5.5
		170	60	4.0	3.7	226	210	−7.0
		170	60	4.0	3.7	226	210	−7.0
		170	93	4.5	4.3	201	195	−2.8
		170	93	4.3	4.3	196	195	−0.4
		170	136	5.1	4.9	191	183	−4.1
		170	119	5.1	4.7	201	187	−6.9
		170	185	5.7	5.5	181	174	−3.9
		170	272	6.5	6.2	172	163	−5.2
		170	298	7.2	6.4	181	161	−11.2
		170	463	7.8	7.4	157	149	−4.9
		170	681	8.8	8.4	147	140	−4.7
E: S50C (annealed) (45)	Notch	177	316	5.9	6.7	144	163	13.2
		177	316	5.9	6.7	144	163	13.2
E: S50C (annealed) (47)	Circum-	177	95	4.4	4.5	196	199	1.5
	ferential	177	379	7.2	7.1	160	158	−1.2
	crack	177	538	7.1	8.0	133	149	11.9
		177	791	8.0	9.1	123	140	13.5
F: S45C (quenched) (29)(30)	Hole	650	37	9.3	8.5	667	604	−9.4
		650	62	10.3	10.1	568	554	−2.5
		650	93	12.4	11.5	559	518	−7.4
		650	117	11.7	12.4	470	499	6.1
G: S45C (quenched and tempered) (29)(30)	Hole	520	37	8.0	7.0	568	502	−11.6
		520	62	9.4	8.4	519	460	−11.3
		520	117	10.5	10.3	421	414	−1.6
		520	202	12.5	12.4	382	378	−1.0
H: S50C (quenched and tempered) (45)	Notch	319	316	8.6	9.9	209	241	15.3
I-1: S50C (quenched and tempered) (45)	Notch	378	316	10.3	11.2	252	273	8.2

Continued

Table 1   (Continued)

Material	Defect	H_v	$\sqrt{area}$ (μm)	ΔK_{th} ($MPa\sqrt{m}$)		σ_w (MPa)		Error (%)
				Expt	Pred.	Expt	Pred.	
I-2: S50C (quenched and	Notch	375	16	4.3	4.1	468	447	−4.4
tempered) (**39**)		375	16	4.4	4.1	478	447	−6.4
		375	63	6.8	6.5	373	355	−4.8
		375	63	6.8	6.5	373	355	−4.8
		375	316	10.3	11.1	252	272	7.6
		375	316	10.3	11.1	252	272	7.6
		375	316	10.3	11.1	252	272	7.6
J: 70/30 brass (**46**)	Notch	70	316	3.6	4.3	87	104	19.5
		70	316	3.6	4.3	87	104	19.5
J: 70/30 brass (**33**)	Hole	70	93	2.6	2.8	118	128	8.4
		70	185	3.4	3.6	108	114	5.6
		70	463	4.9	4.8	98	98	−0.3
K: Aluminum alloy	Hole	114	74	3.0	3.2	152	164	7.6
(2017–T4) (**33**)		114	93	3.3	3.5	147	158	7.2
		114	185	3.9	4.4	123	140	14.1
L: Stainless steel	Hole	355	37	7.4	5.2	530	373	−29.7
(SUS603) (**34**)		355	74	8.7	6.6	441	332	−24.7
		355	93	9.8	7.1	441	320	−27.5
		355	139	11.7	8.1	432	299	−30.8
		355	185	11.7	8.9	373	285	−23.6
M: Stainless steel	Hole	244	93	6.7	5.4	304	245	−19.4
(YUS170) (**34**)		244	139	8.0	6.2	294	229	−22.1
		244	185	8.3	6.8	265	218	−17.6
N: Maraging steel (**35**)	Vickers hardness indentation	720	19	6.9	7.4	686	736	7.3
	Hole	720	37	9.5	9.2	677	659	−2.7
		720	93	11.7	12.5	530	566	6.7
		720	185	13.8	15.8	441	504	14.3
	Notch	720	63	10.0	11.0	546	603	10.4
		720	95	10.2	12.6	454	563	24.0

S10C: ~0.10% carbon steel
S30C: ~0.30% carbon steel
S35C: ~0.35% carbon steel
S45C: ~0.45% carbon steel
S50C: ~0.50% carbon steel

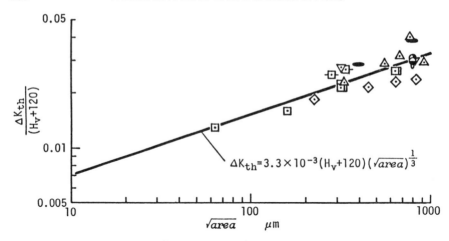

Fig 5   Relationship between $\Delta K_{th}/(H_v + 120)$ and $\sqrt{area}$. Most of the H_v values are estimated by equation (10). See Table 2 for symbol references

Applications

After obtaining equation (8), the data of other references (9)(47)(57)(58)–(64) were investigated. Few of them present the hardness of materials. Therefore, H_v was estimated by using the relationship between H_v and H_B (ASTM E-48-43-T) and an empirical formula which is thought to hold between ultimate tensile strength and H_B, i.e.

$$\sigma_U \cong 0.36(9.8 \times H_B) \qquad (10)$$

Figure 5 and Table 2 provide details of the comparison between the prediction by equation (8) and the experimental data. Although these references lack the data for very small values of $\sqrt{area}$, it may be concluded that equation (8) predicts ΔK_{th} very well for cracked or notched specimens having $\sqrt{area}$ less than 1000 μm.

Table 2   A collection of data from the different studies used in Fig. 5

Material	Defect	Researchers
◗ 0.12% C steel ● 0.53% C steel	Fatigue crack	Isibasi and Uryu (58)
△ SF60	Artificial crack	Ouchida and Kusumoto (59)
▽ Mild steel	Fatigue crack	Frost (9)
⊡ S20C	Circumferential notch	Nisitani (57)(61)
-⊡- S20C	Drilled hole	Nisitani and Kage (60), Hayashi et al. (62)
0 S25C	Fatigue crack	Awatani and Matsunami (63)
➖ S35C	Fatigue crack	Kobayashi and Nakazawa (64)
◇ Eutectoid steel	Circumferential notch	Kobayashi and Nakazawa (47)

The limitation of the applicability of equations (8) and (9)

As seen in Fig. 4, equations (8) and (9) can be applied to small defects or cracks within some range of the value of $\sqrt{area}$. Although the upper limit of the size of $\sqrt{area}$ for the application of these equations is uncertain at present, it may be approximately 1000 μm. The lower limit is dependent on material properties and microstructures. In experiments we have a finite value of the fatigue limit σ_w for specimens containing no defects or cracks. Theoretically, in this case, $\sqrt{area} = 0$ and accordingly $\sigma_w = \infty$. But this never occurs, because cracks along slip bands or grain boundaries nucleate as a result of reversed slip in grains, that is, $\sqrt{area}$ is not zero, and accordingly the fatigue limit of defect-free specimens σ_{w_0} becomes finite. Therefore, as discussed in previous studies (28)–(30), the lower limit of $\sqrt{area}$ for the application of equations (8) and (9) is related to the maximum length, l_0, of non-propagating cracks observed in unnotched (defect-free) specimens. It follows that even if a specimen contains small defects or cracks before fatigue tests, and if the fatigue limit σ_w predicted from the value of $\sqrt{area}$ and equation (9) is greater than σ_{w_0}, the value of σ_w is never measured because such defects do not lower the fatigue strength of the specimen and they are virtually harmless. When we know the value of σ_{w_0} in advance, the lower limit of $\sqrt{area}$ can be determined from equation (9). When σ_{w_0} is unknown, its approximate value can be estimated by the empirical formula

$$\sigma_{w_0} \cong 0.5\sigma_U \cong 1.6H_v \tag{11}$$

where σ_U is the ultimate tensile strength in MPa. So far, it has been said that equation (11) is not necessarily applicable to high-strength or hard steels (65). However, such a conclusion was based on experiments in which the original site of fatigue fracture – whether it was slip bands or defects – was not identified exactly. Murakami and Endo (29)(30) showed on the basis of careful investigations of fracture origins that equation (11) is applicable to hard steels when defects are not the cause of fatigue fracture.

Concluding remarks

The dependence of ΔK_{th} on crack size and material properties was investigated on more than ten materials and microstructures. The value of ΔK_{th} of all materials were unified with one geometrical together with one material parameter.

The geometrical parameter is the square root of the area which is occupied by projecting the defects or cracks onto the plane normal to the maximum tensile stress. It was found that

$$\Delta K_{th} \propto (\sqrt{area})^{1/3}$$

The material parameter to unify data is the Vickers hardness value, H_v. The influence of microstructural variables and the difference of materials can be unified by the following equation

$$\Delta K_{th} \propto (H_v + 120)$$

The constant of 120 in the above equation reflects the experimental fact that relatively large non-propagating cracks are likely to be observed in low strength metals in comparison with hard metals. Combining the above two equations, we find by experiment that

$$\Delta K_{th} = 3.3 \times 10^{-3}(H_v + 120)(\sqrt{area})^{1/3}$$

for threshold stress intensity factor ranges at a stress ratio, R, equal to -1, and

$$\sigma_w = 1.43(H_v + 120)/(\sqrt{area})^{1/6}$$

for the fatigue limit, where the units of the quantities in these equations are MPa$\sqrt{m}$ for ΔK_{th}, MPa for σ_w, and μm for $\sqrt{area}$.

Although the upper limit of $\sqrt{area}$ for the application of the above equations is uncertain at present, it may be approximately 1000 μm. The lower limit can be estimated from the fatigue strength of unnotched (defect-free) specimens or from the length of the maximum non-propagating crack which is observed at the fatigue limit of an unnotched specimen.

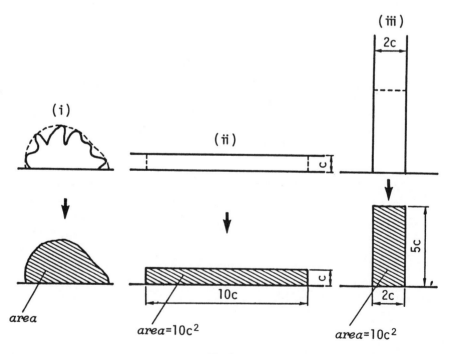

Fig 6

Appendices

Appendix 1

Figure 6 gives rules for the estimation of $\sqrt{area}$ for irregularly shaped cracks and very slender cracks (**24**) (**25**). (See also (**20**) (**66**).)

The long shallow and very deep cracks shown in Fig. 6 as (ii) and (iii), respectively, must be bounded by a maximum length of $10c$ and $5c$, respectively, for the estimation of effective crack area.

Appendix 2

Figure 7 shows the relationship between the maximum stress intensity factor K_{Imax} and $\sqrt{area}$ for surface cracks (elastic analysis) (**24**)(**25**). (See also (**20**)(**66**).)

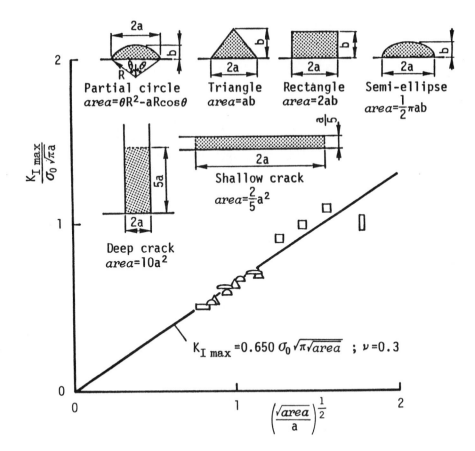

Fig 7

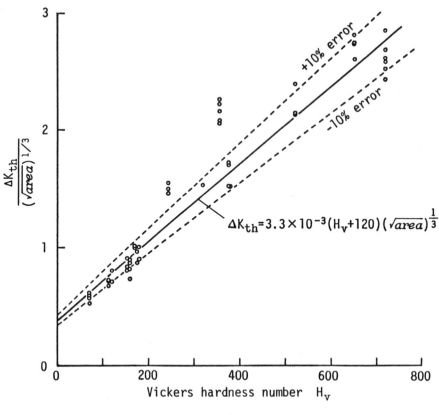

Fig 8

Appendix 3

Figure 8 shows the dependence of $\Delta K_{th}/(\sqrt{area})^{1/3}$ on the material parameter (H_v) for various materials.

References

(1) FROST, N. E., POOK, L. P., and DENTON, K. (1971) A fracture mechanics analysis of fatigue crack growth data for various materials, *Engng. Fracture Mech.*, **3**, 109–126.
(2) KOBAYASHI, H. and NAKAZAWA, H. (1972) A stress criterion for fatigue crack propagation in metals, *Proc. 1st Int. Conf. Mech. Behavior Mater.*, Kyoto, Japan, pp. 199–208.
(3) KITAGAWA, H. and TAKAHASHI, S. (1976) Applicability of fracture mechanics to very small cracks or the cracks in the early stage, *Proc. 2nd Int. Conf. Mech. Behavior. Mater.*, Boston, USA, pp. 627–631.
(4) ROMANOV, O. N., SIMINKOVICH, V. N., and TKACK, A. N. (1981) Near-threshold short fatigue crack growth, *Fatigue Thresholds: Fundamentals and Engineering Applications*, Vol. II (Edited by Bäcklund, J., Blom, A. F., and Beevers, C. J.), pp. 799–807.

(5) LIES, B. N., KANNINEN, M. F., HOPPER, A. T., AHMAD, J., and BREOK, D. (1983) A critical review of the short fatigue crack in fatigue, Battelle's Columbus Laboratories, Report No. AFWAL-TR-83-4019, pp. 1–135.

(6) SURESH, S. and RITCHIE, R. O. (1983) The propagation of short fatigue cracks, Report No. UCB/RP/83/1014, pp. 1–70.

(7) KIKUKAWA, M., JONO, M., TANAKA, K., and TAKATANI, M. (1976) Measurement of fatigue crack propagation and crack closure at low stress intensity level by unloading elastic compliance method, J. Soc. Mater. Sci. Japan, 25, 899–903.

(8) BYRNE, J. and DUGGAN, T. V. (1981) Near-threshold short fatigue crack growth, Fatigue Thresholds: Fundamentals and Engineering Applications, Vol. II (Edited by Bäcklund, J., Blom, A. F., and Beevers, C. J.), pp. 759–775.

(9) FROST, N. E. (1959) A relation between the critical alternating propagation stress and crack length for mild steel, Proc. Instn mech. Engrs, 173, 811–835.

(10) EL HADDAD, M. H., SMITH, K. N., and TOPPER, T. H. (1979) Fatigue crack propagation of short cracks, Trans Am. Soc. Test. Mater., 101, 42–46.

(11) CUMMINGS, H. N., STULEN, F. B., and SCHULTE, W. C. (1958) Tentative fatigue strength reduction factors for silicate-type inclusions in high-strength steels, Proc. Am. Soc. Test. Mater., 58, 505–514.

(12) FUJIWARA, T. and FUKUI, S. (1964) The effect of non-metallic inclusions on the fatigue life of ball-bearing steel, Electr. Furnace Steel, 35, 170–177.

(13) RAMSEY, P. W. and KEDZIE, D. P. (1957) Prot fatigue study of an aircraft steel in the ultra high strength range, J. Metals, 401–407.

(14) HYLER, W. S., TARASOV, L. P., and FAVOR, R. J. (1958) Distribution of fatigue failures in flat hardened steel test bars, Proc. Am. Soc. Test. Mater., 58, 540–551.

(15) JOHNSON, R. F. and SEWELL, J. F. (1960) The bearing properties of 1% C–Cr steel as influenced by steelmaking practice, J. Iron Steel Inst., 196, 414–444.

(16) DUCKWORTH, W. E. and INESON, E. (1963) The effects of externally introduced alumina particles on the fatigue life of En24 steel, Iron Steel Inst. Sp. Rep., 77, 87–103.

(17) KAWADA, Y., NAKAZAWA, H., and KODAMA, S. (1963) The effects of the shapes and the distributions of inclusions on the fatigue strength of bearing steels in rotary bending, Trans Japan Soc. mech. Engrs, 29, 1674–1683.

(18) UHRUS, L. O. (1963) Through-hardening steels for ball bearings – effect of inclusions on endurance, Iron Steel Inst. Sp. Rep., 77, 104–109.

(19) DE KAZINCZY, F. (1970) Effect of small defects on the fatigue properties of medium-strength cast steel, J. Iron Steel Inst., 208, 851–855.

(20) MURAKAMI, Y. and ENDO, M. (1983) Quantitative evaluation of fatigue strength of metals containing various small defects or cracks, Engng Fracture mech., 17, 1–15.

(21) HAMMOUDA, M. M. and MILLER, K. J. (1979) Elastic–plastic fracture mechanics analyses of notches, ASTM STP 668, pp. 703–719.

(22) SMITH, R. A. and MILLER, K. J. (1978) Prediction of fatigue regimes in notched components, Int. J. Mech. Sci., 20, 201–206.

(23) MURAKAMI, Y. and NEMAT-NASSER, S. (1983) Growth and stability of interacting surface flaws of arbitrary shape, Engng Fracture Mech., 17, 193–210.

(24) MURAKAMI, Y. and ISIDA, M. (1985) Analysis of an arbitrarily shaped surface crack and stress field at crack front near surface, Trans Japan Soc. mech. Engrs, 51, 1050–1056.

(25) MURAKAMI, Y. (1985) Analysis of stress intensity factors of modes I, II and III for inclined surface cracks of arbitrary shape, Engng Fracture Mech., 22, 101–114.

(26) NISITANI, H. and MURAKAMI, Y. (1973) Effect of spheroidal graphite on bending and torsional fatigue strength of nodular cast iron, Science of Machines, 25, 543–546.

(27) NISITANI, H. and KAWANO, K. (1971) Correlation between the fatigue limit of a material with defects and its non-propagating crack: some considerations based on the bending or torsional fatigue of the specimen with a diametrical hole, Trans Japan Soc. mech. Engrs, 37, 1492–1496.

(28) MURAKAMI, Y. and ENDO, T. (1980) Effects of small defects on fatigue strength of metals, Int. J. Fatigue, 2, 23–30.

(29) MURAKAMI, Y. and ENDO, T. (1981) The effects of small defects on the fatigue strength of hard steels, Fatigue 81 (Warwick University, UK), pp. 431–440.

(30) MURAKAMI, Y., KAWANO, H., and ENDO, T. (1979) Effect of micro-holes of fatigue strength (the 2nd report: effect of micro-holes of 40 μm–200 μm in diameter on the fatigue strength of quenched or quenched and tempered 0.46% carbon steel), *Trans Japan Soc. mech. Engrs*, **45**, 1479–1489.

(31) MURAKAMI, Y. (1984) Size effect of ΔK_{th}: ΔK_{th} of small surface cracks as the function of crack area, *Prelim. Proc. 61st Annual Meeting of Japan Soc. Mech. Engrs*. No. 840–2, pp. 1–3.

(32) NISITANI, H. and KAGE, M. (1973) Rotating bending fatigue of electropolished specimens with transverse holes – observation of slip bands and non-propagating cracks near the holes, *Trans Japan Soc. mech. Engrs*, **39**, 2005–2012.

(33) MURAKAMI, Y., TAZUNOKI, Y., and ENDO, T. (1984) Existence of the coaxing effect and effects of small artificial holes on fatigue strength of an aluminum alloy and 70–30 brass, *Metall. Trans.*, **15A**, 2029–2038.

(34) NISHIDA, S. (1985) private communication.

(35) MURAKAMI, Y., KIYOTA, T., ENOMOTO, K., and ABE, M. (1985) Effects of artificial small defects on the fatigue strength of a maraging steel, *Prelim. Proc. 34th Annual Meeting of the Soc. Mater. Sci. Japan*, pp. 148–150.

(36) OBA, H., MURAKAMI, Y., and ENDO, T. (1983) Effects of artificial small holes on fatigue strength of notched specimens, *Trans Japan Soc. mech. Engrs*, **49**, 901–910.

(37) NISITANI, H. and NISHIDA, S. (1970) Change in surface states and incipient fatigue cracks in electro-polished low carbon steel (plain and notched specimens) under rotating bending stress, *Bull. Japan Soc. mech. Engrs*, **13**, 961–967.

(38) NISITANI, H. and MURAKAMI, Y. (1969) Torsional fatigue and bending fatigue of electropolished low carbon steel specimens, *Bull. Japan Soc. mech. Engrs*, **13**, 325–333.

(39) NISITANI, H. (1972) Correlation between notch sensitivity of a material and its non-propagating crack, under rotating bending stress, *Proc. 1st Int. Conf. Mech. Behavior of Mater.*, Kyoto, Japan, pp. 312–322.

(40) NISITANI, H. and KAWANO, K. (1968) Non-propagating crack and crack strength of shafts with a shoulder fillet subjected to rotary bending, *Proc. 11th Japan Congr. Mater. Research – Metallic Mater.*, Japan, pp. 49–51.

(41) KOBAYASHI, H. and NAKAZAWA, H. (1969) The effects of notch depth on the initiation, propagation and non-propagation of fatigue cracks, *Trans Japan Soc. mech. Engrs*, **35**, 1856–1863.

(42) NISITANI, H. and ENDO, M. (1985) Fatigue strength of carbon steel specimens having an extremely shallow notch, *Engng Fracture Mech.*, **21**, 215–227.

(43) NISITANI, H. and ENDO, M. (1985) Unifying treatment of notch effects in fatigue, *Trans Japan Soc. mech. Engrs*, **51**, 784–789.

(44) NISITANI, H. and ENDO, M. (1984) Unified treatment of deep and shallow notches in rotating bending fatigue, *ASTM STP*, submitted for publication.

(45) NISITANI, H. and CHISHIRO, I. (1974) Non-propagating micro-cracks of plain specimens and fatigue notch sensitivity in annealed or heat-treated 0.5% C steel. *Trans Japan Soc. mech. Engrs*, **40**, 41–52.

(46) NISITANI, H. and OKAZAKA, K. (1973) Effect of mean stress on fatigue strength, crack strength and notch radius at branch point under repeated axial stress, *Trans Japan Soc. mech. Engrs*, **39**, 49–59.

(47) KOBAYASHI, H. and NAKAZAWA, H. (1970) On the alternating stress required to propagate a fatigue crack in carbon steel (continued report), *Trans Japan Soc. mech. Engrs*, **36**, 1789–1798.

(48) MURAKAMI, Y. and MATSUDA, K., in preparation.

(49) ISIBASI, T. (1960) Fatigue strength of notched stainless steel specimens, *Proc. 3rd Japan Congr. Test Mater*, Japan, pp. 24–26.

(50) OUCHIDA, H. and ANDO, S. (1964) The fatigue strength of notched specimens at low temperatures, *Trans, Japan Soc. mech. Engrs*, **30**, 52–58.

(51) OGURA, K., MIYOSHI, Y., and NISHIKAWA, I. (1983) Fatigue crack growth and closure at notch root of SUS304 stainless steel, *Proc. 26th Japan Congr. Mater. Research – Metallic Mater.*, Japan, pp. 91–96.

(52) KAWAI, S. and KASAI, K. (1985) Considerations of allowable stress of corrosion fatigue (focused on the influence of pitting), *Fatigue Fracture Engng Mater. Structures*, **8**, 115–127.

(53) ISIBASI, T. (1948) On the fatigue limits of notched specimens, *Memo. Fac. Engng. Kyushu University*, **11**, 1–31.
(54) PETERSON, R. E. (1953) *Stress Concentration Design Factors* (John Wiley, New York), p. 8.
(55) SIEBEL, E. and STIELER, M. (1955) Ungleichförmige spannungsverteilung bei schwingender beanspruchung, *Zeitschrift VDI*, **97**, 121–126.
(56) HEYWOOD, R. B. (1962) *Designing against fatigue* (Chapman & Hall, London), p. 95.
(57) NISITANI, H. (1968) Effects of size on the fatigue limit and the branch point in rotary bending tests of carbon steel specimens, *Bull. Japan Soc. mech. Engrs*, **11**, 947–957.
(58) ISIBASI, T. and URYU, T. (1953) Fatigue strength of carbon steel bars with round-crack, *Rep. Res. Inst. Appl. Mech.*, *Kyushu University*, **2**, 65–74.
(59) OUCHIDA, H. and KUSUMOTO, S. (1954) Fatigue strength of carbon steel bars with artificial cracks, *Trans Japan Soc. mech. Engrs*, **20**, 739–745.
(60) NISITANI, H. and KAGE, M. (1977) Effect of annealing or change in stress level on the condition for propagation of non-propagating fatigue cracks, *Trans Japan Soc. mech. Engrs*, **43**, 398–406.
(61) NISITANI, H. (1968) *Prelim. Proc. 46th Annual Meeting Japan Soc. Mech. Engrs*, Japan, No. 198, pp. 37–40.
(62) HAYASHI, I., SUZUKI, T., and MORITA, J. (1982) Mechanism of non-propagating crack of notched low carbon steel specimen near the stage of rotating bending fatigue limit, *Proc. 25th Japan Congr. Mater. Research – Metallic Mater.*, Japan, pp. 80–86.
(63) AWATANI, J. and MATSUNAMI, K. (1977) On non-propagating fatigue cracks in specimens with cracks, *J. Soc. Mater. Sci. Japan*, **26**, 343–347.
(64) KOBAYASHI, H. and NAKAZAWA, H. (1967) On the alternating stress required to propagate a fatigue crack in carbon steels, *Trans Japan Soc. mech. Engrs*, **33**, 1529–1534.
(65) GARWOOD, M. F., ZURBERG, H. H., and ERIKSON, M. A. (1951) Correlation of laboratory test and service performance. *Interpretation of tests and correlation with service* (American Society of Metals), pp. 1–77.
(66) ENDO, M. and MURAKAMI, Y. (1985) Fatigue strength of defect and cracked materials – discussion based on the crack area, *Prelim. 38th Annual Meeting of Japan Soc. Mech. Engrs. Kyushu*, Japan, No. 858–1, pp. 63–65.

O. Oni* and C. Bathias*

Fatigue Crack Growth in Micro-notched Specimens of High Strength Steels

REFERENCE Oni, O. and Bathias, C., **Fatigue Crack Growth in Micro-notched Specimens of High Strength Steels**, *The Behaviour of Short Fatigue Cracks*, EGF Pub. 1 (Edited by K. J. Miller and E. R. de los Rios) 1986, Mechanical Engineering Publications, London, pp. 295–307.

ABSTRACT Small fatigue crack formation and growth, from single-edge micro-notched specimens of high strength low alloy martensitic steels, were studied under pure bending and constant amplitude loading conditions, employing a carefully calibrated pulsed direct current electric potential technique.

It was observed that small cracks formed sporadically along the notch root, and that their growth rate was different from long through-cracks. The growth rate of the small cracks fell to a minimum before rising again, to join with the long through-crack growth curve determined on conventional specimens. The observed short crack depth varies from a few microns to approximately 900 microns from the notch root.

Introduction

Very recently the study of short or small part-through fatigue cracks has attracted considerable attention, not only because of its academic interest but also because of its practical implications in design, as they are generally the origin of most structural failures.

Observations made during fatigue experiments have shown that this type of crack propagates at rates different from long through-cracks subjected to the same nominal crack driving force (1)(2) and, furthermore, they can also grow under stress intensity conditions that are below the threshold stress intensity factor range, ΔK_{th}, determined from long through-cracks in conventional specimens (3)(4).

The principal reasons advanced to explain these observations include plasticity effects, crack closure phenomena, the breakdown of linear elastic fracture mechanics analysis in the short crack region, non-equilibrium crack profiles, and the decrease in stress concentration factor with distance from a notch root (3)(5).

In this paper we report part of our current experimental observations on the formation and the growth of small part-through fatigue cracks, emanating from single edge micro-notched (SEN) bend specimens of high yield strength martensitic steels. The pulsed direct current (d.c.) electric potential technique was used for the detection and the continuous measurement of crack growth. The results on the growth of the part-through fatigue cracks, were compared with those of long through-cracks determined from conventional compact tension specimens.

* Laboratoire de Mécanique – UA, 849 du CNRS, Université de Technologie de Compiègne, B.P. 233, 60206 – Compiègne, France.

Table 1   Percentage weight composition of the materials

Steel type	C	Si	Mn	S	P	Cr	Ni	Mo	Cu	Al
16 NCD 13	0.16	0.28	0.48	<0.03	<0.09	0.96	3.14	0.23	0.10	—
30 NCD 16	0.27	0.22	0.38	0.005	0.005	1.30	3.79	0.44	0.16	0.006

Table 2   Heat treatments and mechanical properties

Material	Heat treatment	0.2% yield strength σ_{ys} (MPa)	Ultimate tensile strength σ_{ult} (MPa)	Elongation (%)
16 NCD 13	Austenized at 825°C for 0.5 h oil quenched and tempered at 190°C for 2 h	1070	1360	13
30 NCD 16	Austenized at 850°C for 0.5 h oil quenched, tempered at 585°C for 1 h	1076	1192	15.3

Experimental procedure

Material

Two similar nickel chromium martensitic steels, types 16 NCD 13 and 30 NCD 16 of the French standards, principally used in the aeronautical industry, were used in the present investigation. Their chemical composition, heat treatments, and mechanical properties are presented in Tables 1 and 2. The average grain size of the steel 30 NCD 16 is about 30 μm while that of 16 NCD 13 is of the order of 25 μm.

Specimen preparation

Single edge notched specimens having nominal dimensions $100 \times 20 \times 10$ mm were machined in the LT direction of the plate (Fig. 1(a)). A small 60 degree V shaped notch, of depth 0.1 mm and root radius $\rho = 45$ μm, was machined across the 10 mm width by electro-discharge machining. The notch was mechanically polished, first with abrasive paper, and then with alumina paste down to 1 μm. Four pairs of potential probes were positioned as shown in Fig. 1(b). Thin copper wire probes were micro-welded to the specimen surface.

The fatigue crack growth tests were carried out in air, on a ± 10 kN Instron fatigue machine at constant amplitude of cycling ($\sigma_{max}/2 = 0.39\sigma_{ys}$, $\sigma_{min}/\sigma_{max} = 0.1$, where σ_{max} is the gross maximum bending stress). The test frequency was 20 Hz and four point bending was applied. The formation of a crack was detected and its growth monitored continuously with a pulsed d.c. electric

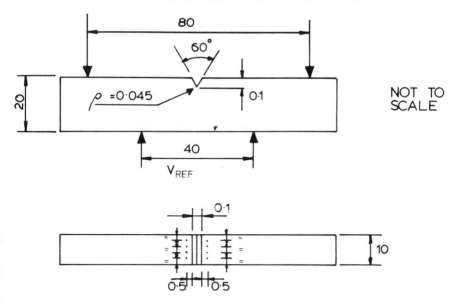

Fig 1 Specimen dimensions and the equi-distance positions of the eight probes near the notch.
All dimensions in millimetres (not to scale)

potential technique developed by Baudin and Policella (**6**). A commutator in
the electric circuit permitted switching from one probe position to another.
Potential drop changes and the number of cycles were continuously recorded.
The moment the crack was considered to have propagated significantly, the test
was stopped and the specimen broken open under static loading, in order to
reveal the crack profile. The fracture surfaces were then observed and the crack
dimensions (surface length and depth) measured in the scanning electron
microscope (SEM).

Results and discussions

Microscopic observations

Experimental observations, confirmed subsequently by the scanning electron
microscope, showed that despite having a continuous small through-thickness
micro-notch in the specimen, several small part-through fatigue cracks were
sporadically formed along the notch root and their formation was not concentrated at a particular point. The four pairs of potential drop probes therefore
facilitated the detection and monitoring of the growth of such cracks at any
point along the notch root. Figure 2 shows some of the SEM micrographs of the
cracks. In some cases several cracks were observed, each at a different stage in
its development (Fig. 2(a)). Regarding the crack shape, the semi-elliptical form

Fig 2   Typical small part-through fatigue cracks emanating from through-thickness micro-notched specimens

was the most common though other shapes were also evident, particularly the quarter ellipse form at specimen corners.

Quantitative microscopic observations revealed that these cracks originated principally from irregularities at the notch root. It is also thought that inclusion pits, formed as a result of the poor adhesion between the aluminium oxide inclusion and the matrix material (7), could also contribute to the sporadic formation of small cracks.

Further experimental tests and observations showed that these cracks grew for a large number of cycles as small part-through cracks before coalescing to become a single through-thickness crack, which eventually led to the catastrophic failure of the specimen.

Calibration of the crack length measuring system

The calibration of the potential drop system was carried out as follows. The variation in electric potentials was measured by the four pairs of probes which were summed and averaged and then normalized (6)(8)(9) with a reference potential, V_{ref}. Here V_{ref} is the potential difference measured at the beginning of a test at a distance of $2W$ from the centre of the specimen, where W is the specimen width.

The normalized averaged potential difference $\Delta V_m/V_{ref}$ was then plotted as a function of the total cracked area measured from the microscopic observations. Figure 3 shows the calibration curve obtained from several specimens cracked under the same experimental conditions. The curve shows that the electrical potential sensitivity to crack growth is the same in the two steels, though agreement was not perfect; this is expected as the electrical potential sensitivity of a material to crack growth depends on the electrical conductivity and the dimensions of the crack formed in the material (8). Since the two materials have similar physical properties, and the dimensions of the cracks formed in the materials were of the same order it is not surprising that their electrical potential sensitivity to crack growth is almost identical.

Crack length and growth analysis

As the objective is to express the average growth rate of these cracks in terms of the stress intensity factor range, the following assumptions were made in order to simplify the analysis.

(1) Linear elastic fracture mechanics analysis is applicable to the short crack problem.
(2) The interaction between cracks is negligible and the biggest of these cracks (in cases where there were several) has the most damaging effect.
(3) The aspect ratio of the crack remains constant.

If these assumptions are valid, then the present results can be analysed by applying the empirical stress intensity factor solutions developed by Raju and

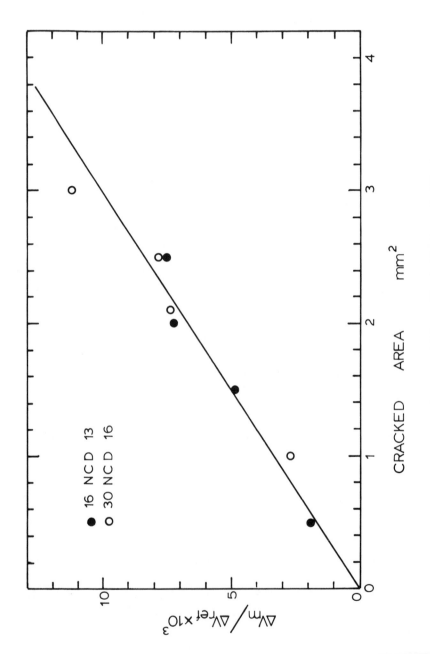

Fig 3   Electrical potential drop calibration for the notch of Fig. 1

Newman (10) for surface crack growth under pure bending conditions. The analysis is as follows.

From the calibration curve we obtain an equation of the form

$$\Delta V_m / V_{ref} = AS \tag{1}$$

further, we know that the area, S, of a semi ellipse is given as

$$S = \pi/2 a_i c_i \tag{2}$$

and, finally, from assumption (3) above we have

$$a/c = a_i/c_i = a_{final}/c_{final} = \text{constant} \tag{3}$$

$$S = \text{crack area}$$
$$a = \text{depth of crack}$$
$$c = \text{half of the total surface length of crack}$$
$$A = \text{constant determined from calibration curve}$$

Equations (1) and (2) are therefore used to calculate the depth and the surface length of the cracks. The minimum detectable crack depth from the notch root measured by the present electrical potential instrumentation was of the order of 30 μm when the crack formed exactly opposite a probe wire, and of 50 μm when it formed in between probe wires. The maximum crack depth measured in this latter case was approximately 900 μm. Furthermore, the calculated crack depths were found to be approximately ± 10 per cent of their actual values in the former case and ± 15 per cent in the latter.

Figure 4 gives crack length values c, in terms of the number of cycles, showing that, at the loading level considered, crack formation takes place earlier in steel 16 NCD 13 than in 30 NCD 16. Detailed experimental results at different loading levels confirm this point. They also show that 30 NCD 16 is characterized by the formation of several more small cracks than 16 NCD 13. The earlier crack formation in the latter is thought to be due to its lower threshold stress intensity factor value of 6 MPa$\sqrt{m}$ for 30 NCD 16 steel.

The ability of 30 NCD 16 steel to form several cracks can be associated with the presence of Al_2O_3 inclusions which, according to a quantitative fatigue crack initiation study carried out by Kunio et al. (7) on a similar martensitic steel, may initiate cracks at inclusion pits formed due to the poor adhesion between this oxide and the matrix. Since these pits are randomly distributed in the matrix, multiple crack initiation sites would be expected. Scanning electron microscope studies of the crack formation zones near the notch root in the two materials provided some evidence of inclusions on the fractured surface of the 30 NCD 16, while in Fig. 5 the crack is seen to initiate at the site of an irregularity on the notch root, even though some inclusions must be present in this material.

Figure 6 shows the relation between the fatigue crack growth rate in the depth direction, as a function of the stress intensity factor range, for the part through- and the long through-cracks in the two materials.

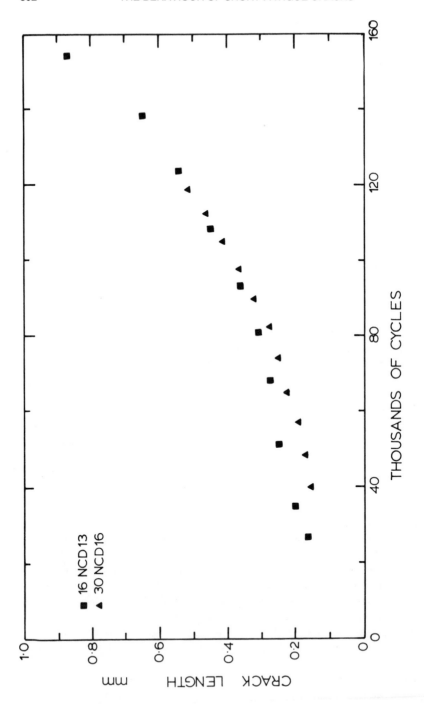

Fig 4 Crack length of a part-through crack in terms of the number of cycles: $R = 0.1$, $f = 20$ Hz

Inclusions

Fig 5   Micro-notch root irregularities

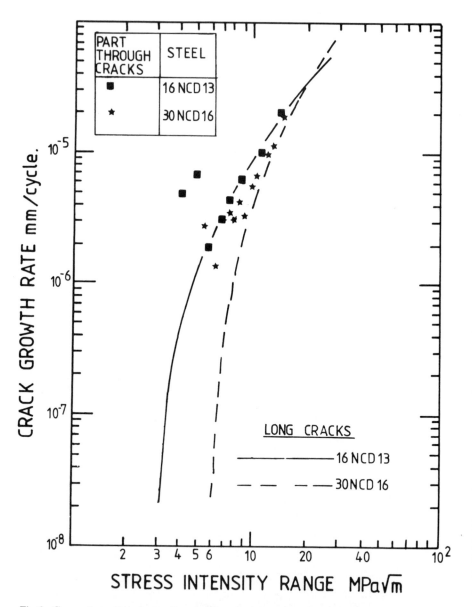

Fig 6 Comparison of crack growth rates in terms of the stress intensity factor range in two steels:
$R = 0.1$, $f = 20$ Hz

The long through-crack

The full line and the dashed line of Fig. 6 represent the long through-crack results ($a > 1$ mm) in 16 NCD 13 and in 30 NCD 16 steels, respectively. The experimental data points were omitted for clarity. The data were obtained using compact tension specimens and the near threshold was determined using the load shedding technique. It was found that the value of the threshold stress intensity factor of the material of steel 30 NCD 16 is nearly double that of steel 16 NCD 13 – a surprising result, as the two materials have identical mechanical properties. This further shows that two materials may have similar mechanical properties but not necessarily similar defect tolerance.

The part-through crack

Although some of the experimental data exhibit scatter, especially at very short crack depth or low stress intensity, Fig. 6 indicates that, at very short crack depth, most experimental points lay outside the long through-crack data, indicating a higher propagation rate of the short crack in comparison with the equivalent long through-crack. There is also an initial dip or minimum growth in the two materials. In the case of 16 NCD 13 steel at this minimum point, the short crack data merges with the long through-crack curve, and this point, according to our analysis, corresponds to a crack depth of about 260 μm. In the case of material 30 NCD 16 the minimum point occurs at a lower crack depth (190 μm); moreover, the merger point of the short crack data with the long crack curve is not at this minimum point, but at a crack depth of about 220 μm. Also at very short crack lengths, the ΔK values for steel 16 NCD 13 were higher than the equivalent value of the ΔK at threshold, while for steel 30 NCD 16 the initial ΔK value for the short crack is approximately the same as its threshold stress intensity factor value.

Figure 7 shows our experimental results on the two materials compared with the work of Fine and Heubaum (5) on the short crack growth behaviour of a micro-notched ($\rho = 25$ μm) single edge notch specimen of a high strength low alloy steel (HSLAVAN 80) in which the crack length was measured with an optical microscope. Figure 7(a) shows the comparison with the steel 16 NCD 13 while Fig. 7(b) shows the comparison with the steel 30 NCD 16. From the Figs it can be seen that our experimental curves have similar shape to those previously reported in that each curve can be said to go through a minimum before rising up again to join the conventional long-crack growth behaviour. In addition, the short-crack growth behaviour in the three materials takes place at a growth rate of the order of 10^{-6} mm/cycle, and the ΔK values at very short crack length are higher than the equivalent ΔK at the threshold of each material.

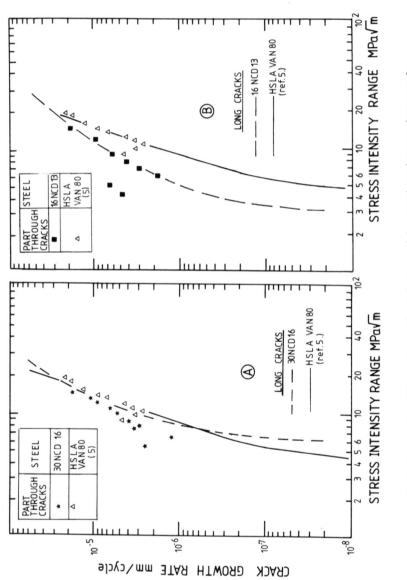

Fig 7 Comparison of the crack growth rate in terms of the stress intensity range in two steels.
(a) 30 NCD 16
(b) 16 NCD 13

Conclusions

Part-through fatigue crack formation and growth were studied in micro-notched specimens of high yield strength low alloy martensitic steels.

Several cracks developed sporadically along the notch root and were of varying shapes and sizes. They originated principally from the irregularity of the notch root. It is also thought that inclusion pits formed as a result of the poor adhesion between aluminium oxide inclusions and the matrix, and these pits were responsible for sporadic crack formation.

The cracks were observed to grow for a large number of cycles as part-through cracks before joining up to form a single through-thickness crack that eventually lead to the failure of the specimen. The growth rate of the several small cracks was also observed to be different from that of equivalent through-cracks measured on conventional compact tension specimens.

The experimental results further confirmed the ability of the direct current electric potential technique to detect and monitor continuously and quantitatively the sub-critical growth of a crack emanating from a small surface defect.

References

(1) GANGLOFF, R. P. (1981) Electrical potential monitoring of crack formation and sub-critical growth from small defects, *Fatigue Engng Mater. Structures*, **4**, 15–33.
(2) NEWMAN, Jr, J. C. (1982) A non linear fracture mechanics approach to the growth of small cracks, Agard Specialist Meeting on Behaviour of Short Cracks, Toronto.
(3) SURESH, S. and RITCHIE, R. O. (1983) The propagation of short fatigue cracks, Report No. RP/83/1014. University of California at Berkeley.
(4) TANAKA, K. and NAKAI, Y. (1983) Propaation and non propagation of short fatigue cracks at a sharp notch, *Fatigue Engng Mater. Structures*, **6**, 315–327.
(5) HEUBAUM, F. and FINE, M. F. (1984) Short fatigue crack growth behaviour in a high strength low alloy steel, *Scripta Met.*, **18**, 1235–1240.
(6) BAUDIN, G. and POLICELLA, H. (1982) A pulsed d.c. P.D. technique applied to three dimensional crack fronts, *Advances in crack length measurements* (Edited by C. J. Beevers), EMAS, Warley, pp. 159–174.
(7) KUNIO, T., SHIMIZY, M., YAMADA, K., SAKURA, K. and YAMAMOTO, T. (1981) The early stage of fatigue crack growth in martensitic steel, *Int. J. Fracture*, **17**, 111–119.
(8) ATKINSON, E. C. and SMITH, M. E. F. (1982) The detection and measurement of non-through cracks using the A.C. potential difference method, National Gas Turbine Establishment Memorandum, M. 81001.
(9) GANGLOFF, R. P. (1982) Electrical potential monitoring of the formation and growth of small fatigue cracks in embritting environments, *Advances in crack length meaurements* (Edited by C. J. Beevers) EMAS, Warley, pp. 174–229.
(10) NEWMAN, Jr, J. C. and RAJU, I. S. (1984) Stress intensity factor equations for cracks in three dimensional finite bodies subjected to tension and bending loads, 19827 NASA Technical Memorandum.

E. Hay* and M. W. Brown†

Initiation and Early Growth of Fatigue Cracks from a Circumferential Notch Loaded in Torsion

REFERENCE Hay, E. and Brown, M. W., **Initiation and Early Growth of Fatigue Cracks from a Circumferential Notch Loaded in Torsion**, *The Behaviour of Short Fatigue Cracks*, EGF Pub. 1 (Edited by K. J. Miller and E. R. de los Rios) 1986, Mechanical Engineering Publications, London, pp. 309–321.

ABSTRACT Fatigue crack growth rates for notched specimens loaded in torsion are compared for various materials, temperatures and specimen sizes, using data from three different publications. It is demonstrated that the mode III stress intensity factor cannot provide a unifying parameter for crack propagation at different temperatures, but a reasonable correlation is obtained by using an elasto-plastic strain intensity factor that characterizes mode III deformation close to the crack tip. An upper bound equation is derived to give conservative estimates of crack growth rate, the strain intensity factor being derived from the ultimate shear stress of the material.

Notation

da/dN	Fatigue crack growth rate
G	Modulus of rigidity
ΔK_{III}	Stress intensity factor range (Mode III)
M_A	Maximum applied torque
M_L	Limit torque
r	Radius from specimen centre line
r_o	Specimen maximum radius
r_n	Specimen minimum radius (to crack tip)
r_p	Radius to edge of plastic zone
τ_y	Shear yield stress
τ_{ult}	Shear ultimate stress
Γ_{III}	Maximum strain intensity factor (Mode III)

Introduction

Under normal circumstances rotating shafts may be considered to transmit a constant torque, but frequently an oscillating component is superimposed on this constant torque and, under conditions of torsional resonance, such oscillations can cause yielding at stress concentrating features which can lead to the formation of fatigue cracks. An example of such a component is shown in Fig. 1, a compressor motor shaft of EN3A approximately 150 mm diameter.

* International Research and Development Co. Limited, Fossway, Newcastle upon Tyne NE6 2YD.
† University of Sheffield, Mechanical Engineering Department, Mappin Street, Sheffield S1 3JD.

Fig 1   Cracking associated with an oil seal collar and keyway

Materials, specimen geometries, test conditions

Results of reversed torsion fatigue tests, carried out in air, from three sources are compared here. In each case the specimen was of circular cross-section with a circumferential vee-notch (Fig. 2). The sources and testing conditions were:

(a)   from Brown, Hay, and Miller (**1**) on 1% Cr–Mo–V steel at ambient temperature and 565°C, specimen diameter 10 mm with a 1 mm deep notch;

(b)   from Tschegg (**2**) on AISI 4340 steel at ambient temperature, specimen diameter 12.7 mm with a 0.631 mm deep notch;

(c)   from Tschegg (**3**) on AISI 4340 steel at ambient temperature, specimen diameter 25.4 mm with a 1.06 mm deep notch;

Tables 1 and 2 show the chemical composition and mechanical properties of the materials considered.

Table 1   Chemical composition (wt %)

	C	Mn	Ni	Cr	Mo	Si	p	S	Cu	V	Remainder
1% Cr–Mo–V	0.24	0.64	0.21	1.02	0.57	0.29	0.016	0.01	—	0.29	Fe
AISI 4340	0.4	0.78	1.77	0.81	0.25	0.26	0.07	0.013	0.14	—	Fe

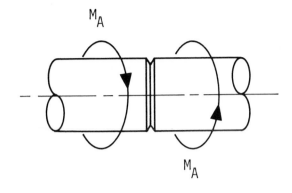

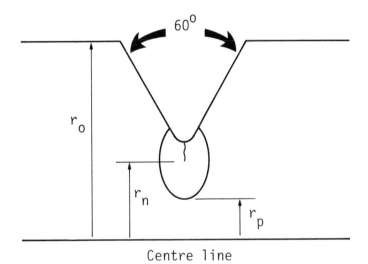

Fig 2   Specimen geometry

Table 2   Mechanical properties

	1% Cr–Mo–V		AISI 4340 Room Temp.
	Room Temp.	565°C	
Ultimate tensile strength (MPa)	805	547	1076
Ultimate shear strength (MPa)	481	261	
Torsion yield strength (MPa)	336	198	552*

* For AISI 4340 Torsion Yield Strength was taken as $1/\sqrt{3}$ times the tensile yield strength.

In the cases compared here a small constant axial tension was applied to the specimen to minimize crack face rubbing and interference. The tests reported in (1) were conducted under constant angular deflection while those in (2) and (3) were conducted under constant quasi-apparent ΔK_{III} conditions, the applied torque being reduced as the fatigue crack depth increased. Fatigue crack depth was monitored in each case using the d.c. potential drop technique.

Fractography

After fatigue testing, a number of specimens were cooled in liquid nitrogen and broken under impact loading. Two distinct fracture surfaces were produced by different test conditions, which have been designated 'factory roof' (Fig. 3) and 'flat' (Fig. 4), in the 1% Cr–Mo–V steel.

The factory roof type of fracture surface was produced by the lower values of applied stress intensity range both at room and elevated temperatures. Under

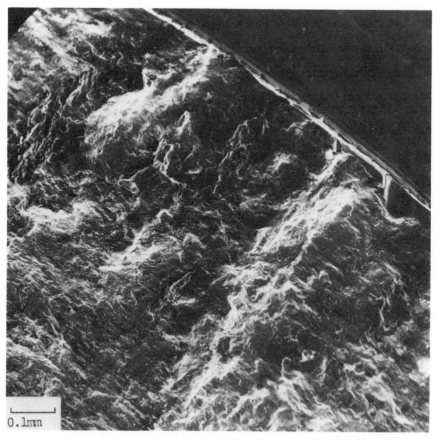

Fig 3   Factory roof fracture surface in 1% Cr–Mo–V steel at 20°C. Initial torque 34.8 Nm

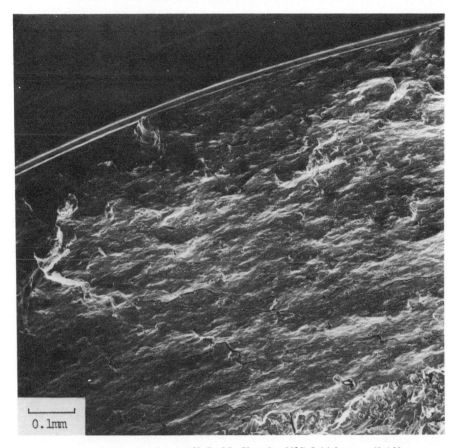

Fig 4   Flat fracture surface in 1% Cr–Mo–V steel at 20°C. Initial torque 60.1 Nm

constant angular deflection conditions (1), several cracks propagated by a mode I mechanism at approximately 45 degrees to the plane of the notch and linked up to give the serrated fracture surface; see also Fig. 1. The cracks propagated radially in these 45 degree planes until the applied stress intensity factor increased to a value where the mode of cracking changed to a mode III type, which gave a flat fracture surface. If the initial loading produced a high value of stress intensity factor range the crack initiated in mode III and a flat fracture surface was produced across the entire cross-section. The change-over in fracture mode occurred at a crack growth rate of 10 nm/cycle at room temperature and 100 nm/cycle at 565°C. The application of axial load did not affect the major features of the fracture surfaces, but did reduce the amount of surface rubbing. This was most evident on the flat fracture surfaces.

For AISI 4340 steel, tests were carried out under apparent constant stress intensity conditions, the applied torque being progressively reduced as the cracks propagated. Fracture surfaces again showed flat faces at high stress intensity values and factory roof facets at low values of stress intensity factor. Where cracks were propagated to a depth of up to 4.5 mm in the range of ΔK_{III} 30 to 60 MPa$\sqrt{m}$ a change from an external flat to an internal section factory roof fracture surface was produced coincident with a reduction in crack growth rate, indicating that there was some crack face rubbing preventing the full effect of the applied loading being felt at the tip of the long crack. This change occurred at a crack depth which depended on the value of ΔK_{III}, but always in the region of a crack growth rate of 100 nm/cycle.

Fatigue crack growth behaviour

The present analysis has been confined to small cracks of depth less than 0.6 mm for the results from (1), and less than 1 mm for results from (2) and (3), since this corresponds to the greatest portion of fatigue lifetime. On a circumferential crack the axial force applied gave a mode I stress intensity factor of less than 3 MPa$\sqrt{m}$, but this force had a significant effect on crack growth behaviour, permitting maximum crack growth rates to be achieved by minimizing crack flank interference. In practical situations such high crack growth rates could be achieved by imposed bending stresses.

The fatigue crack growth rates, at room and elevated temperature, with an applied axial tension, are shown versus ΔK_{III} in Fig. 5. The results from different sources show agreement for the ambient temperature tests, but the results of tests carried out at 565°C show a different scatter band. Thus a single expression may not be used to describe fatigue crack growth behaviour in terms of ΔK_{III} at both temperatures.

The fatigue crack growth data of (1) were plotted against Γ_{III}, a plastic strain intensity factor based on the work of Walsh and MacKenzie (4) and Nayeb-Hashemi *et al.* (5)

$$\Gamma_{III} = \lim_{r \to r_n} \{\gamma(r_n - r)\}$$

$$= \frac{\tau_y}{G} \left(\frac{r_n}{r_p}\right)^2 (r_n - r_p) \tag{1}$$

where r, r_p, and r_n are defined in Figure 2. Here, $(r_n - r_p)$ is the plastic zone size, which is given by (5)

$$\frac{r_p}{r_n} = 1 - qM^2 - M^4(1 - q - 1.5873(1 - M)^{1/3}) - 3.17(M^5 - M^6)$$

where

$$q = \frac{(16/9)\{(r_o/r_n) - 1\}}{1 + (64/9)\{(r_o/r_n) - 1\}}$$

and

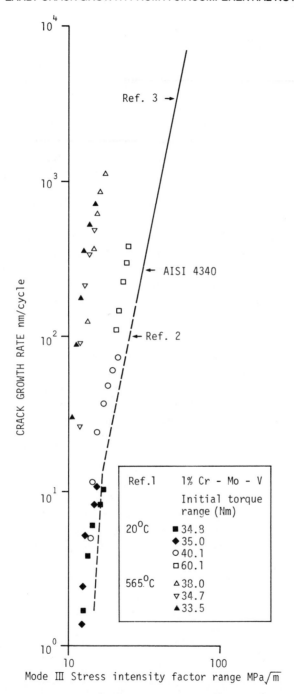

Fig 5   Crack growth rates versus stress intensity factor range

$$M = M_A/M_L$$

where M_A = maximum applied torque and M_L = limit torque = $(2/3)\pi\tau_y r_n^3$.

These equations assume elastic–perfectly plastic behaviour. To avoid the additional complications of strain hardening effects in the real material, the monotonic yield stress, τ_y, was replaced by the ultimate shear stress, τ_{ult}, of the material, since this gives a better estimate of cyclic flow stress. Comparison of the results at ambient and elevated temperature, Fig. 6, shows that there is no clear separation of the two sets of results, the curves from different tests falling into one band for the 1% Cr–Mo–V steel. The arrows in Fig. 6 indicate that the tests, carried out under constant angular deflection, produced increasing crack growth rates. The results from (2) and (3) were recalculated so that fatigue crack growth rates could be plotted as a function of Γ_{III}. A value for ultimate shear stress was used assuming that the ultimate shear stress was $1/\sqrt{3}$ of the value of the ultimate tensile stress. Figure 7 shows the fatigue crack growth rates versus Γ_{III} for these results on AISI 4340 steel. Here also there is no discernible distinction between the two sets of results due to a specimen size effect. The arrows in Fig. 7 show that the crack growth rates decreased under constant ΔK_{III} test conditions.

Discussion

Figure 8 shows the results of tests conducted under angular deflection control (increasing fatigue crack growth rates) and those of tests conducted under constant ΔK_{III} conditions (decreasing fatigue crack growth rates). No clear separation exists for the various conditions considered namely (a) two temperatures for one material, (b) two materials at one temperature, and (c) three specimen sizes. Thus the strain intensity factor gives a parameter which achieves a unification of results which is not possible using elastic stress intensity factors.

Figure 8 may be divided into three regions according to the type of fatigue cracking produced. At crack growth rate less than 10 nm/cycle, region A, the cracks form a factory roof type of surface; at crack growth rates greater than 100 nm/cycle, region C, flat fracture faces are produced; and between these, region B, is a transition region where a change occurs between factory roof and flat surface morphologies.

From Fig. 8 it is clear that some crack face interference effects are still present despite the application of axial tension, but an upper bound may be set from which a conservative estimate of early fatigue crack growth from a notch can be made. The extent of crack face interference, or crack closure, will be difficult to quantify since the actual morphology of individual fracture surfaces close to the crack tip will dictate both the relevant closure stress and the amount of crack face sliding subsequent to closure. Variation in closure conditions will be particularly great in the case of factory roof cracking, since the depth of individual 45 degree facets can vary widely from one crack to another. Thus the

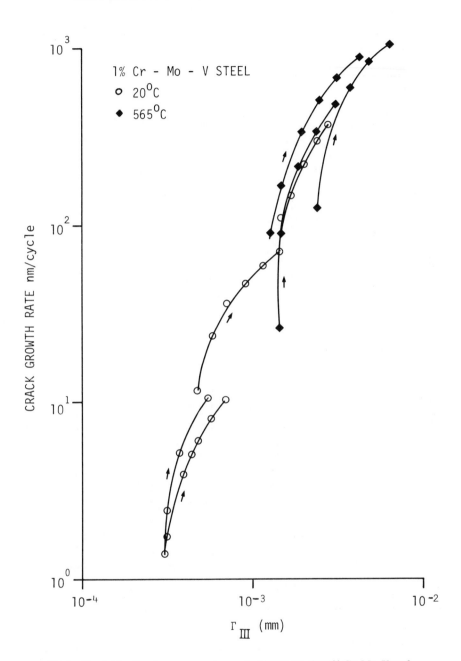

Fig 6   Crack growth rate versus maximum strain intensity for 1% Cr–Mo–V steel

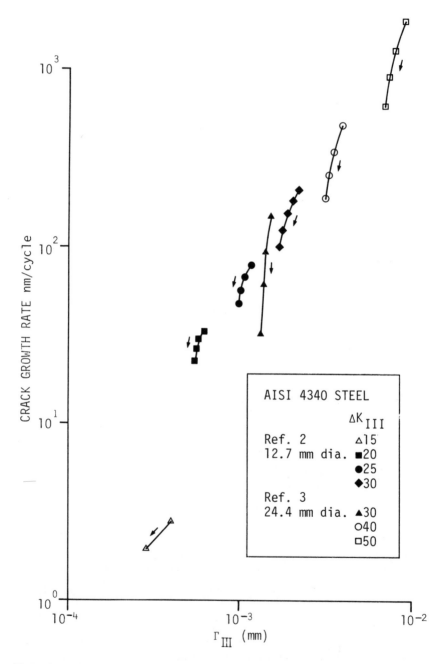

Fig 7   Crack growth rate versus maximum strain intensity factor for AISI 4340 steel (2)(3) where
ΔK_{III} is given in MPa$\sqrt{m}$

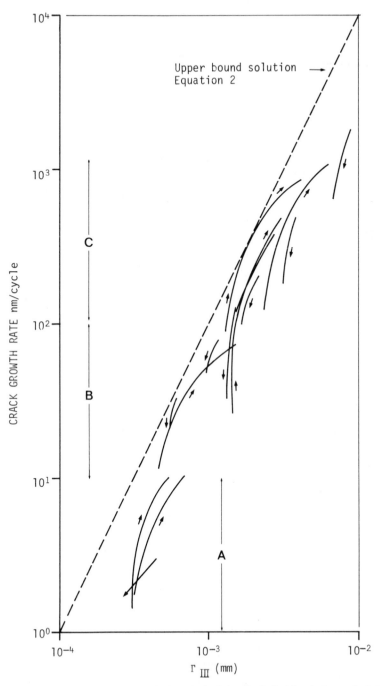

Fig 8   Crack growth rate versus maximum strain intensity factor for tests conducted under angular deflection control and constant stress intensity factor range

use of an upper bound solution which embraces closure effects, provides an expedient method of quantifying mode III crack propagation in components experiencing cyclic torsion.

The upper bound solution indicated in Fig. 8 may be described by

$$\frac{da}{dN} = 0.1\Gamma_{III}^2 \tag{2}$$

where da/dN is in μm/cycle and Γ_{III} is in μm.

This equation can only be used with the cyclic strain intensity factor amplitude, Γ_{III}, which itself is calculated from equation (1) using the ultimate shear stress as the cyclic flow stress, τ_y, and the applied torque amplitude, M_A. The curves in Fig. 8 suggests the possible presence of a threshold value for Γ_{III} in the region of 0.2 to 0.3 μm, below which equation (2) would not apply. However, the change to a mode I dominant type of cracking at low growth rates suggests that failure by mode I crack growth (see Fig. 1) can be expected, unless stresses are sufficiently low to keep below the mode I threshold (1). This apparent mode I threshold for factory roof fracture surfaces will also be highly susceptible to crack closure effects (6).

Conclusions

For circumferentially notched members loaded in reversed torsion, with a small axial tension applied, the following conclusions may be drawn.

(1) The stress intensity factor does not give a unifying parameter which will describe fatigue crack growth at different temperatures.
(2) The strain intensity factor (Γ_{III}) unifies mode III fatigue crack growth results from different materials, temperatures and specimen geometries.
(3) An upper bound equation may be used to give a conservative estimate of fatigue crack growth rates occurring above a threshold value of Γ_{III}.

Acknowledgements

The authors are indebted to E. K. Tschegg whose detailed experimental work has made the analysis possible. The authors are grateful to the Central Electricity Generating Board for provision of test material and for sponsorship of the project through a research fellowship and an SERC/CEGB Case Award.

References

(1) BROWN, M. W., HAY, E., and MILLER, K. J. (1985) Fatigue at notches subjected to reversed torsion and static axial loads, *Fatigue Fracture Engng Mater. Structures*, **8**, 243–258.
(2) TSCHEGG, E. K. (1983) Mode III and Mode I fatigue crack propagation behaviour under torsional loading, *J. Mater. Sci.* **18**, 1604–1614.
(3) TSCHEGG, E. K. (1982) A contribution to Mode III fatigue crack propagation, *Mater. Sci. Engng*, **54**, 127–136.
(4) WALSH, J. B. and MacKENZIE, A. C. (1959) Elastic–plastic torsion of a circumferentially notched round bar, *J. Mech. Phys Solids*, **7**, 247–257.

(5) NAYEB-HASHEMI, H., McCLINTOCK, F. A., and RITCHIE, R. O. (1982) Effect of friction and high torque on fatigue crack propagation in Mode III, *Met. Trans A*, **13A**, 2197–2204.

(6) TSCHEGG, E. K. (1983) The influence of the static I load mode and *R* ratio on Mode III fatigue crack growth behaviour in mild steel, *Mater. Sci. Engng*, **59**, 127–137.

D. L. DuQuesnay,* T. H. Topper,* and M. T. Yu*

The Effect of Notch Radius on the Fatigue Notch Factor and the Propagation of Short Cracks

REFERENCE DuQuesnay, D. L., Topper, T. H., and Yu, M. T., **The Effect of Notch Radius on the Fatigue Notch Factor and the Propagation of Short Cracks**, *The Behaviour of Short Fatigue Cracks*, EGF Pub. 1(Edited by K. J. Miller and E. R. de los Rios) 1986, Mechanical Engineering Publications, London, pp. 323–335

ABSTRACT Strength reductions in fatigue due to sharp notches are less severe than indicated by elastic stress analysis, even when notch root strains are nominally elastic. This investigation evaluates the reduction in fatigue concentration for central circular notches in thin plate specimens of a 2024-T351 aluminum alloy and a SAE 1045 steel. The specimens were tested in uniaxial, fully reversed stressing ($R = -1$) until fracture, using a closed-looped servo-controlled electrohydraulic testing system. Successively smaller diameter notches: 3 mm, 1 mm, 0.5 mm, and 0.24 mm, were tested at different stress levels to establish the stress–life curves. Smooth specimen stress–life data was obtained from previous investigations. The fatigue notch factor was observed to decrease with decreasing notch radius from its theoretical maximum of 3 to values approaching its minimum of 1. The observed fatigue notch factors were in agreement with the predictions of short crack fracture mechanics.

Introduction

The majority of fatigue failures in engineering components occur due to local stress concentrations at notches which promote crack initiation and subsequent crack propagation. An elastic notch solution defines the theoretical stress concentration factor, K_t, as the ratio of the local stress at the notch root, σ_{notch}, to the nominal stress in the component, σ_{nom}.

$$K_t = \frac{\sigma_{notch}}{\sigma_{nom}} \tag{1}$$

The theoretical stress concentration factor depends on the geometry of the component, the geometry of the notch and the type of loading. It is well established that $K_t = 3$ for a central circular notch in a plate of infinite width subjected to pure uniaxial tension/compression(**1**)(**2**).

Strength reductions in fatigue due to sharp notches are generally less severe than is indicated by the elastic solution, even when notch root strains are nominally elastic. Therefore, K_t is replaced by the fatigue notch factor which is the ratio of the unnotched fatigue strength of the material to the notched

* Department of Civil Engineering, University of Waterloo, Waterloo, Ontario, Canada N2L 3G1.

component fatigue strength at a given fatigue life. In this paper the fatigue notch factor is defined at the fatigue limit as:

$$K_f = \frac{\Delta\sigma_e}{\Delta S_{fat}} \tag{2}$$

where $\Delta\sigma_e$ is the fatigue limit stress range of an unnotched specimen, and ΔS_{fat} is the nominal fatigue limit stress range on the net area of a notched specimen. Several empirical relationships (1)–(5) have been suggested for the prediction of K_f in the general form

$$K_f = \text{fn}\ (K_t,\ \rho,\ \alpha) \tag{3}$$

where ρ is the notch root radius and α is a material constant. It has been known for some time that the reduced fatigue concentration at sharp notches is associated with the existence of non-propagating cracks (6)–(10). Several investigators (11)–(16) have developed analytical expressions for the fatigue notch factor based on traditional fracture mechanics. However, such expressions appear to be valid only when the size of the notch plastic zone and the length of the non-propagating crack are small relative to the depth of the notch.

Topper et al. and El Haddad et al. (17)–(22) showed that a modified form of fracture mechanics that they had developed for short cracks could predict both the geometries in which non-propagating cracks would appear and the fatigue notch factors for these geometries. In their analysis notches are divided into two groups: 'blunt' notches, which have radii greater than a critical radius, ρ_{cr}, and 'sharp' notches, which have radii less than ρ_{cr}. The critical radius is defined by a length parameter, l_o, which was thought to be a material constant. They also showed that for blunt notches the fatigue strength depends on the resistance to crack initiation and that $K_f \approx K_t$, while for sharp notches the fatigue strength depends on the resistance to crack propagation to a critical length, equal to $\sqrt{(c/l_o)}$, and they suggested the equation

$$K_f = \frac{1}{F}\ \{1 + \sqrt{(c/l_o)}\} \tag{4}$$

where F is a geometric constant of the order of unity and c is the notch depth. This expression appears to have general validity.

The objectives of the present investigation are to determine the manner in which the fatigue notch factor varies with the root radius of sharp circular notches, and to determine whether short crack fracture mechanics can predict the fatigue notch factors for these notches.

Materials, equipment, and methods

The materials used in this investigation were a 2024-T351 aluminum alloy and a SAE 1045 steel. The chemical compositions and the mechanical properties of

Table 1   Chemical compositions (% by weight)

			2024-T351 aluminum alloy				
Si	Fe	Cu	Mn	Mg	Cr	Zn	Ti
0.50	0.50	4.35	0.60	1.50	0.10	0.25	0.15

		SAE 1045 steel			
C	Si	Mn	P	S	Fe
0.46	0.17	0.81	0.027	0.023	Remainder

Table 2   Mechanical properties

	2024-T351 aluminum alloy	SAE 1045 steel
Elastic modulus (MPa)	72 400	203 500
Yield stress (MPa) (0.2% offset)	356.5	471.6
Ultimate tensile strength (MPa)	466.1	744.7
True fracture stress (MPa)	623.3	1046.0
Hardness (HB)	—	235
Fatigue limit ($R = -1$, MPa)	124.0	303.0
Threshold stress intensity factor (MPa$\sqrt{m}$)	3.52	6.93

these metals are given in Table 1 and Table 2, respectively. Both materials were tested in the as-received condition.

Flat plate specimens, with the geometry and dimensions illustrated in Fig. 1, were machined from rolled plates of the materials such that the loading axis was parallel to the final direction of rolling. Circular notches were then drilled at the centre of the plate specimens. Notch diameters of 3 mm, 1 mm, 0.5 mm, and 0.24 mm were used. The surface of each specimen was hand polished to remove any sharp edges surrounding the notch. The geometry and dimensions of each notch was then checked by means of a travelling microscope with a resolution of 0.025 mm.

All tests were performed in a laboratory environment at room temperature (23°C) using a closed-loop servo-controlled electrohydraulic testing system. The loading frame was equipped with a liquid metal gripping system which ensured proper alignment of the specimen. Previous investigations using this type of specimen and gripping system (23)(24) have shown that the amount of bending induced by high compressive loading is negligible.

The specimens were subjected to a constant amplitude sinusoidal loading waveform (load control) until fracture occurred. The typical test frequencies ranged between 1 Hz and 60 Hz depending on the applied load amplitude. The long life tests on the aluminium alloy were performed at frequencies between 100 Hz and 120 Hz. Unnotched specimen stress–life data was obtained from previous investigations (25)–(27).

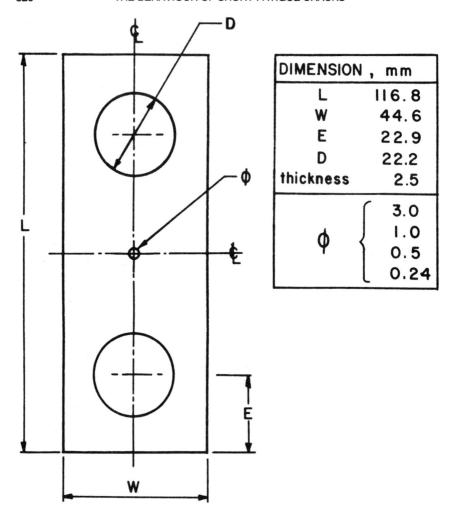

DIMENSION , mm	
L	116.8
W	44.6
E	22.9
D	22.2
thickness	2.5
φ	3.0
	1.0
	0.5
	0.24

Fig 1   Notched plate specimen geometry and dimensions (not to scale)

Results

The observed stress–life data for the notched specimens are given in Fig. 2(a) for the 2024-T351 aluminum alloy and in Fig. 2(b) for the SAE 1045 steel. The unnotched specimen curves are also shown in those figures. In general, the fatigue strength is observed to decrease as the notch radius is increased.

The fatigue notch factors, K_f, for the 2024-T351 aluminum alloy were computed using the fatigue strengths at 2×10^7 cycles since the notched specimens of this material did not exhibit a distinct endurance limit. The endurance limit stresses were used to compute the K_f values for the SAE 1045

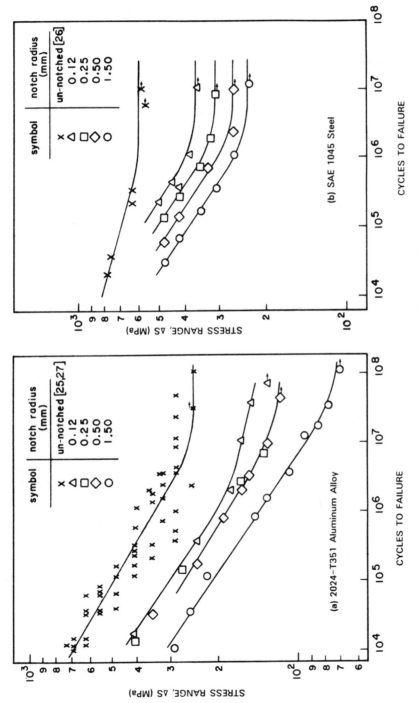

Fig 2 Fully reversed constant amplitude stress-life curves

Table 3   Theoretical and observed stress concentration factors

	2024-T351 aluminum alloy		
Notch radius (mm)	ΔS_{fat} (MPa)	K_t	K_f
Unnotched	248	1.00	1.00
0.12	160	3.00	1.55
0.25	124	2.96	2.00
0.50	124	2.94	2.00
1.50	90	2.82	2.76
	SAE 1045 steel		
Notch radius (mm)	ΔS_{fat} (MPa)	K_t	K_f
Unnotched	608	1.00	1.00
0.12	360	3.00	1.69
0.25	310	2.96	1.96
0.50	276	2.94	2.17
1.50	248	2.82	2.45

steel. The observed fatigue notch factors and the theoretical elastic stress concentration factor for each notch size, corrected to account for the effects of finite specimen width (2), are given in Table 3.

The results for the 2024-T351 aluminum alloy (Table 3) show that the fatigue notch factor decreases from a value of 2.76 for a notch radius of 1.5 mm to a value of 2.00 for notch radii of 0.50 mm and 0.25 mm, and decreases further to a value of 1.55 for a notch radius of 0.12 mm. The data in Fig. 2(a), obtained for the 0.50 mm and 0.25 mm radii notches, are not significantly different, and consequently these notch radii have the same value for K_f. Larger K_f values would be obtained for this material if the fatigue limit was defined at a life greater than 2×10^7 cycles. The blunt notch ($\rho = 1.50$ mm) has an observed K_f of about 3.2 at a life of 10^8 cycles, which is greater than the theoretical maximum value ($K_t = 2.82$) for this notch.

The results for the SAE 1045 steel given in Table 3 show that the fatigue notch factor decreases from a value of 2.45 for a notch radius of 1.5 mm to values of 2.17 and 1.96 for notch radii of 0.50 mm and 0.25 mm, respectively. The value of K_f decreases further to a value of 1.69 for a notch radius of 0.12 mm.

Discussion

Short crack fracture mechanics

Topper *et al.* and El Haddad *et al.* (17)–(19) postulated the following relationship for the stress intensity factor, ΔK, for short cracks in notches when the material behaviour is nominally elastic

$$\Delta K = K' \Delta S \sqrt{\{\pi(l_o + l)\}} \tag{5}$$

where K' is the elastic stress concentration at the crack tip and is a function of the crack length (28), ΔS is the nominal stress range in the component, l is the length of the crack, and l_o is a length parameter given by the expression

$$l_o = \frac{1}{\pi}\left(\frac{\Delta K_{th}}{\Delta \sigma_e}\right)^2 \tag{6}$$

where ΔK_{th} is the threshold stress intensity factor range. At any crack length the condition for continued propagation is that the local stress intensity factor exceeds the threshold value. This condition also implies that the local stress range at the crack tip exceeds the threshold stress range

$$K' \Delta S > \Delta \sigma_{th} \tag{7}$$

The threshold stress range is given in terms of local parameters as:

$$\Delta \sigma_{th} = \frac{\Delta K_{th}}{F\sqrt{\{\pi(l_o + l)\}}} \tag{8}$$

In the case of a centrally notched plate specimen subjected to constant amplitude fully reversed stressing in the elastic range, the crack propagates through a diminishing stress field depicted in Fig. 3. At every increment of length while the crack is short the magnitude of both the maximum and minimum stresses at the crack tip are decreasing. Recent work by Topper, Au, and Yu(23)–(25) has shown that the threshold stress intensity factor, ΔK_{th}, increases as the compressive portion of the stress cycle, σ_{min}, decreases. In addition, closure gradually builds up behind the crack tip as the crack length increases. When the crack has just initiated ($l \approx 0$) no closure exists because there is no material behind the crack tip. As the crack propagates to a few grain diameters in length, the material behind the tip becomes plastically deformed. However, the presence of a locally high cyclic compressive stress tends to retard the build up of closure (29). Steady-state conditions will prevail when the crack tip is remote form the notch.

It is apparent that the length parameter, l_o, defined by equation (6), is a function of the stress range and the minimum stress, and is not a constant for a given material. This length parameter may be thought of as a measure of the rate at which closure builds up and ΔK_{th} increases to its steady-state value.

Figure 4 illustrates the variation of the threshold stress, $\Delta \sigma_{th}$, with crack length, l, for a crack in a notch. The threshold stress varies between the bounds given by the material fatigue limit, $\Delta \sigma_e$, and the local threshold stress intensity factor, ΔK_{th}. The local threshold stress intensity factor varies from a lower bound, ΔK_{th} at initiation, which represents the 'no-closure' condition, to an upper bound, ΔK_{th} steady-state, which represents the 'full-closure' condition. The actual path followed by the threshold stress may not be identical to the path given by equation (8) using only the steady-state value of l_o. However, if the

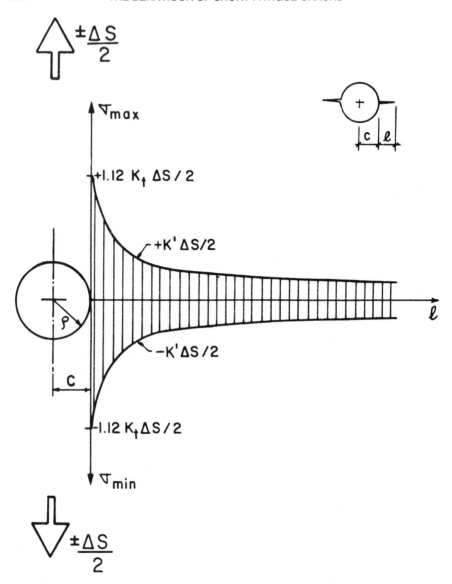

Fig 3 The diminishing stress field at the tip of a crack in a notch

distance over which closure builds up is not appreciably different on either of these paths, then the steady-state value of l_o should provide a suitable approximation for characterizing short crack behaviour in a notch. The region where non-propagating cracks are possible is indicated by the hatched area in Fig. 4. A crack will become non-propagating at any crack length for which the inequality (7) is not satisfied.

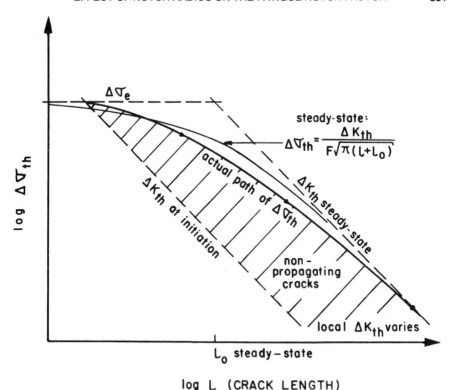

Fig 4   The variation of the threshold stress with crack length

Prediction of the fatigue notch factor

Equation (4) may be modified for central circular notches by assuming the notch to be small compared with the specimen width, and using an approximate value of 1.12 for the geometric factor, F. Hence

$$K_f = 0.89\{1 + \sqrt{(\rho/l_o)}\} \tag{9}$$
$$(1.0 \leq K_f \leq K_t)$$

The critical radius, ρ_{cr}, which divides sharp and blunt notches can be approximated, using the value of F given above, as

$$\rho_{cr} \approx 5l_o \tag{10}$$

Topper and El Haddad (**17**) produced satisfactory results by using the steady-state value of l_o for notched mild steel specimens. In this investigation, the steady-state value of l_o were also used to predict the variation of K_f with ρ for the 2024-T351 aluminum alloy and the SAE 1045 steel.

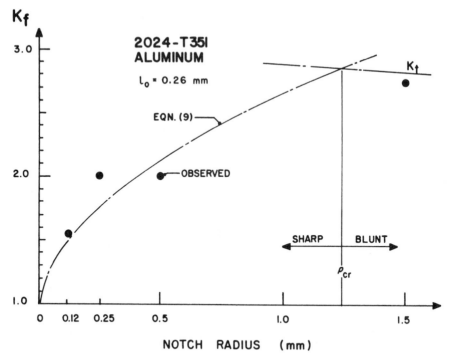

Fig 5 Predicted and observed variation of K_f with notch radius for 2024-T351 aluminum alloy

The predicted and the observed data are shown together in Figs 5 and 6 for the two materials. These figures show a reasonable agreement between the predicted and the observed values of K_f for the sharp notches. The results suggest that the steady-state l_o is an appropriate parameter for use in short crack fracture mechanics applied to constant amplitude stressing at notches. The difference between the observed K_f and the predicted K_t for the 1.5 mm blunt notch in the SAE 1045 steel is due to cyclic plasticity which is exhibited by this material at the high stresses present at the notch root.

The above results show that short crack fracture mechanics may be useful for analysing 'notch-size' effects in fatigue. This is significant from a practical engineering viewpoint, despite the fact that no physical interpretation of the parameter l_o has yet been determined. The method is relatively simple to apply and provides reasonable predictions of notch strength reductions for a range of sizes, including very small flaws, which occur in real components.

In variable amplitude loading the analysis becomes more complex. Both ΔK_{th} and the closure level will be varied by the load history and l_o may not be a useful parameter. However, some finite length term, perhaps a grain diameter, is needed in order for an engineer to apply the short crack fracture mechanics analysis. In this case, it may be useful to think of l_o as a fictitious length term

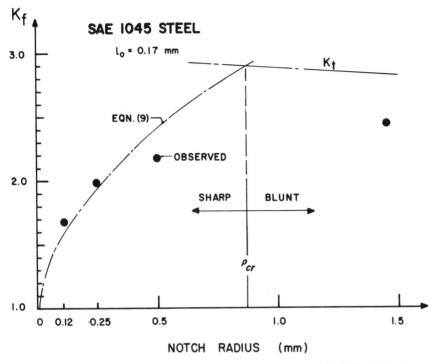

Fig 6 Predicted and observed variation of K_f with notch radius for SAE 1045 steel

used to give a smooth transition from traditional crack initiation conditions to fracture mechanics growth calculations.

Conclusions

(1)   The fatigue notch factor, calculated at a fatigue limit of 2×10^7 cycles, decreases from close to its theoretical maximum value of 3 to values approaching its minimum of 1 as the notch root radius is decreased.

(2)   Short crack fracture mechanics gives reasonable predictions of the observed values of the fatigue notch factor for the sharp circular notches.

(3)   The steady-state value of the length parameter, l_o, is appropriate for use in the equations of short crack fracture mechanics applied to constant amplitude loading of thin notched plates.

Acknowledgements

The authors gratefully acknowledge GKN Technology Ltd, the Ford Automobile Company Ltd, and the Natural Sciences and Engineering Research Council of Canada (grant No. A1694) for financial support of this research.

References

(1) NEUBER, H. (1946) *Theory of notch stresses* (J. S. Edwards, Ann Arbor, Michigan).

(2) PETERSON, R. E. (1974) *Stress concentration factors* (John Wiley, New York).

(3) KUHN, P. and HARDRATH, H. F. (1952) An engineering method for estimating notch-size-effect in fatigue tests of steel, NACA Technical note 2805.

(4) SIEBEL, E. and STIELER, M. (1955) *Z. Ver. Deutsch. Ing.*, **97**, 121.

(5) HEYWOOD, R. B. (1962) *Designing against fatigue* (Chapman and Hall, London).

(6) FENNER, A. J., OWEN, N. B., and PHILLIPS, C. E. (1951) A note on the fatigue crack regarded as a stress raiser, *Engineering*, **171**, 637–638.

(7) FROST, N. E. (1955) Crack formation and stress concentration effects in direct stress fatigue, *The Engineer*, **200**, 464–467; 501–503.

(8) FROST, N. E. (1959) A relation between the critical alternating propagation stress and crack length for mild steel, *Proc. Instn mech. Engrs*, **173**, 811–827.

(9) FROST, N. E. (1960) Notch effects and the critical alternating stress required to propagate a crack in an aluminum alloy subject to fatigue loading, *J. mech. Engng Sci.*, **2**, 109–119.

(10) FROST, N. E., MARSH, K. J., and POOK, L. P. (1974) *Metal fatigue* (Clarendon Press, Oxford).

(11) FROST, N. E., POOK, L. P., and DENTON, K. (1971) A fracture mechanics analysis of fatigue crack growth data for various metals, *Engng Fracture Mech.*, **3**, 109–126.

(12) SMITH, R. A. and MILLER, K. J. (1977) Prediction of fatigue regimes in notched components, *J. Mech. Sci.*, **20**, 201–206.

(13) SMITH, R. A. and MILLER, K. J. (1977) Fatigue cracks at Notches, *J. Mech. Sci.*, **19**, 11–22.

(14) HAMMOUDA, M. M., SMITH, R. A., and MILLER, K. J. (1979) Elastic–plastic fracture mechanics for initiation and propagation of notch fatigue cracks, *Fatigue Engng mater. Structures*, **2**, 139–154.

(15) DOWLING, N. E. (1979) Notched member fatigue life predictions combining crack initiation and propagation, *Fatigue Engng Mater. Structures*, **2**, 129–138.

(16) DOWLING, N. E. (1979) Fatigue at notches and the local strain and fracture mechanics approaches, *Fracture mechanics, ASTM STP 677*, pp. 247–273.

(17) TOPPER, T. H. and EL HADDAD, M. H. (1982) Fatigue strength prediction of notches based on fracture mechanics, *Fatigue thresholds* (EMAS, Warley, UK), Vol. 2, pp. 777–797.

(18) EL HADDAD, M. H., SMITH, K. N., and TOPPER, T. H. (1979) A strain based intensity factor solution for short fatigue cracks initiating from notches, *Facture Mechanics, ASTM STP 677*, pp. 274–289.

(19) EL HADDAD, M. H., TOPPER, T. H., and SMITH, K. N. (1979) Prediction of non-propagating cracks, *Engng Fract. Mech.*, **11**, 573–584.

(20) EL HADDAD, M. H., SMITH, K. N., AND TOPPER, T. H. (1979) Fatigue crack propagation of short cracks, *J. Engng Mater. Technol.*, **101**, 42–46.

(21) TOPPER, T. H. and EL HADDAD, M. H. (1979) Fracture mechanics analysis for short fatigue cracks, *Canadian Metallurgical Q.*, **18**, 207–213.

(22) EL HADDAD, M. H., TOPPER, T. H., and TOPPER, T. N. (1981) Fatigue life predictions of smooth and notched specimens based on fracture mechanics, *J. Engng Mater. Technol.*, **103**, 91–96.

(23) YU, M. T. and TOPPER, T. H. (1985) The effects of material strength, stress ratio and compressive overload on the threshold behavior of a SAE 1045 steel, *J. Engng Mater. Technol.*, **107**, 19–25.

(24) YU, M. T., TOPPER, T. H., and AU, P. (1984) The effects of stress ratio, compressive load and underload on the threshold behavior of a 2024-T351 aluminum alloy, *Fatigue 84* (University of Birmingham, UK), Vol. 1, pp. 179–186).

(25) AU, P. (1982) *The effect of compressive loads on fatigue crack growth behavior*, MASc thesis, University of Waterloo, Ontario, Canada.

(26) YU, M. T., TOPPER, T. H., and DUQUESNAY, D. L. (1985) The effect of microstructure on the fatigue behavior of a SAE 1045 steel, *Proceedings of the International Symposium on Microstructure and Mechanical Behavior of Materials*, Xian, China.

(27) ILLG, W. (1956) Fatigue tests on notched and unnotched sheet specimens of 2024-T3 and 7075-T6 aluminum alloys and of SAE 4130 steel with special consideration of the life range from 2 to 10,000 cycles, National Advisory Committee For Aeronautics, TN 3866, Langley Field, Virginia, USA.

(28)   BOWIE, O. L. (1956) Analysis of an infinite plate containing radial cracks originating from the boundary of an internal hole, *J. Math. Phys*, **35**, 60–71.

(29)   YU, M. T., TOPPER, T. H., DUQUESNAY, D. L., and POMPETZKI, M. A. (1985) The fatigue crack growth threshold and crack opening of a mild steel, *ASTM J. Testing Evaluation Mater.*, in press.

I. W. Hussey,* J. Byrne,* and T. V. Duggan*

Behaviour of Small Fatigue Cracks at Blunt Notches in Aero-Engine Alloys

REFERENCE Hussey, I. W., Byrne, J., and Duggan, T. V., **Behaviour of Small Fatigue Cracks at Blunt Notches in Aero-Engine Alloys**, *The Behaviour of Short Fatigue Cracks*, EGF Pub. 1 (Edited by K. J. Miller and E. R. de los Rios) 1986, Mechanical Engineering Publications, London, pp. 337–351.

ABSTRACT This investigation studies the influence of notch stress fields on the near-threshold and mid-range FCG behaviour of small cracks in Waspaloy and Nimonic 105. Experimental estimates of K for cracks within the notch influence have been obtained using the principle of stress intensity similitude for equivalent growth rates. Fatigue crack growth data has been generated for the two nickel-base alloys at stress ratios of 0.1 and 0.5. It has been established that in common with other nickel-based alloys near-threshold FCG is strongly dependent on the stress ratio.

Cracks at the root of circular notches were generated by drilling holes behind the tips of long cracks in compact tension specimens, after achieving the required stress intensity condition. The results obtained are compared with various proposals for defining the stress intensity factor range of a crack in a notch, including finite element and the boundary between long/short crack empirical methods. It is concluded that ΔK values estimated by the finite element and empirical methods, with the exception of the Cameron and Smith based solution, yield a conservative approach to life prediction. It is confirmed that a non-propagating crack condition will not develop at a blunt notch under bulk elastic loading conditions due to a continuous increase in stress intensity factor as the crack grows.

Notation

k_T	Elastic stress concentration factor
σ_o	Nominal stress
a_n	Crack length measured from the notch root
δ	Extent of notch stress field measured from the notch root
ΔP	Load range
Y	Compliance function
Y_n	Compliance function based on length $(a_n + D)$
B	Specimen thickness
W	Distance from load line to rear of specimen
D	Notch depth
ρ	Notch radius
ΔK	Stress intensity factor range
ΔK_{eff}	Effective stress intensity factor range
da/dN	Fatigue crack growth rate

* Department of Mechanical Engineering, Portsmouth Polytechnic, Anglesea Building, Anglesea Road, Portsmouth, England.

Introduction

The initiation and growth of fatigue cracks at notches is of great practical importance in engineering components and structures. It is well known that the majority of fatigue failures occur at some form of geometric discontinuity (1)(2), and of necessity most components contain such features. A recent survey of serious aircraft accidents involving fatigue fracture clearly underlines the importance of stress concentrators at crack initiation sites (3).

The fatigue life of many engineering structures is spent initiating cracks at stress concentrators and the subsequent propagation of one dominant macro-crack through the highly stressed region of the notch. The conventional philosophy employed in fatigue design at stress concentrators involves the local stress–strain concept (2). This approach essentially represents designing against crack initiation. For the purposes of this paper, crack initiation is defined as the processes of local plastic deformation, micro-crack initiation, and subsequent growth to form an 'engineering-size crack'. If structures and components, particularly welded and rivetted components, contain inherent defects, their fatigue lives may depend on the propagation of these flaws, since their initiation period is either very small or non-existent.

In these circumstances, the fatigue integrity of the component or structure should be evaluated using the so-called defect tolerance philosophy as used in the military specification MIL-A-83444 (4). Briefly, this involves assessment of the fatigue life spent propagating a fictitious flaw (estimated from non-destructive testing) to some pre-defined failure condition. Estimates of fatigue life are obtained using a fracture mechanics based methodology and a knowledge of the following factors: (a) crack size (length, shape, etc.); (b) service loads; (c) stress intensity factor; (d) material properties, e.g., K_{IC}, fatigue crack growth data.

Often the determination of K requires recourse to numerical methods which are costly and time consuming. As an alternative, the application of approximate K solutions is considered in this study, which is concerned with the growth of mode I, through-thickness fatigue cracks in blunt notch elastic stress fields. Through-thickness cracks are investigated since they represent the most fatigue-critical conditions which can occur. The behaviour of such cracks has been studied here and in an earlier paper (5) using what effectively is a blunt notched compact type (CT) specimen geometry, see Fig. 1. Two nickel based aero-engine alloys are considered, Waspaloy and Nimonic 105. The combination of high strength associated with these alloys, and the low stress concentration factors associated with the CT specimen configuration, are the principal reasons for bulk elastic conditions being maintained.

Approximate stress intensity solutions

A pre-requisite for assessing the fatigue life for a crack at a notch where linear elastic fracture mechanics (LEFM) is applicable, is a knowledge of the stress

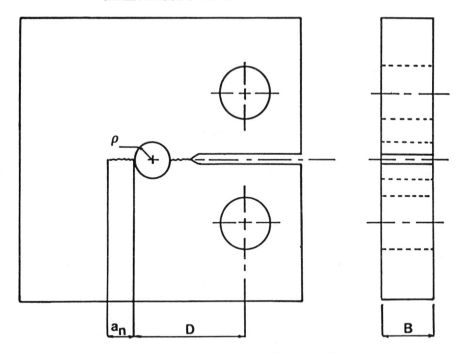

Fig 1   Keyhole specimen design with machined hole

intensity factor. When a crack is small compared to the notch size it behaves in a similar way to a crack growing from a free surface, with the FCG rate being dominated by the local stress-strain field. In these circumstances the stress intensity factor may be defined as

$$K = 1.12k_T\sigma_o\sqrt{(\pi a_n)} \tag{1}$$

where 1.12 is the free-edge correction factor. The product $k_T\sigma_o$ represents the maximum elastic stress at the notch root. This solution (6) yields a pessimistic estimate of K for all crack lengths since no account is taken of the stress gradient which exists at the notch root.

Errors in K at small crack lengths can be reduced by using the stress distribution or the average stress without a flaw instead of the peak stress at the notch. Both of these approximations suffer from two disadvantages. Firstly, a detailed knowledge of the stress distribution is necessary, and secondly, non-conservative estimates of K might be obtained (6).

When the crack is remote from any notch influence the FCG rate is controlled by the bulk stresses. The stress intensity may be estimated using the 'long crack' solution which considers the notch to be part of the crack. For the blunt notched CT specimen geometries this engineering solution corresponds to the standard long crack solution (7)

$$K = \frac{P}{B(W)^{1/2}} Y \tag{2}$$

where Y represents the specimen compliance function. If the full notch depth is assumed to contribute to crack length for small cracks a very pessimistic estimate of K is obtained.

Dowling (8) and Rooke (6) proposed that a K solution for a crack at a notch can be satisfactorily obtained by combining the two approximate limiting K solutions. The transition in behaviour between the limiting solutions may be optimized to ensure that errors in K are minimized. In practice a transition occurs between notch stress field dominated 'small crack' behaviour and the bulk stress dominated 'long crack' behaviour.

Smith and Miller (9) proposed an empirical K solution for a crack at a notch. This solution is based on comparing FCG rates for a fatigue crack of length l growing in an unnotched specimen with a fatigue crack of length a_n growing from a notch under identical bulk stresses. Where identical growth rates exist for the two cases the same crack tip driving force must also exist, however defined, i.e., whether using K or some other parameter. Therefore, in situations where LEFM is applicable, the same stress intensity factor, K, must apply.

For a crack within the notch influence, Smith and Miller proposed that

$$K = \{1 + 7.62(D/\rho)^{1/2}\}^{1/2} \sigma_o (\pi a_n)^{1/2} \tag{3}$$

where the extent of the notch stress/strain field based on this approximate K solution is given by

$$\delta = 0.13\sqrt{(D\rho)} \tag{4}$$

Later work conducted by Cameron and Smith (10) suggested that a better approximation for the extent of the notch field is given by

$$\delta = 0.21\sqrt{(D\rho)} \tag{5}$$

which, following the method of Smith and Miller (9), yields a K solution given by

$$K = \{1 + 4.672(D/\rho)^{1/2}\}^{1/2} \sigma_o (\pi a_n)^{1/2} \tag{6}$$

Application of these empirical K solutions to compact tension specimen geometries, requires an approach utilized by Duggan (11). This analysis yields a general expression of the form

$$K = (P/BW)(a_n + D)^{1/2} Y_n \tag{7}$$

The modified compliance function Y_n, based on the length $(a_n + D)$, is given by

$$Y_n = 29.6 - 185.5(a_n + D)/W + 655.7\{(a_n + D)/W\}^2$$
$$- 1017\{(a_n + D)\}^3 + 638.9\{(a_n + D)/W\}^4 \tag{8}$$

It follows that the modified K solution (11) for the CT specimen configuration gives, for Smith and Miller (9)

$$K = (P/BW)\{a_n + 7.69(D/\rho)^{1/2}a_n\}^{1/2}Y_n \tag{9}$$

and after the work of Cameron and Smith (10) for the extent of the notch stress field

$$K = (P/BW)\{a_n + 4.672(D/\rho)^{1/2}a_n\}^{1/2}Y_n \tag{10}$$

Experimental studies

In order to assess the influence of blunt notches on the behaviour of through-thickness fatigue cracks in both the near-threshold and mid-range FCG regimes, cracks were generated by pre-cracking sharp notched CT specimens. This was followed by a subsequent load shedding procedure which involved load reductions of less than 10 per cent, reducing to 5 per cent when the required crack length and stress intensity condition were approached. The crack was grown a distance of at least three times the size of the previous forward crack tip plastic zone (plane stress) before the next load reduction. After obtaining the required conditions, a hole was then drilled behind the crack tip to leave a short remnant crack. Subsequent testing was performed under constant load conditions. An Instron 1603 electromagnetic resonance machine operating between 70 and 90 Hz was used for the near-threshold programme, whilst for the mid-range FCG tests, an Instron 8032 servo hydraulic test machine operating at 10 Hz was used.

Experimental estimations of the stress intensity of a crack within the influence of a notch were obtained using the principle of stress intensity similitude for equivalent crack growth rates at the same stress ratio, R. Where appropriate, the effective stress intensity range, ΔK_{eff}, was determined by ascertaining the onset of crack opening/closure loads. Crack closure was measured using both the d.c. potential drop and backface strain compliance change methods (12). Crack length was measured using the d.c. potential drop technique supported by single stage acetate film replicas.

Introducing blunt holes by machining is likely to result in some strain hardening of the material and residual stresses in the vicinity of the bore. In order to offset such effects, these specimens were stress relieved at 600°C for 1 hour in a vacuum of less than 10^{-6} torr.

Materials

Two materials were used in this study, Waspaloy and Nimonic 105. These materials are nickel–cobalt–chromium alloys used for high temperature applications, which include gas turbine discs, blading, flame tubes, and casings, due to their resistance to oxidation and creep.

Waspaloy

Waspaloy CT specimens were machined from the diaphragm of an RB211-22HP turbine disc. The nominal chemical composition of the material (per cent weight) was 19 Cr, 13.5 Co, 4.3 Mo, 3 Ti, 1.3 Al, 0.08 C, and Ni (balance). The mechanical properties measured at 20°C were as follows: 0.2 per cent proof stress, 923 MPa; tensile strength, 1300 MPa; reduction in area, 25 per cent; elastic modulus, 215.7 GPa.

Nimonic 105

The CT specimens of Nimonic 105 were machined from 45mm diameter wrought bar. The nominal chemical composition of the material (per cent weight) was 20 Co, 15 Cr, 5 Mo, 4.7 Al, 1.3 Ti, 0.13 C, and Ni (balance). The mechanical properties measured at 20°C were as follows: 0.2 per cent proof stress, 888 MPa; tensile strength, 1256 MPa; reduction in area, 30 per cent; elastic modulus, 200 GPa.

Results

Base line, long crack FCG data was generated under constant load amplitude conditions for the two alloys, at two specific stress ratios, $R = 0.1$ and $R = 0.5$. Crack closure was assessed using the experimental techniques previously stated. Figure 2 gives the mid-range FCG results obtained for Nimonic 105, where only a slight R dependence was found. The data was generated on the servo-hydraulic machine operating at 10 Hz using sharp vee notched CT specimens.

Figure 3 shows the FCG results for Waspaloy, for which a strong dependence on R was found for near-threshold FCG but as with Nimonic 105 much reduced for the mid-range. This data was obtained on the resonance machine over the frequency range of 70–90 Hz.

The results for the various notch geometries were obtained at the same stress ratio of $R = 0.5$. The purpose of using this high stress ratio is to minimize the effects of crack closure. Four tests are considered in this paper, two on Waspaloy in the near-threshold regime and two in the mid-range regime, one for each material. The FCG data obtained from these tests was processed using the three point secant method. In the near-threshold regime, where FCG rates demonstrate increased scatter, an incremental seven point polynomial method was used (**13**). For comparative purposes the FCG data were converted to values of dimensionless ΔK obtained from stress intensity similitude and plotted against crack length, a_n, Figs 4–7. The dimensionless ΔK parameter ($\Delta K_{eff} BW^{1/2}/\Delta P_{eff}$) refers to the specimen compliance. In all cases values of effective stress intensity range, ΔK_{eff}, were used and the similitude principle was applied for identical R values. Each set of data pertaining to a particular notch geometry, material and FCG regime, has been plotted twice in subsequent figures. Firstly, against empirically derived solutions (i.e., that due to Smith and Miller (**9**) and that based on the work of Cameron and Smith (**10**) for the extent of the notch stress field) and secondly against finite element predictions (**14**)(**15**). In both cases the limiting K solutions represented by the small crack solution and the long crack solution are included.

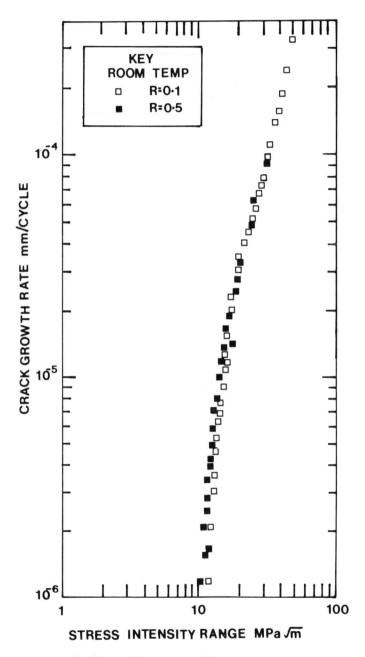

Fig 2   Fatigue crack growth data for Nimonic 105

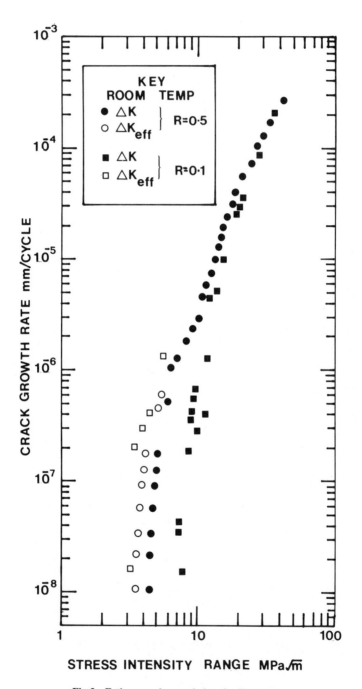

Fig 3   Fatigue crack growth data for Waspaloy

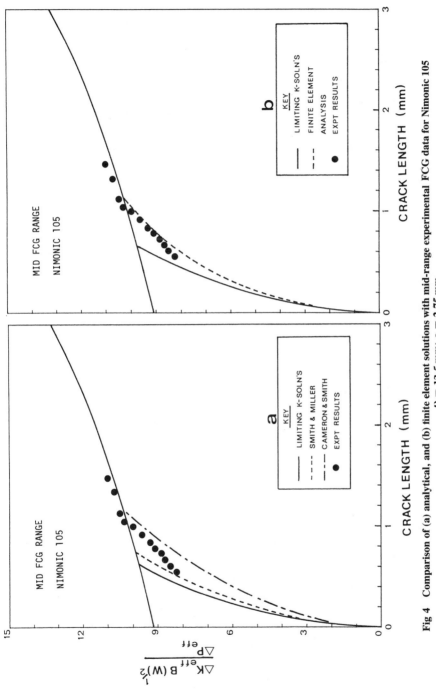

Fig 4 Comparison of (a) analytical, and (b) finite element solutions with mid-range experimental FCG data for Nimonic 105
$D = 12.5$ mm; $\rho = 2.75$ mm

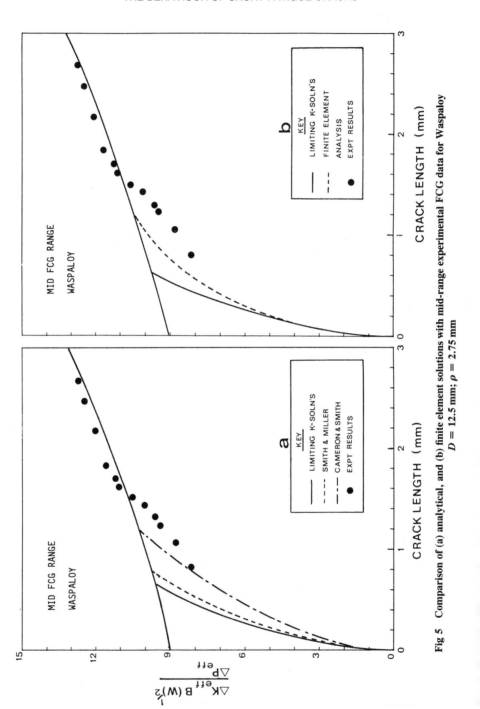

Fig 5   Comparison of (a) analytical, and (b) finite element solutions with mid-range experimental FCG data for Waspaloy
$D = 12.5$ mm; $\rho = 2.75$ mm

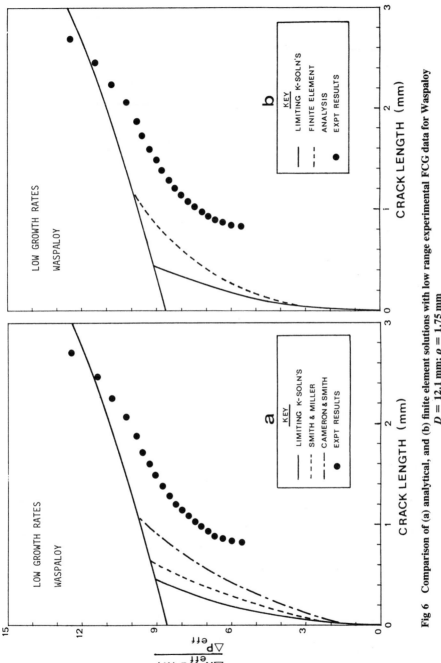

Fig 6 Comparison of (a) analytical, and (b) finite element solutions with low range experimental FCG data for Waspaloy
$D = 12.1$ mm; $\rho = 1.75$ mm

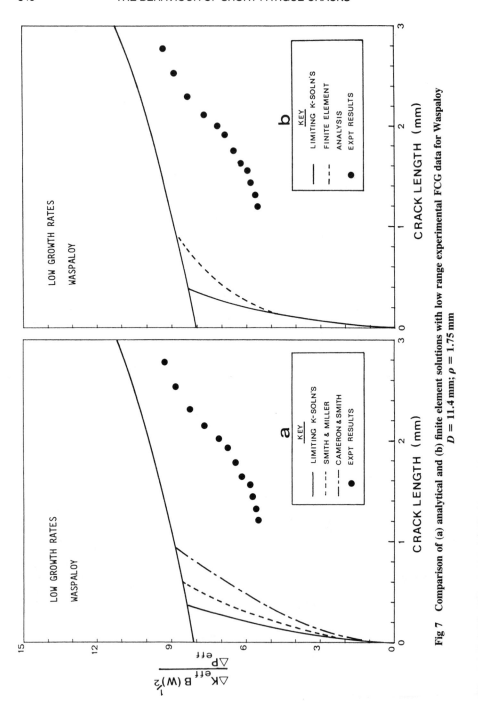

Fig 7 Comparison of (a) analytical and (b) finite element solutions with low range experimental FCG data for Waspaloy
$D = 11.4$ mm; $\rho = 1.75$ mm

Discussion

In order to preserve the validity of applying the principle of stress intensity similitude it must firstly be shown that residual stresses introduced by drilling do not affect the FCG rates and secondly that LEFM is not invalidated by the generation of plasticity at the notch root. The removal of possible residual stresses by stress relieving is considered precautionary rather than essential. Work conducted by Forsyth (**16**) on RR58 extrusions has revealed that drilling and reaming holes produces deformed material; the surface only being affected to depths of between 40 and 50 μm, an order of magnitude smaller than the initial crack lengths considered in this study, which were between 0.3 and 0.9 mm. To ensure that elastic conditions were obtained at the notch root, the maximum notch root stresses were estimated for each specimen using k_T values predicted from finite element analyses (**14**)(**15**). In no case has the maximum stress at the notch root exceeded 70 per cent of the 0.2 per cent proof stress.

Blunt notch FCG results

The mid-range results obtained for Nimonic 105 plotted against the empirical solutions, Fig. 4(a), show that the K solution based on the work of Cameron and Smith (**10**) can yield non-conservative estimates of K and slightly over-predict the crack length over which there exists a notch effect. The same results correlate very closely with the finite element prediction for the extent of the notch influence, Fig. 4(b).

The result obtained for Waspaloy in the mid-range, Fig. 5, support the experimental observations found for Nimonic 105, i.e., a reasonably close correlation with both the finite element prediction of the extent of the notch field and that found more simply and cheaply by the Cameron and Smith based solution. The experimental results later coincide with the long crack solution. Figure 5(a) shows that both the empirical solutions gave conservative estimates of K when compared with the experimental results.

The results for Waspaloy (Figs 6 and 7) obtained in the near-threshold regime indicate considerably lower values of ΔK than expected from the various K solutions. The attenuation of the experimental K values in this regime may in fact be accounted for by the observed crack branching, a common feature in nickel-based alloys in the near-threshold regime. A further contributory factor may be incurred by prolonged cycling at a fatigue threshold condition. It is possible that some form of hysterisis may occur, e.g., by crack tip blunting, requiring higher loads to be applied to cause further crack propagation than was necessary prior to reaching a threshold condition. A further possibility is that the methods used to monitor closure may not be sufficiently sensitive at $R = 0.5$.

General observations

The experimental results clearly confirm that a non-propagating crack condition will not develop at a blunt notch under elastic conditions since a continuous increase in K accompanies crack growth. The short crack and the empirical Smith and Miller solution (9) both provide consistently conservative estimates of stress intensity at small crack lengths. At longer crack lengths the long crack solution provides an accurate stress intensity characterization of the crack outside the influence of the notch. In certain cases the experimental results provide higher K solutions than theoretical long crack solutions. This may result from inherent inaccuracies in the compliance function.

In practical situations where more complex component configurations exist than those considered here, the Smith and Miller approximate solution may prove difficult to manipulate and the finite element method may be too expensive. In these situations the small crack solution which yields only a marginally more pessimistic estimate of K may be used.

Conclusions

For through-cracks at blunt notches under elastic conditions.

(1) The results obtained from both Waspaloy and Nimonic 105, for mid-range FCG rates, are reasonably consistent. They show that the extent of the notch influence for various geometries compares favourably with both the finite element solutions and the simple Cameron and Smith based solution.

(2) The Nimonic 105 results show that the Cameron and Smith based solution can yield non-conservative estimates of the stress intensity factor in the small crack growth phase.

(3) The Waspaloy results obtained in the near-threshold regime, however, infer much lower values of stress intensity factor than those predicted from any of the available K solutions.

(4) The attenuation of the inferred stress intensity values obtained for a crack in the near-threshold regime emanating from a notch in Waspaloy may be caused by:
   (i) crack tip branching reducing the effective stress intensity;
   (ii) retardation of FCG rates after threshold load shedding as a result of crack tip blunting;
   (iii) crack closure effects.

(5) The experimental results confirm that a non-propagating crack will not develop at a blunt notch since a continuous increase in the stress intensity factor accompanies crack growth.

(6) The use of the long crack solution in conjunction with the small crack approximations (with the possible exception of the Cameron and Smith

based solution) has been shown to be a safe method of estimating K for both near-threshold and mid-range fatigue crack growth regimes.

Acknowledgements

Provision of specimens and technical support by the Materials Engineering Department of Rolls-Royce Limited, Derby, is gratefully acknowledged. The authors are also indebted to Dr A. C. Pickard for providing finite element stress intensity solutions.

References

(1) DUGGAN, T. V. and BYRNE, J. (1979) *Fatigue as a design criterion* (Macmillan Press, London).

(2) FUCHS, H. O. and STEPHENS, R. I. (1980) *Metal fatigue in engineering* (John Wiley, New York).

(3) CAMPBELL, G. S. and LAHEY, R. (1984) A survey of serious aircraft accidents involving fatigue fracture, *Int. J. Fatigue*, **6**, 25–30.

(4) *Airplane Damage Tolerance Requirements*, Military Specification MIL-A-83444 USAF, 1974.

(5) HUSSEY, I. W., BYRNE, J., and DUGGAN, T. V. (1984) The influence of notch stress field on the fatigue crack growth threshold condition, Proceedings of the 2nd International Conference on Fatigue and Fatigue Thresholds, pp. 807–816.

(6) ROOKE, D. P. (1980) Asymptotic stress intensity factors for fatigue crack growth calculations, *Int. J. Fatigue*, **2**, 69–75.

(7) ASTM Standard E-399-83 (1983) Standard test method for plane-strain fracture toughness of metallic materials, *ASTM Book of Standards* (ASTM, Philadelphia), Part 3, pp. 518–551.

(8) DOWLING, N. E. (1979) *Fatigue at notches and the local strain and fracture mechanics approaches, ASTM STP 677*, pp. 243–273.

(9) SMITH, R. A. and MILLER, K. J. (1977) Fatigue cracks at notches, *Int. J. Mech. Sci.*, **19**, 11–22.

(10) CAMERON, A. D. and SMITH, R. A. (1981) Upper and lower bounds for the lengths of non-propagating cracks, *Int. J. Fatigue*, **3**, 9–15.

(11) DUGGAN, T. V. (1982) Influences of notch geometry on fatigue thresholds, *Fatigue Thresholds, Fundamentals and Engineering Applications* (Chameleon Press), Vol. II, pp. 809–826.

(12) BEEVERS, C. J. (Editor) (1980) *The measurement of crack length and shape during fracture and fatigue* (Chameleon Press), pp. 28–68.

(13) ASTM Standard E647-83 (1983) Standard test method for constant-load-amplitude fatigue crack growth rates above 10^{-8} m/cycle, *ASTM Book of Standards* (ASTM, Philadelphia), Part 3, pp. 710–730.

(14) PICKARD, A. C. (1981) Stress and stress intensity and current flow of the Derby keyhole specimen RLH 5295, Rolls-Royce Limited (Derby Aero-Engine Div.), Aero Stress Report, ASR 99144.

(15) TERRY, C. S. (1984) Stress and stress intensity analysis of four different geometries of keyhole specimens, Rolls-Royce Limited (Derby Aero-Engine Div.), Stress Method Report, SMR 99007.

(16) FORSYTH, P. J. E. (1972) Microstuctural changes that drilling and reaming can cause in the bore holes in DTD 5014 (RR58 extrusions), *Aircraft Engng*, 20–23.

J. Foth,* R. Marissen,† K.-H. Trautmann,† and H. Nowack†

Short Crack Phenomena in a High Strength Aluminium Alloy and Some Analytical Tools for Their Prediction

REFERENCE Foth, J., Marissen, R., Trautmann, K.-H., and Nowack, H., **Short Crack Phenomena in a High Strength Aluminium Alloy and Some Analytical Tools for Their Prediction**, *The Behaviour of Short Fatigue Cracks*, EGF Pub. 1 (Edited by K. J. Miller and E. R. de los Rios) 1986, Mechanical Engineering Publications, London, pp. 353–368.

ABSTRACT Short cracks show in many aspects a different behaviour to that predicted on the basis of long crack properties. This is especially true for microstructurally short cracks. As the cracks become longer they may grow through a regime denoted as mechanically short cracks. The present investigation presents results of the mechanically short crack regime in unnotched and notched Al 2024-T3 specimens and proposes a rational fracture-mechanics based approach for semielliptical surface cracks within an arbitrary stress field.

For a moderate plastic environment for short cracks in a notch field the notch root stress and strain ratios attain a special significance. Calculations based on the stress ratio, R_σ, described the lower bound of the short crack data. Predictions based on the strain ratio were more conservative.

Microstructurally short cracks showed a behaviour which could not be covered by the described crack growth evaluation method.

Nomenclature

a	Crack depth
c	Crack length on the notch surface
f	Notch depth
da/dN	Crack propagation rate
K	Stress intensity factor
ΔK	Cyclic stress intensity factor
K_{CC}	Stress intensity factor of an embedded circular crack in a stress field without a gradient
K_{EC}	Stress intensity for a semi-elliptical crack in a specimen without a notch
K_{NCC}	Stress intensity factor of an embedded circular crack in a stress field with a stress gradient
K_{NEC}	Stress intensity factor for a semi-elliptical crack at a notch
K_t	Stress concentration factor
K_x	Stress intensity due to an incremental stress $\sigma(x)$
l	Crack length from the notch root

* *Presently* Project Engineer, Fatigue and Fracture Mechanics, IABG, D-8012 Ottobrunn, FRG; *Formerly* Researcher, Fatigue Branch, DFVLR, D-5000 Cologne 90, FRG.
† Fatigue Branch, DFVLR, D-5000 Cologne 90, FRG.

N	Cycle number
r, φ	Polar coordinates
R	Stress ratio
R_ε	Strain ratio at the notch root
R_σ	Stress ratio at the notch root
S	Nominal stress
ε	Local strain
ε_{max}	Maximum local strain
ε_{min}	Minimum local strain
σ	Local stress
$\sigma(x) \cdot dx$	Wedge force at location x
σ_{max}	Maximum local stress
σ_{min}	Minimum local stress
ρ	Notch radius

Introduction

Short crack investigations have become a major subject of research work during the past 8–10 years. Several general trends of short crack behaviour have been observed. A survey is given in Fig. 1 (**1**). A classification of the short crack regime may be performed by a subdivision into two characteristic areas. At the beginning the size of the cracks correlates to microstructural features (such as grain size, length of slip bands, etc.). These cracks are denoted as micro-structurally short cracks (see Fig. 1). As the cracks grow larger and reach higher da/dN vs ΔK values the microstructural influences become less significant and the mechanical environment of the cracks becomes more important. Cracks of the later category are denoted as mechanically short cracks (**2**).

When considering short cracks, for example in engineering design, analytical predictions have to be available. Based on the results of experimental investigations the present paper outlines some analytical tools for the representation and prediction of short crack behaviour.

Experimental investigations

In some previous publications (**3**)–(**5**) the results of an extensive experimental investigation of crack initiation and propagation behaviour of Al 2024-T3 specimens were given. Experimental details and the special mapping procedure which had been applied for the registration and reconstruction of the crack initiation, short crack and macro-crack behaviour were also described in the referenced publications. An overview of the results is given in Figs 2–5.

Figure 2 shows the specimen types. Unnotched and notched specimens were considered. In Fig. 3 the crack propagation behaviour and the microstructural features are presented as observed at the notch root of the $K_t = 3.3$ specimen (compare Fig. 2). The cracks always initiated at hard intermetallic inclusions

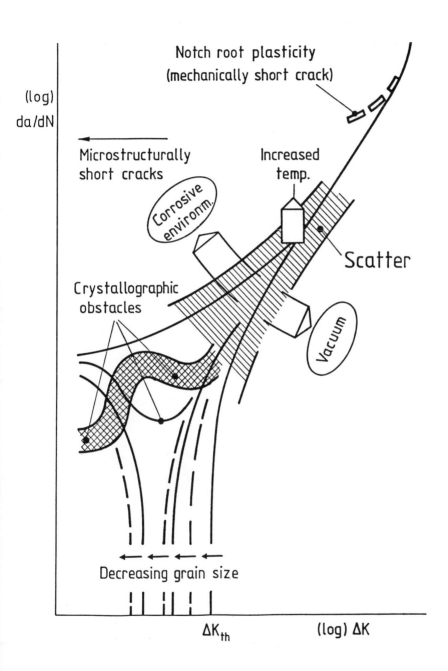

Fig 1   Characteristics of short cracks and transition to long crack behaviour

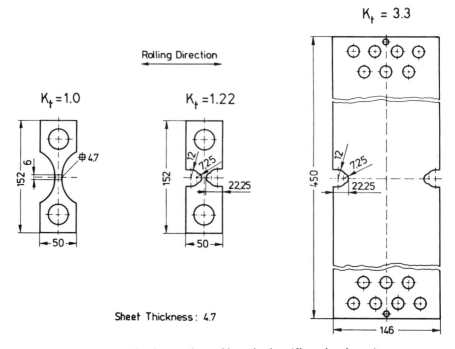

Fig 2   Specimens for the experimental investigations (dimensions in mm)

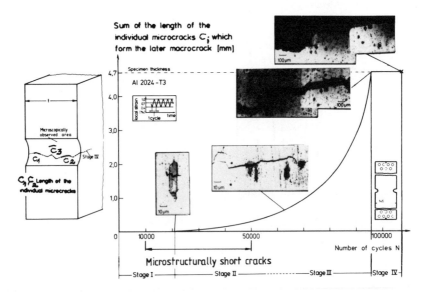

Fig 3   Stages of crack initiation and crack propagation at notched specimens, and the micro-structurally short crack regime

for this material, and the period for cracks to attain a size at the notch surface of 50–100 μm is marked as stage I in the figure. For the loading level applied in the study cracks initiated at more than one location in the notch. The cracks grew at first independently of each other (stage II in Fig. 3), and some of them started to coalesce after some while to form the later macro-crack (stage III in Fig. 3). The macro-crack stage, when cracks become visible at the side surface of the notched specimen, is denoted as stage IV (not that on the ordinate of Fig. 3 the sum of the lengths only of those cracks which later formed the macro-crack is considered). Specially indicated in Fig. 3 is that range of lifetime where microstructurally short cracks are present.

Figure 4 shows the corresponding crack propagation behaviour of all three types of specimens used in this study. It can be clearly seen that the initiation of short cracks occurred at a similar number of cycles, but that this equivalence disappeared as the cracks grew larger. In the unnotched specimens the cracks rapidly increased in size, whereas for the notched specimens the cracks grew more slowly (due to the stress gradient behind the notch).

It has to be pointed out that the trends in Fig. 4 are subject to a very large scatter in experimental data. In order to visualize the scatter the test data for the notched specimen with $K_t = 3.3$ are given in Fig. 5 in the form of a da/dN vs crack length plot. It can be seen that the scatter band is very large at the beginning of growth, decreases in width as the cracks become larger, and develops into the growth direction as observed for long cracks as the crack length increases.

In the following a fracture-mechanics based interpretation of the crack propagation behaviour will be performed.

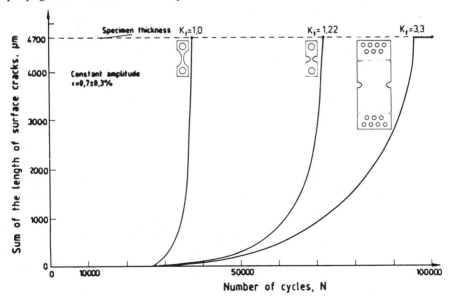

Fig 4   Crack initiation and propagation behaviour for the specimens shown in Fig. 2

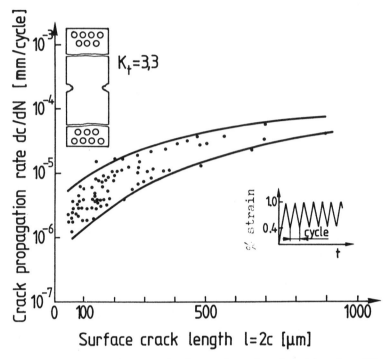

Fig 5   **Scatter band of the crack propagation data as observed on the notch surface of** $K_t = 3.3$
specimens

Evaluation of stress intensity factors

For a fracture mechanics based description of the propagation behaviour of
short cracks, which are usually half or quarter elliptical, suitable stress intensity
factors have to be available. For unnotched configurations such stress intensity
factors have already been derived (**6**). In the case of notched specimens,
solutions are available from the literature mainly for two-dimensional
(through-thickness) problems (**7**)(**8**). Three-dimensional problems have also
been considered by systematic finite element analyses and solutions are given
either in the form of tabular solutions or as analytical expressions which are
fitted to the tabular solutions (**9**). So far only cracks at unnotched specimens
and at specimens with circular notches have been treated. In the following a
procedure will be derived which may be applied to cases which are not covered
by the present literature.

Basically, stress intensity factors for cracks at notches depend on the notch
stress field (without a crack) and the crack length. The notch stress field is
described by ρ, the notch root radius, by f, the notch depth, and by the applied
stresses, S. That means that, altogether, four variables become relevant for the
stress intensity in a notch field (up to a maximum distance from the notch of

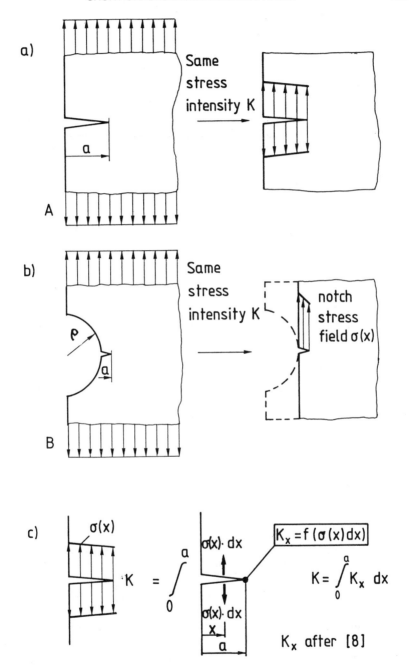

Fig 6   Evaluation of stress intensity factors in unnotched and notched configurations (two-dimensional case)

$\approx \rho/2$). Several authors have shown that stress intensity factors based on the described four variables yielded satisfactory results for two-dimensional cases (10)(11). However, practical solutions have not yet been established for the three-dimensional case; this is now proposed and the working principle is initially demonstrated, for the sake of simplicity, by a two-dimensional case.

Figure 6(a) shows how elementary fracture mechanics can be used to calculate the stress intensity factor for a semi-infinite plate containing an edge crack. Based on the wedge force model the remote stress has to be applied along the flanks of the crack. For a crack at a notch an analogous procedure is utilized (see Fig. 6(b)). This approximate solution remains satisfactory as long as the crack does not become too long as compared to the notch root radius, ρ. The calculation procedure for the stress intensity is given in Fig. 6(c). The notch stress field (without a crack) is split up into finite increments and individual wedge forces $\sigma(x) \cdot \mathrm{d}x$ can be calculated. Each individual wedge force leads to a stress intensity factor $K_x = \mathrm{f}\{\sigma(x) \cdot \mathrm{d}x\}$ at the tip of the crack. The K_x values are then integrated over the entire crack length, a, to give the actual stress intensity factor at the crack tip. Values of K_x can be taken, for example, from stress intensity factor handbooks (7)(8).

The analogous procedure for the three-dimensional case is demonstrated in Fig. 7. Elementary solutions as mentioned in connection with Fig. 6(c) for the two-dimensional case are not yet available for three-dimensional problems. That is the reason why, for half or quarter elliptical crack shapes, a circular embedded crack in an infinite body is here taken as a basis. As mentioned before, the notch stress field (without a crack) has to be known and this has been calculated for the $K_t = 3.3$ specimen (which is the example here) by a finite element analysis (2)(12). The notch field is applied symmetrically to the circular embedded crack (see Fig. 7). Wedge forces are now calculated from the notch stress field for the whole area of the crack. Together with the stress intensity solution for the crack area (which is taken, for example, from a stress intensity handbook) individual stress intensity contributions at the crack tip can be evaluated. As demonstrated for the two-dimensional case in Fig. 6(c), an integration of the individual stress intensity contributions is performed (double integrals) and leads to the stress intensity factors, for example at A and B in Fig. 7 for the embedded crack, which are of special interest.

A similarity approach is now introduced for the evaluation of the stress intensity factor for a semi-elliptical surface crack at notches, K_{NEC} (see Fig. 8). This basis is a K solution for a semi-elliptical surface crack in an unnotched body, K_{EC} (see Fig. 8), which is given in the literature (6). It is assumed that the relative 'difference' between the K value of a semi-elliptical surface crack at a notch and of a semi-elliptical surface crack in an unnotched body is the same as the 'difference' between an embedded circular crack in an infinite body with the notch stress field and the same type of crack within a homogeneous stress distribution, K_{cc}. This leads to the approach presented in Fig. 8.

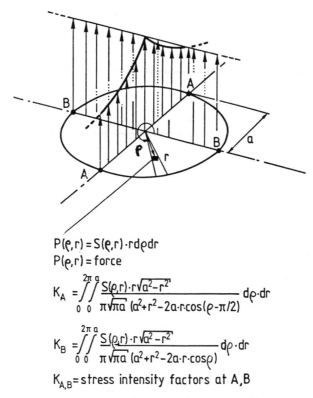

$$P(\varrho,r) = S(\varrho,r) \cdot r \, d\varrho \, dr$$
$$P(\varrho,r) = \text{force}$$

$$K_A = \int\limits_{0}^{2\pi}\int\limits_{0}^{a} \frac{S(\varrho,r) \cdot r \sqrt{a^2 - r^2}}{\pi \sqrt{\pi a} \ (a^2 + r^2 - 2a \cdot r \cdot \cos(\varrho - \pi/2))} \, d\varrho \cdot dr$$

$$K_B = \int\limits_{0}^{2\pi}\int\limits_{0}^{a} \frac{S(\varrho,r) \cdot r \sqrt{a^2 - r^2}}{\pi \sqrt{\pi a} \ (a^2 + r^2 - 2a \cdot r \cdot \cos\varrho)} \, d\varrho \cdot dr$$

$$K_{A,B} = \text{stress intensity factors at A,B}$$

Fig 7   Stress intensity factors at A and B for an embedded three-dimensional crack

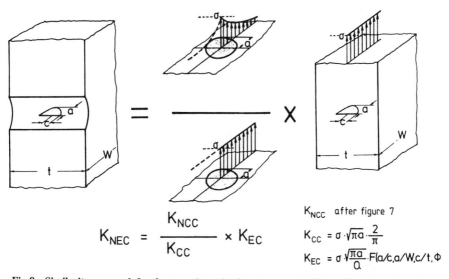

$$K_{NEC} = \frac{K_{NCC}}{K_{CC}} \times K_{EC}$$

K_{NCC} after figure 7

$$K_{CC} = \sigma \cdot \sqrt{\pi a} \cdot \frac{2}{\pi}$$

$$K_{EC} = \sigma \sqrt{\frac{\pi a}{Q}} \cdot F(a/c, a/W, c/t, \Phi)$$

Fig 8   Similarity approach for the stress intensity factor at a notch (three-dimensional case)

Evaluation of the crack growth behaviour

The ΔK solution as proposed in the previous section can immediately be applied to calculate the crack propagation rate on the basis of da/dN vs ΔK data from long cracks as long as no plasticity occurs at the notch tip. If, however, plasticity is present, a correction has to be performed. As shown in Fig. 9 for the notch specimen with K_t = 3.3 (for which a finite element calculation had been performed for a nominal loading of the specimen of 100 ± 60 MPa) plasticity occurs during the first uploading of the specimen and a redistribution of the stresses takes place. Afterwards the cyclic stresses behave predominantly elastically (because about twice the yield stress is available for further cyclic

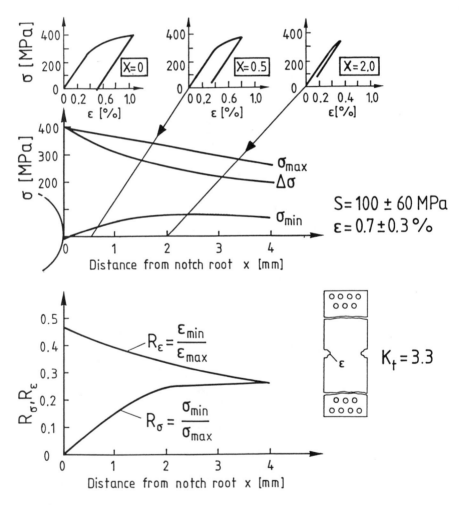

Fig 9 Behaviour of stresses and strains (without a crack) as a function of the distance from the notch for the K_t = 3.3 specimen, and behaviour of the stress ratio, R_σ, and the strain ratio, R_ε

loading due to the Masing criterion). This is shown in the insert of Fig. 9 on the left for a distance from the notch root of $x = 0$. At some distance from the notch root the amount of plasticity caused during uploading decreases.

From the local stress–strain paths in the other inserts of Fig. 9 local stress and

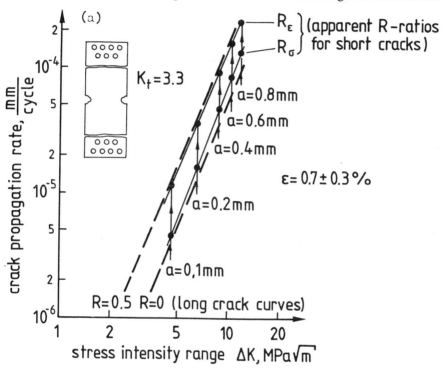

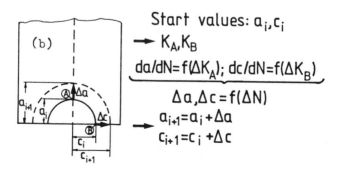

Fig 10   Crack growth prediction on the basis of the apparent R ratios, R_σ and R_ε for short cracks.

(a) Evaluation of apparent R ratios for short cracks

(b) crack growth prediction (schematic)

local strain ratios ($R_\sigma = \sigma_{min}/\sigma_{max}$ and $R_\varepsilon = \varepsilon_{min}/\varepsilon_{max}$, respectively) can always be calculated for cyclic loading. As shown in the lower part of Fig. 9 the parameter R_ε decreases while R_σ increases as the distance from the notch root increases; both R_σ and R_ε approach the nominal stress ratio for very large x values.

It is well established that the crack propagation behaviour of long cracks depends on the R ratio of the nominal stresses. As shown in Fig. 9, local stress and local strain ratios occur which change as a function of the distance from the notch root. It seems reasonable to take the local stress ratio for crack propagation calculations for short cracks at notches. Another possibility is to take the strain ratio, R_ε (3)–(5). In Fig. 10(a) the apparent R ratios, R_σ and R_ε, are shown on the basic (long crack) da/dN vs ΔK curves. These modified curves are taken as a basis for the consideration of the propagation behaviour of the short cracks. In Fig. 10(b) a crack growth calculation scheme for semi-elliptical surface cracks is outlined. If this procedure is applied (together with the modified data in Fig. 10(a)) to calculate the crack propagation rates for the conditions as were present in the experiments indicated in Fig. 5, the behaviour in Fig. 11(a) can be observed. From the figure it can be seen that neither the R_σ based nor the R_ε based crack propagation calculations represent the experimental short crack behaviour exactly. At shorter crack lengths the R_ε based predictions yield better results. The R_σ based calculations always follow the lower bound of the scatter band of the test data. At longer crack lengths the R_ε based predictions overestimate the actual crack rates. This may be due to the fact that several short cracks may coalesce and form a flat elliptical surface crack. This leads to lower crack propagation rates along the specimen surface as compared to a situation, where the shape of the crack has already reached a stabilized condition.

From an engineering standpoint the more conservative R_ε-based predictions may be preferred. It is also important to note that at very short crack lengths both types of prediction tend to fail. The R_ε approach is essentially similar to the 'strain based stress intensity' concept.

The given trends are further supported by Fig. 11(b), where the experimental results and the crack propagation predictions are compared for the unnotched specimen type of Fig. 2.

So far we have mainly considered mechanically short cracks. In a recent study the behaviour of microstructurally short cracks has been investigated, as well (13) and these results are given in Fig. 12. It can be seen that for very short cracks the observed crack propagation behaviour is very different from that of long cracks. But the hatched area, which represents the results from the tests with mechanically short cracks on the specimen types of Fig. 2, deviates from the long crack curve into the zone of lower da/dN vs ΔK values as well.

The present study considers the behaviour of predominantly mechanically short cracks. If Fig. 12 is considered, where the da/dN vs ΔK data of microstructurally short cracks are also shown, a representation of the behaviour seems to be possible if the long crack ΔK threshold value is shifted into the direction of

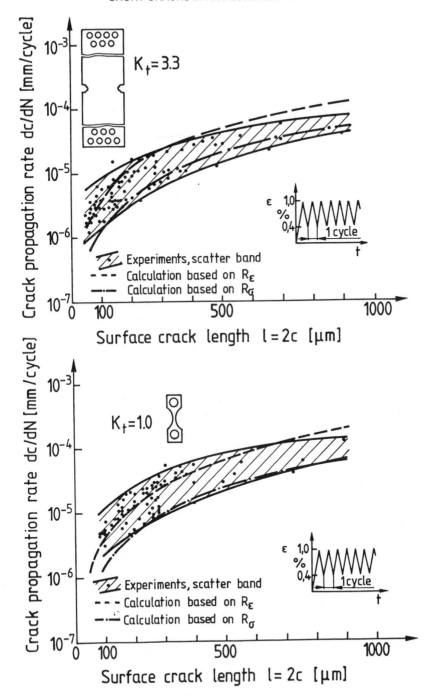

Fig 11 Crack growth calculations based on R_σ and R_ε for the experimental data from Fig. 5 and for the unnotched specimen in Fig. 2

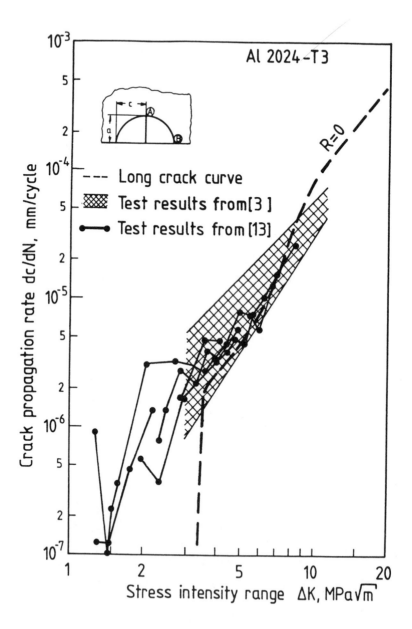

Fig 12 Representation of the behaviour of microstructurally short cracks and of mechanically short cracks in a dc/dN vs ΔK representation

$\Delta K = 0$. Similar trends have also been recognized in references (14) and (15). In (1) a more detailed discussion of the threshold behaviour of short cracks is presented.

Conclusions

On the basis of the results of an extensive experimental program on unnotched and notched Al 2024-T3 specimens a fracture mechanics based evaluation of short crack behaviour was performed. The trends of the observed short crack behaviour are represented in an adequate manner on the basis of a stress intensity concept which incorporated the notch root strain ratio.

Remaining limitations in this concept may be overcome by an investigation into the influence of prior deformation on material behaviour.

Acknowledgements

The support of the Deutsche Forschungsgemeinschaft (DFG) for this investigation is gratefully acknowledged. The authors also thank Mrs Schmidt and Mr Hermanns for their help in preparing the manuscript.

References

(1) NOWACK, H., MARISSEN, R., TRAUTMANN, K. H., and FOTH, J. (1986) Significance of the short crack problem as a function of fatigue life range, *International Conference on Fatigue of Engineering Materials and Structures*, Sheffield 15–19 September 1986.

(2) SCHIJVE, J. (1984) The practical and theoretical significance of small cracks. An Evaluation, *Fatigue 84* (2nd International Conference on Fatigue and Fatigue Thresholds), (EMAS, Warley), pp. 751–772.

(3) FOTH, J. (1984) Einfluß von einstufigen und nicht-einstufigen Schwingbelastungen auf Rißbildung und Mikrorißausbreitung in Al 2024-T3, DFVLR-Rep. No. FB 84-29, DFVLR, D-5000 Cologne 90.

(4) FOTH, J., MARISSEN, R., NOWACK, H., and LÜTJERING, G. (1984) A fracture mechanics based description of the propagation behaviour of small cracks at notches (Edited by L. Faria), *Proceedings of ECF-5*, Lisbon, pp. 135–143.

(5) FOTH, J., MARISSEN, R., NOWACK, H., and LÜTJERING, G. (1984) Fatigue crack initiation and microcrack propagation in notched and unnotched aluminium 2024-T3 specimens, Proceedings of the ICAS Symposium, Toulouse, France, pp. 791–801.

(6) NEWMAN, J. C., Jr., and RAJU, I. S. (1981) An empirical stress intensity factor equation for the surface crack, *Engng Fracture Mech.*, **15**, 185–192.

(7) ROOKE, D. P. and CARTWRIGHT, D. J. (1976) *Compendium of stress intensity factors* (HMSO, London).

(8) TADA, H., PARIS, P. C., and IRWIN, G. R. (1973) The stress analysis of cracks handbook, Del Research Corporation, St. Louis, Missouri.

(9) NEWMAN, J. C., Jr. and RAJU, I. S. (1983) Stress-intensity factor equations for cracks in three-dimensional finite bodies. *Fracture Mechanics: Fourteenth Symposium – Volume I: Theory and Analysis, ASTM STP 791* (Edited by J. C. Lewis and G. Sines), pp. I-238–I-265.

(10) SCHIJVE, J. (1982) The stress intensity factor of small cracks at notches, *Fatigue Engng Mater. Structures*, **5**, 77–90.

(11) JAERGEUS, H. Å. (1978) A simple formula for the stress intensity factors of cracks in side notches, *Int. J. Fracture*, **14**, R113–R114.

(12) OTT, W. (1982) Elastisch-plastische FE-Analyse an Kerben, Ber. Bd. Betriebsfestigkeit von Leichtbauwerkstoffen, DFVLR, D-5000 Cologne 90.

(13) FOTH, J. (1986) AGARD collaborative effort on short cracks, IABG Techn. Rep. TF-V, to be published.
(14) SOCIE, D. F., HUA, C. T., and WORTHEM, D. W. (1985) Mixed mode small crack growth, *Int. J. Fatigue Fracture Engng Mater. Structures*, To be published.
(15) HEITMANN, H. H., VEHOFF, H., and NEUMANN, P. (1984) Life prediction for random load fatigue based on the growth of microcracks, Proceedings 6th International Conference on Fracture (ICF 6), New Delhi, India.

P. Lalor,* H. Sehitoglu,* and R. C. McClung*

Mechanics Aspects of Small Crack Growth from Notches – the Role of Crack Closure

REFERENCE Lalor, P., Sehitoglu, H., and McClung, R. C., **Mechanics Aspects of Small Crack Growth from Notches – the Role of Crack Closure**, *The Behaviour of Short Fatigue Cracks*, EGF Pub. 1 (Edited by K. J. Miller and E. R. de los Rios) 1986, Mechanical Engineering Publications, London, pp. 369–386.

ABSTRACT The concept of fatigue crack closure has come to be one of the most effective tools in explaining the anomalous propagation behaviour of small cracks. A two-dimensional, plane stress, elasto-plastic finite element model is employed to predict the opening and closure behaviour of small fatigue cracks emanating from blunt notches. The model incorporates changing boundary conditions associated with the intermittent contact and separation of the crack faces and allows cyclic crack extension through the mesh. Crack opening and closure levels, near crack stress fields, and crack tip plastic zone sizes are determined as a function of crack length. The model predicts crack opening and closure levels to increase and then stabilize as the crack grows away from the notch. The influence of load level on small crack closure behaviour is also considered. Predictions correlate with, and give insight to, experimental results.

Nomenclature

c	Half notch width
$d\mu$	Scalar used in Ziegler's rule
E	Modulus of elasticity
$\dot{e}_{ij}^p$	Plastic strain rate tensor
E_p	Plastic modulus
f	von Mises yield function
k	Stiffness of truss element
K	Stress intensity factor
K_{min}, K_{max}	Minimum and maximum stress intensity
$\Delta K, \Delta K_{eff}$	Range in stress intensity factor, and effective stress intensity factor
l	Crack length measured from the notch
l/c	Crack length normalized by half notch width
R	Stress ratio, minimum stress/maximum stress
r_p	Notch plastic zone size
$S, \Delta S$	Applied stress level and stress range
S_a, S_{max}	Stress amplitude and maximum applied stress level
S_{clos}, S_{open}	Applied stress level at crack closure and at crack opening
S_{ij}, S_{ij}^c	Deviatoric stress and back stress tensor
$w, \Delta w$	Maximum plastic and reversed plastic zone size
x, y	Coordinate axes at crack tip
X, Y	Global coordinate axes located at the centre of notch (FEM)

* Department of Mechanical and Industrial Engineering, University of Illinois at Urbana-Champaign, 1206 West Green Street, Urbana IL 61801, USA.

σ_o	Yield stress in tension
σ_{ij}	Stress tensor
σ_{yy}	y component of local (notch and/or crack tip) stress
δ_{ij}	Kronecker delta (identity tensor)

Introduction

It is now well known that a crack may be closed during a significant portion of a fatigue cycle (1). This crack closure is a result of the plastically deformed material ahead of the crack tip and is most significant under plane stress conditions. As the crack extends, this material is unloaded and forms residual displacements on the crack surfaces. These residual displacements cause the crack surfaces to contact earlier than expected during unloading. Once contact is attained, further unloading results in residual compressive stresses in the wake of the crack tip. In order for the crack to re-open, these stresses must be overcome by the applied load. Based on experimental and analytical results (2)(3), it is expected that crack opening and closure levels approach a steady-state value with increasing crack length. However, a majority of a component's fatigue life may well be consumed in the small crack growth regime where opening and closure levels are continuously changing. It is, therefore, desirable to understand the influence of load level, geometry, crack length, and material microstructure on the behaviour of these small cracks. The effects of crack length and applied load on opening and closure levels are addressed in this study.

Analytical studies (4)–(6) based on assumed residual displacement fields have been performed to predict crack opening loads. Estimates of the compressive residual stresses behind the crack at minimum load have been used to establish the crack opening loads necessary to overcome them. These models were based on the Dugdale (strip-yield) model (7) modified to leave plastically deformed material in the wake of the advancing crack. Many assumptions are inherent in these analytical methods.

For physically small cracks, especially small cracks growing in the vicinity of a notch plastic zone where plastic zone size is not small compared to crack size, linear elastic fracture mechanics (LEFM) concepts fall short in characterizing crack tip stress–strain behaviour. Finite element methods (FEM) have been applied to crack closure studies to account for both notch and crack tip plasticity (8)–(10). Elasto-plastic material behaviour and geometry can be accurately accounted for with finite element methods. In the present work, small fatigue crack growth from a blunt notch (circular hole) is modeled with FEM and the results are compared to analytical and experimental studies performed by Sehitoglu (2)(3). Near crack stress fields determined from the FEM study add much support to the understanding of small crack growth behaviour.

Several factors thought to influence small crack growth behaviour are not included in the present work. Grain size is critical to small crack growth since deceleration and non-propagation of small cracks may be attributed to blockage of slip by the grain boundaries (**11**). In the present work the crack was allowed to grow perpendicular to the principal stress direction with no effect of shear loading on crack surfaces. Specifically, tortuosity of the crack path observed near the threshold region may influence crack growth rates. The crack growth rate behaviour reported here, however, is considered to be outside the regime in question.

The purpose of the present study is to:

(a) simulate the crack opening and closure behaviour of the 1070 steel using elasto-plastic finite element methods. (This is the steel used in previously reported experimental studies (**2**)(**3**).)

(b) determine stress (including residual stress) fields in the vicinity of growing cracks;

(c) analyse the influence of the notch inelastic field on crack closure;

(d) determine the influence of crack length and applied load level on crack opening and closure levels and compare the results with experiments (**2**)(**3**).

Analysis

A small fatigue crack growing in a notched plate (Fig. 1) subjected to completely reversed ($R = -1$) cyclic loading is analysed here. The mesh is composed of four-noded isoparametric quadrilateral elements with a 2×2 integration rule. Nodal coordinates are updated after application of each load increment. Material properties incorporated in the model were those of a 1070 wheel steel (class U). Experimental results on this material are tabulated in references (**2**) and (**3**). A replica technique with a scanning electron microscope has been used to determine crack tip opening and closure levels in these references. The stress–strain curve of the material is shown in Fig. 2. Monotonic and cyclic curves at 20°C coincide for this material.

The plasticity theory incorporated in the model employs a von Mises yield surface and uses Ziegler's modification to Prager's hardening rule (**12**) for translation of the yield surface. This hardening rule was chosen to account for the material anisotropy (Bauschinger effect) associated with reversed yielding (kinematic hardening) in the vicinity of a fatigue crack. The von Mises yield condition is given by (repeated indices imply summation)

$$f = \tfrac{3}{2}(S_{ij} - S_{ij}^c)(S_{ij} - S_{ij}^c) - \sigma_o^2 = 0 \tag{1}$$

where σ_o is the yield stress in tension and S_{ij} are the stress deviators

$$S_{ij} = \sigma_{ij} - \tfrac{1}{3}\delta_{ij}\sigma_{kk} \tag{2}$$

S_{ij}^c represents the total translation of the centre of the yield surface in deviatoric stress space due to work hardening and is often termed the deviatoric back

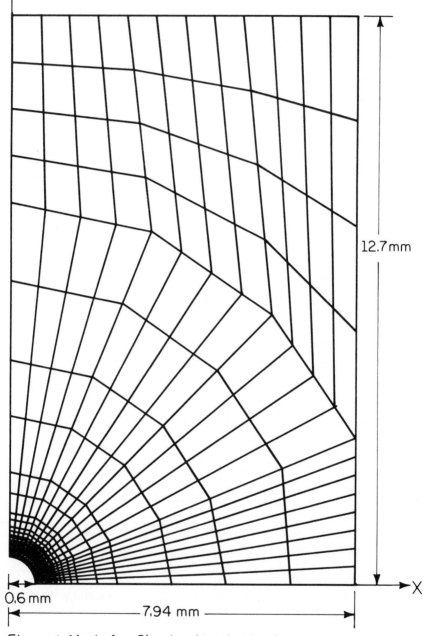

Element Mesh for Circular Notch Member

Fig 1 Element mesh for centre notched member

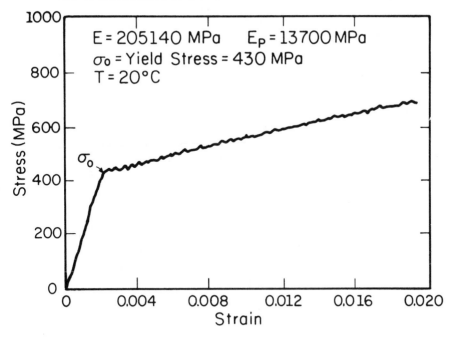

Fig 2   Experimental stress–strain relation for 1070 wheel steel (Class U)

stress tensor. The flow rule and Ziegler's kinematic hardening rule are represented, respectively, by

$$\dot{e}_{ij}^{p} = G \, \frac{\partial f}{\partial S_{ij}} \frac{\partial f}{\partial S_{kl}} \, \dot{S}_{kl} \tag{3}$$

and

$$\dot{S}_{ij}^{c} = d\mu(S_{ij} - S_{ij}^{c}) \tag{4}$$

Here, G is a scalar (constant) and $d\mu$ is a scalar determined from the consistency condition $df = 0$. The term $\dot{e}_{ij}^{p}$ represents increments of the plastic strain tensor. A state of plane stress is modeled and for simplicity a bilinear representation of the stress–strain curve is used. This simplification is reasonable for the material of interest (Fig. 2). The yield stress, σ_{o}, is 430 MPa.

The procedure which permits cyclic crack extension and variable boundary conditions associated with the intermittent contact and separation of the crack surfaces during the loading history is illustrated in Fig. 3. Truss elements of variable stiffness are attached to the nodes along the crack face. The stiffness of a truss is set to an extremely high value ($k \to \infty$) when that portion of the crack is closed. When a portion of the crack satisfies the condition for opening, the stiffness of the associated truss is set to zero ($k = 0$) thus allowing the crack to open. (Note that the crack surfaces are free to move in the x direction.)

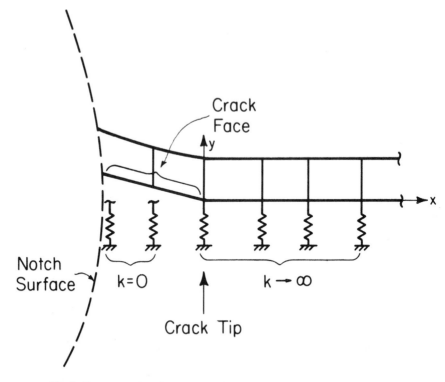

Fig 3   Representation of control of boundary conditions along the crack face

In other studies (**8**) of crack closure using the finite element method, a section of crack was permitted to open ($k = 0$) when the applied loading caused the y displacement of that node along the crack surface to become positive. This displacement appears to be dependent on the finite stiffness value assigned to the spring holding the crack surface closed. Although most results showed good agreement with experiments, this criterion gave closure levels higher than opening levels, which does not correspond to experimental results (**2**)(**3**).

The criteria for crack opening implemented in this study provides the correct trends in opening and closure levels. It has been postulated that in order for a fatigue crack to open, the applied loading must overcome the compressive residual stresses left in the wake of the crack tip (**1**). Here, the σ_{yy} components in the crack surface elements are monitored throughout the loading portion of the history. When σ_{yy} for a crack surface material point changes from compressive to zero to slightly tensile, the stiffness of the associated truss element is changed to zero, thereby allowing that portion of the crack to open. Crack opening level, S_{open}, is defined as the applied load level that overcomes all compressive residual stresses in the wake of the crack tip. At this time, crack propagation is once again assumed to be possible.

A section of crack is said to be closed ($k \to \infty$) when the y component of the coordinates of the particular node being considered along the crack face returns to its original ($y = 0$) position. These nodal coordinates are monitored throughout the unloading portion of the history. In this study, crack closure level, S_{clos}, is defined as the applied load level at which the node directly behind the crack tip first closes thus eliminating the possibility of further crack propagation. Similarly, the first contact of the crack tip region has been defined as the experimental S_{clos}. An alternate definition of S_{clos} based on complete closure of the crack surfaces is also possible but yields S_{clos} levels which are lower than the above case.

Crack extension is achieved by arbitrarily choosing to release the present crack tip node ($k = 0$) when the peak of each load cycle is reached. Upon doing so, the load is redistributed and the new crack tip is located in front of the node just released. The increment of crack extension is thus equal to the side length of a crack surface element (0.028 mm). Presently, this crack extension criteria is arbitrary but could readily be modified to represent a crack-growth law based on, for instance, crack tip opening displacements.

The procedure described above was implemented to propagate a small fatigue crack from initiation ($l/c = 0$) well into the region of long crack behaviour ($l/c = 1.5$) where $c = 0.6$ mm (half notch width). The crack grew out of the notch in a plate (Fig. 1) subjected to $R = -1$ loading with an applied stress amplitude of $S_a = 207$ MPa. Then, the case of $S_a = 310$ MPa was considered. In order to reduce computation time, the crack length was extended by two element side lengths in the fine region of the mesh or by one element side length in the coarser region of the mesh at the top of each load cycle. The crack growth increment was then 0.056 mm per cycle. If a slightly larger or smaller increment in crack extension per cycle were used, the results for the same total crack lengths were not affected provided that the increment of crack extension did not exceed the size of the crack tip plastic zone.

The loading was carried out in small increments in order to monitor the crack surface elements stresses and nodal coordinates as criteria for determining opening and closure levels. Plots of numerical results are superimposed on experimental results where applicable.

Results

Crack opening and closure levels, S_{open} and S_{clos}, determined with FEM, are plotted as a function of normalized crack length in Fig. 4. These results are indicated with symbols F* and F, respectively. Results consistently indicate crack opening levels to be higher than closure levels. There exists a favourable agreement among opening and closure levels obtained with the experimental (Exp.) and analytical methods (A*, A) shown in Fig. 4. Note that closure levels predicted by FEM compare better with experimental values than did the closure levels predicted by the analytical method (based on the Dugdale model)

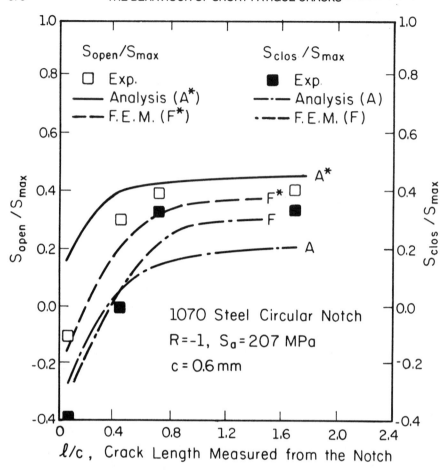

Fig 4   Three different techniques to determine crack opening and closure levels for a crack growing from a circular notch

This difference is due to the fact that both the experimental and the FEM techniques considered closure at the crack tip but the analytical closure levels were calculated based on closure over the entire crack length (2). The experimental crack opening and closure results were obtained by a replicating technique and viewing the replicas in the scanning electron microscope (3).

The effective stress range during which the crack is open and capable of propagating is defined as

$$\Delta SK_{eff} = S_{max} - S_{open} \tag{5}$$

The corresponding effective stress intensity range, ΔK_{eff}, has been used to intepret crack growth rate behaviour (1)–(3). In Fig. 5, ΔK_{eff} is used to indicate that changing crack opening level with increasing crack length and its influence

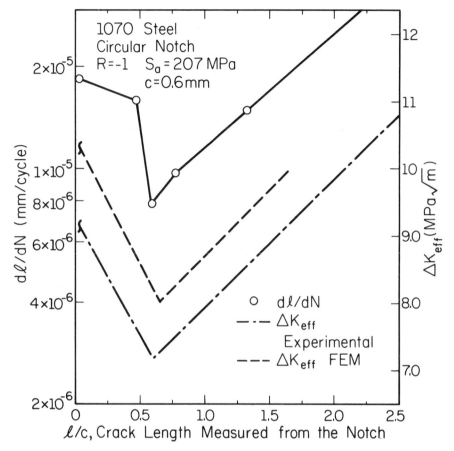

Fig 5 Changing effective stress intensity range as a function of crack length and its influence on crack growth rate

on the experimentally measured crack growth rate. In this case ΔK_{eff} was computed using ΔK (equations (5a) and (5b) of reference (**2**) modified with experimental and FEM opening levels (S_{open}), respectively. The experimentally and numerically (FEM) computed ΔK_{eff} levels agree within 10 per cent for most cases. Crack growth rates are seen to decrease with increasing crack length then increase again. Closure effects appear to account for this behaviour as shown in Fig. 5; note that the ΔK_{eff} curves in Fig. 5 are not extended to $\Delta K_{eff} = 0$.

Use of the finite element method in modelling fatigue crack growth problems gives access to some parameters that are not as readily obtained with other methods. Two of these parameters, near crack stress fields and crack tip plastic zones for the hardening material considered, are discussed below.

Figure 6 presents plots of the near tip stress fields for different crack lengths.

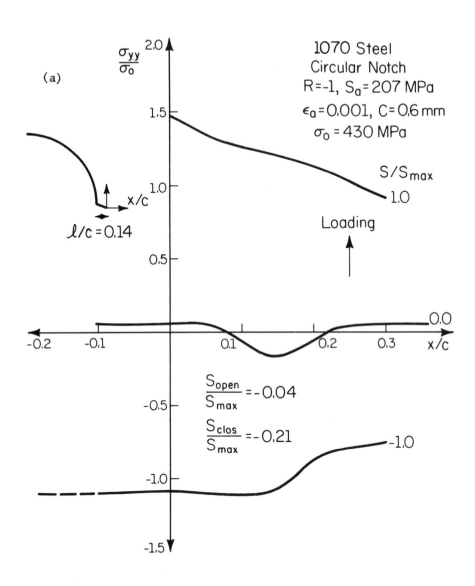

Fig 6   Near crack stress fields during different stages of loading
(a) *l/c* = 0.14   (b) *l/c* = 0.6   (c) *l/c* = 0.8

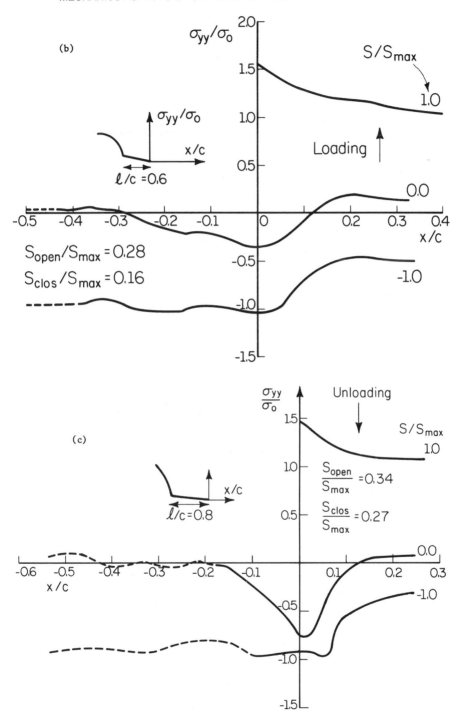

Three different stress fields are plotted on each curve representing different stages of the cycle; loading in Fig. 6(a) and (b), unloading in Fig. 6(c). The vertical axis is the stress field in the y direction normalized by the uniaxial yield stress, σ_o. On the horizontal axis, the location $x/c = 0$ denotes the crack tip; the position $x/c = -l/c$ corresponds to the notch surface. The shorter crack (Fig. 6(a)) is open during a larger portion of a loading cycle than the longer crack in Fig. 6(b). This occurs due to the extent of closure as a function of crack length. Note that due to the influence of the notch stress field, the magnitudes of the compressive residual stresses at minimum load ($S/S_{max} = -1.0$) are approximately equal for both crack lengths. Stress fields for a fatigue crack during different stages of the unloading portion of a cycle are shown in Fig. 6(c). The onset and evolution of the compressive residual stresses in the wake of the crack tip can be seen after contact of the crack surfaces is observed.

The stress fields created by the presence of the notch also appear to affect small crack behaviour (13)–(15). The uncracked notch root material stress–strain response is shown in Fig. 7 to indicate the plastic strain field the small

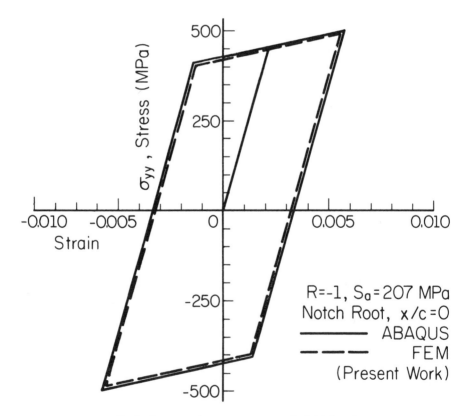

Fig 7   Stress–strain response at uncracked notch root

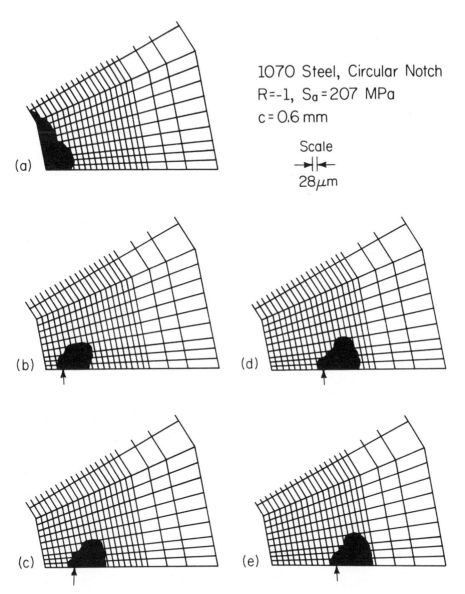

1070 Steel, Circular Notch
R=-1, S_a=207 MPa
c=0.6 mm

Scale
→|←
28 μm

(a)

(b)

(d)

(c)

(e)

Fig 8 Notch and current crack tip plastic zone profiles at a maximum load for a series of crack lengths. Corresponding experimental crack growth rates are supplied below. Arrow indicates location of crack tip

(a) $l/c = 0.0$
(b) $l/c = 0.138$ $da/dN = 1.77 \times 10^{-5}$ mm/cycle
(c) $l/c = 0.232$ $da/dN = 1.74 \times 10^{-5}$ mm/cycle
(d) $l/c = 0.418$ $da/dN = 1.62 \times 10^{-5}$ mm/cycle
(e) $l/c = 0.512$ $da/dN = 1.25 \times 10^{-5}$ mm/cycle

crack will be initially embedded in. A comparison of results is also made with the ABAQUS commercial FEM code. The present work uses a plasticity theory and constitutive law integration method similar to those in the ABAQUS code. The integration scheme is a mean normal, tangent stiffness method with radial return and subincrementation. For a discussion of these, refer to (16)–(19).

The notch plastic zone profile and crack tip plastic zone profile are shown for a series of crack lengths in Fig. 8. Examination of experimentally observed growth rates for this problem (Fig. 5) reveals the approximate crack length where growth rates reach a minimum value. This length appears to be at about $l/c = 0.60$ where c is half notch width. FEM results indicate the notch plastic zone to extend out to only about $l/c = 0.24$ for $S_a = 207$ MPa. Clearly, small crack behaviour continues at crack lengths well past the notch plastic zone size.

The ratio of the reversed crack tip plastic zone size, Δw, to the maximum crack tip plastic zone size, w, is influenced by the notch plastic zone as seen in Fig. 9. There exists a correspondence of this ratio with closure level (2). The opening and closure stresses increase with decreasing $\Delta w/w$ ratios. The rate of decrease in $\Delta w/w$ appears to be highest for $0 \geq l/c \geq r_p/c$ where r_p is the plastic zone of the notch.

The level of applied load also greatly influences crack closure behaviour for the $R = -1$ loading considered. Figure 10 is a plot of opening and closure levels versus normalized crack length for two different applied stress amplitudes. These results were generated with FEM. Although the opening and closure levels for $S_a = 310$ MPa have not been experimentally verified, the qualitative trends do agree with observed data (20). Both crack opening and closure levels are seen to be lower at higher applied stress amplitudes. Also note that the difference between crack closure level and opening level increases with higher applied stress amplitude. This has been observed experimentally (20).

Discussion of results

Experimental evidence (2)–(4)(13)–(15) has shown that physically small fatigue cracks grow at rates higher than expected based on the trends of long crack data. More specifically, in the notched member simulated here, growth rates for very small cracks were found to decrease to a minimum then increase with increasing crack length (Fig. 5). It has been proposed that crack closure may have a significant influence on the explanation of small crack growth behaviour. The FEM simulation described here accurately modeled the plastically deformed material ahead of a growing crack and thereby predicted crack opening and closure levels that are in good agreement with experimentally observed results.

The crack opening and closing levels increase progressively and then stabilize as the crack grows away from the notch (Fig. 4). Thus, the extent of closure appears to be related to physical crack size. The level of closure is strongly dictated by the material in the wake of the crack tip. In general, a smaller crack

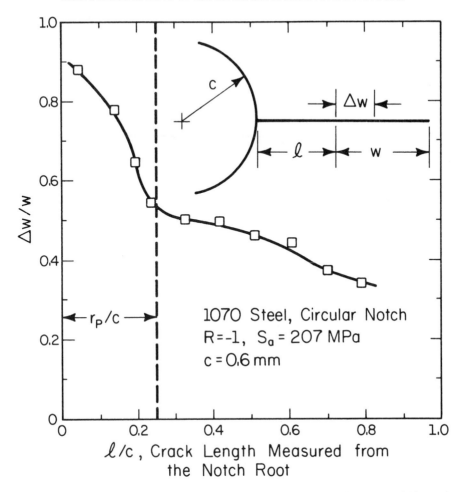

Fig 9   Ratio of reversed crack tip plastic zone to maximum crack tip plastic zone, $\Delta w/w$, for a crack growing from a circular notch

with a limited wake is less likely to experience closure effects than a long crack with a significant amount of plastically deformed material in its wake. Examining the stress fields for a relatively small and a long crack, Fig. 6(a) and 6(b), respectively, revealed that the longer crack has a larger region of compressive residual stresses that must be overcome by the applied load than that of the smaller crack. Since the small crack is embedded in the notch plastic zone, the magnitudes of the compressive residual stresses and the ensuing maximum tensile stresses for it are about the same as those for the longer crack. Therefore, the magnitude of these stresses seems less significant than the length of the wake in this argument.

It should be pointed out here that other factors also influence closure as

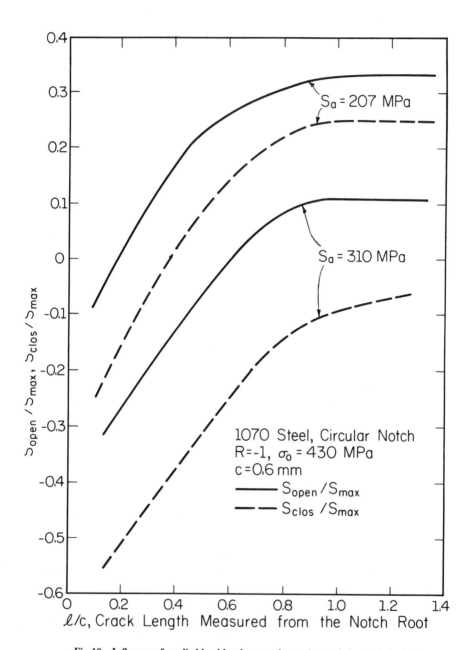

Fig 10   Influence of applied load level on crack opening and closure behaviour

experimental evidence in some studies did not observe closure levels to vary significantly with crack length (20). The effect of crack length on closure levels is more dominant for $R < 0$ than for $R \geq 0$ loading as indicated in other studies (4). Crack opening and closure levels were reported to be fairly independent of crack length for $R = 0$ in (2). This apparent lack of crack length influence may be explained by the results of an FEM analysis on the mesh in Fig. 1 run under $R = 0$ conditions. It was found that crack face contact did not extend completely behind the crack at minimum applied load for $R = 0$ loading for the same crack lengths considered. This is contrary to the $R = -1$ case. Therefore in the $R = 0$ case, the contact zone behind the crack for two different crack sizes may be equal, resulting in similar opening levels (2)(21).

Results of the numerical prediction of crack opening and closure levels can be applied to partly explain the anomalous propagation behaviour of small cracks. The ΔK_{eff} calculated over the range of crack lengths considered is illustrated in Fig. 5. Note the qualitative similitude between the dl/dN and ΔK_{eff} versus l/c curves. Since the extent of crack closure is minimal at small crack lengths (Fig. 4), the small cracks measured experience an initially high ΔK_{eff} that translates into a high growth rate. As the crack grows, the stress level required to open the crack allowing further propagation increases. This accounts for the decreasing growth rates. Eventually crack opening levels stabilize and one observes the small crack growth curve to merge with the long crack data.

Conclusions

A detailed simulation of a small fatigue crack growing from a notch has been carried out and results compare favourably with experiments. Admittedly, numerous factors thought to influence small crack behaviour have been omitted from the present analysis. Nevertheless the work supports some important conclusions.

(1) Crack closure levels for cracks emanating from blunt notches are a function of crack length for $R = -1$ loading. This dependency of crack closure level on crack length corresponds with observed propagation behaviour for both small and long cracks.

(2) Opening levels are consistently higher than closure levels for all crack lengths considered. This difference becomes more pronounced at higher applied stress amplitudes.

(3) Finite element method results correspond favourably with both experimental and analytical work. The FEM also provides other information (i.e., near crack stress fields and plastic zones for hardening material and for cyclic anisotropy) which can aid in the understanding of small crack behaviour.

(4) Crack opening and closure levels change with applied load level for $R = -1$ loading. The FEM satisfactorily predicts this dependence. Further work is needed in this area.

Acknowledgements

This research was funded by the Fracture Control Program, College of Engineering, University of Illinois at Urbana-Champaign.

References

(1) ELBER, W. (1971) The significance of fatigue crack closure, *Damage tolerance in aircraft structures, ASTM STP 486* (American Society for Testing and Materials), pp. 230–242.

(2) SEHITOGLU, H. (1985) Crack opening and closure in fatigue, *Engng Fracture Mech.*, **21**, 329–339.

(3) SEHITOGLU, H. (1985) Characterization of crack closure, *ASTM 16th Symposium on Fracture Mechanics, ASTM STP 868*, Colombus, Ohio (American Society for Testing and Materials), pp. 361–380.

(4) NEWMAN, J. C., Jr. (1982) A nonlinear fracture mechanics approach to growth of small cracks, Paper presented at the AGARD Meeting on Behaviour of Short Cracks in Airframe Components, Toronto, Canada.

(5) FUHRING, H. and SEEGER, T. (1979) Dugdale crack closure analysis of fatigue cracks under constant amplitude loading, *Engng Fracture Mech.*, **22**, 99–122.

(6) DILL, H. D. and SAFF, C. R. (1976) Spectrum crack growth prediction based on crack surface displacement and contact analysis, *ASTM STP 595* (American Society for Testing and Materials), pp. 306–319.

(7) DUGDALE, D. S. (1960) Yielding of steel sheet containing slits, *J. Mech. Phys. Solids*, **8**, 100–104.

(8) NEWMAN, J. C., Jr. (1976) A finite element analysis of fatigue crack closure, *Mechanics of Crack Growth, ASTM STP 590* (American Society for Testing and materials), pp. 281–301.

(9) SOCIE, D. F. (1973) Prediction of fatigue crack growth in notched members under variable amplitude loading histories, *Engng Fracture Mech.*, **9**, 849–865.

(10) OHJI, K., OGURA, K., and OHKUBA, Y. (1975) Cyclic analysis of a propagating crack and its correlation with fatigue crack growth, *Eng. Fracture Mech.*, **17**, 457–464.

(11) MORRIS, W. L., JAMES, M. R., and BUCK, O. (1981) Growth rate models for short surface cracks in Al 2219-T851, *Metall Trans*, **12A**, 57–64.

(12) ZIEGLER, H. (1959) A modification of Prager's hardening rule, *Q. Appl. Math.*, **XVII**, 55–65.

(13) SEHITOGLU, H. (1983) Fatigue life prediction of notched members based on local strain and elastic plastic fracture mechanics concepts, *Engng Fracture Mech.*, **18**, 609–621.

(14) HAMMOUDA, M. M., SMITH, R. A., and MILLER, K. J. Elastic–plastic fracture-mechanics for initiation and *R* of notch fatigue cracks, *Fatigue Engng Mater. Structures*, **2**, 139–154.

(15) HUDAK, S. J., Jr. (1981) Small crack behaviour and the prediction of fatigue life, *J. Engng Mater. Technol.*, **103**, 26–35.

(16) ABAQUS, *Theory Manual*, Version 4, 1982 (Hibbit, Karlson, and Sorenson, Providence, RI).

(17) RICE, J. R. and TRACEY, D. M. (1973) Computational fracture mechanics, *Numerical and computer methods in structural mechanics* (Edited by S. J. Fenves, N. Perrone, A. R. Robinson, and W. C. Schnobrich), (Academic Press, New York), pp. 585–623.

(18) KRIEG, R. D. and KRIEG, D. B. (1977) Accuracies of numerical solution methods for the elastic–perfectly plastic model, *J. Pressure Vessel Technol.*, **99**, 510–515.

(19) SCHREYER, H. L., KULAK, R. L., and KRAMER, J. M. (1979) Accurate numerical solutions for elastic–plastic models, *J. Pressure Vessel Technol.*, **101**, 226–234.

(20) McCLUNG, R. C. and SEHITOGLU, H. (1986) Closure behavior of short cracks under high strain fatigue histories, *Int. Symp. on Fatigue Crack Closure*, 1–2 May, Charleston, SC, USA.

(21) BUDIANSKY, B. and HUTCHINSON, J. W. (1978) Analysis of closure in fatigue crack growth, *J. Appl. Mech.*, **45**, 267–276.

Yves Verreman,* Jean-Paul Baïlon,† and Jacques Masounave*

Fatigue Short Crack Propagation and Plasticity-Induced Crack Closure at the Toe of a Fillet Welded Joint

REFERENCE Verreman, Y., Bailon, J.P., and Masounave, J., **Fatigue Short Crack Propagation and Plasticity-Induced Crack Closure at the Toe of a Fillet Welded Joint**, *The Behaviour of Short Fatigue Cracks*, EGF Pub. 1 (Edited by K. J. Miller and E. R. de los Rios) 1986, Mechanical Engineering Publications, London, pp. 387–404.

ABSTRACT During fatigue tests performed at constant load amplitude on stress-relieved fillet welded joints, the crack initiated from the weld toe stress concentrator, which is similar to a V-notched specimen in plane strain.

A highly sensitive crack monitoring system was developed for these tests and short cracks as small as 10–20 μm could be measured, as well as crack opening load levels. Under fully reversed loading, the fatigue fracture could be subdivided into three successive periods.

(1) No crack length greater than 20 μm was detected during the initial portion of life.
(2) A rapid crack growth was recorded, together with an absence of crack closure. The crack growth rate decreased and remained nearly constant, during which time the crack opening level increased and then tended to a stabilized positive value.
(3) When the crack opening level was stabilized, the crack growth rate invariably increased until final fracture; this corresponding to typical long crack behaviour.

Short crack propagation behaviour (period 2) is well correlated with long crack behaviour on the basis of the effective stress intensity factor range. Crack lengths, after stabilization of the crack opening level, increase with nominal applied stress, and are well correlated with notch cyclic plastic zone sizes computed by finite element analysis. Reasons why short crack behaviour is exclusively controlled by notch plasticity are discussed.

Under zero-to-tension loading the short crack effect is barely observed because of a smaller crack closure variation.

Nomenclature

a	Crack length
a_{pz}	Notch plastic zone extent
da/dN	Crack growth rate
h	Height of gauge centre from fracture plane
K	Stress intensity factor
K_{min}, K_{max}	Minimum, maximum value of K during a loading cycle
R	S_{min}/S_{max}
S	Nominal stress
S_{min}, S_{max}	Minimum, maximum value of S during a loading cycle

* Industrial Materials Research Institute, National Research Council Canada, 75, De Mortagne, Boucherville, Québec, Canada J4B 6Y4.
† Department of Metallurgy Engineering, Ecole Polytechnique, University of Montreal, C.P. 6079 Succursale A, Montréal, Québec, Canada H3C 3A7.

S_{op}	Crack opening stress level
α	Singularity of a V-notch stress field
ΔK	Stress intensity factor range
ΔK_{th}	Conventional ΔK threshold
$\Delta K_{eff,th}$	Effective ΔK threshold
θ	Weld toe angle
σ_{yc}	Cyclic yield stress

Introduction

Many experimental investigations have clearly demonstrated that linear elastic fracture mechanics (LEFM), i.e., the ΔK parameter, cannot account for the behaviour of short fatigue cracks. Whatever the initiation site geometry (notch root or smooth surface), the following 'anomalies' of short cracks are often reported: (i) faster growth than long cracks at the same ΔK value; (ii) growth below the conventional long crack threshold, ΔK_{th}; and (iii) short cracks can grow at a decreasing rate, and possibly arrest; so-called non-propagating cracks are observed near the endurance limit level.

This non-LEFM propagation behaviour was first studied for cracks emanating from notches, and it was proposed that the notch plasticity is responsible (1)(2), since a newborn crack, which is completely surrounded by the notch plastic zone, exceeds the conditions of LEFM analyses and such a crack can be defined as a 'mechanically short crack' (3). The notch plasticity influence was well substantiated by Leis (4); he found, for a variety of notches, materials and nominal stress levels, a 'one-to-one correspondence' (in log–log coordinates) between the plastic zone extent and the crack length at the transition to LEFM behaviour. However, the comparison, which covered the range 50 μm–10 mm, revealed about half a decade of scatter. Moreover, experimental data were limited to fully reversed loading ($R = -1$) and little data is available for $R \neq -1$ when monotonic and cyclic plastic zones are not of the same size (5).

Ohji *et al.* (6) and, later, Newman (7) performed numerical analyses in order to simulate the plasticity-induced closure of short cracks growing at notches under fully reversed loading (a 60 degree V-notch and a circular notch, respectively). Both analyses revealed an initial transient variation of the crack opening level: S_{op} starts from $S_{min} = -S_{max}$ at very short crack lengths, rapidly increases as the crack tip moves away from the notch root, and finally tends to a stabilized slightly positive value given by $S_{op}/S_{max} \approx 0.10$–$0.20$. Such variation explains qualitatively the short crack propagation behaviour (5). For example, although K_{max} increases with crack length, the effective crack driving force, ΔK_{eff} decreases because of the rapid and important increase of S_{op}; which causes short cracks to arrest at low nominal stresses when ΔK_{eff} decreases below an effective threshold $\Delta K_{eff,th}$ (6). The transient variation was reported to be less pronounced for a larger radius at the V-notch root for which case ΔK_{eff} continuously increased. This is also in general agreement with experiments which show that the growth rate only decreases if the geometry is severe enough (**2**).

Although previous analyses show that plasticity-induced crack closure can explain the short crack propagation behaviour, they do not show if it is due to notch plasticity. Moreover, crack closure values have not yet been experimentally calibrated or verified since conventional techniques for monitoring crack opening (compliance, potential drop) are inappropriate to short cracks for which high resolution is needed (**8**).

A number of studies performed within the last years have shown that short crack growth at smooth surfaces presents the same anomalies as mentioned above. However, a fundamental difference is that the crack growth rate decreases, and possible arrests, are due to microstructural barriers (e.g., grain boundaries) during the initial crystallographic cracking stage, or at a transition to non-crystallographic cracking (**9**). This 'microstructurally short crack' behaviour cannot be explained by continuum mechanics, especially since it appears to be characterized by transient retardation(s) and re-acceleration(s) from a 'mean', and possibly a continuously increasing growth rate. El Haddad *et al.* have reported that such a 'less anomalous' short crack growth could be successfully described by a mechanical parameter (**2**). Although multiple microstructural interactions are sometimes reported, in most cases such retardation(s) take place at the very first grain boundary, while transition to LEFM behaviour often occurs at a larger scale (**9**)(**10**).

Several arguments have been put forward to explain the existence of a mechanical component in short crack behaviour at smooth surfaces; for example: (i) the loss of the inverse square root singularity (**11**); (ii) closure-based arguments which includes an initial absence of Elber's mechanism in the crack wake (**12**), or of rugosity-induced crack closure (**8**) which can be considered as a microstructural potential effect; (iii) an initial absence of a surrounding elastic medium in which the crack tip plastic zone is fully contained and constrained (**11**).

Regarding the specific problem of short cracks growing at notches, an exclusive control by notch plasticity appears somewhat doubtful when considering the above microstructural and mechanical effects. Furthermore, the experimental and theoretical information previously mentioned does not allow an exact evaluation of notch plasticity effects on short crack propagation and closure. However, the present investigation shows that short crack behaviour is exclusively controlled by notch plasticity when some conditions are fulfilled.

Experimental conditions

This paper presents some results of a project of the fatigue of automatic welded joints (**13**). A cruciform welded joint configuration was selected (Fig. 1). Fatigue tests were performed at constant amplitude loading in the X direction ($R = -1$ and R = 0). The crack propagated from one weld toe in the Y–Z

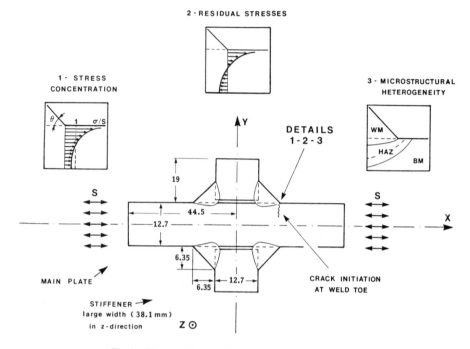

Fig 1   The cruciform welded joint (dimensions in mm)

plane as shown in Fig. 1; the three other weld toes were machined to give a smooth geometry. The major specimen dimensions were; main plate and stiffener thickness, 12.7 mm and specimen width, 38.1 mm.

The base metal was ASTM A36 steel and the weld metal AWS-CSA E70S-7 steel. Welding was performed by an automatic MIG process with gravity in the Y direction. This gave the weld toe an ideal V-shaped geometry, as schematically drawn in Fig. 1, and a straight weld bead in the Z direction. The weld toe stress concentration fulfils three conditions: (i) severity – the curvature radius at the weld toe apex is about 50 μm (or less); (ii) triaxiality, due to severity and large dimensions in the Z direction which impose a plane strain state; (iii) uniformity in the Z direction.

Fatigue at the weld toe is also related to the residual stress field and the heterogeneous microstructure due to welding. The residual stress influence (13)(14) is beyond the scope of this paper, which is only concerned with the stress-relieved state of the welded joint obtained after an appropriate heat treatment. The microstructure grain size of the heat affected zone is generally fine ($\simeq 5$ μm), while the unaffected base metal (Fe and Fe_3C), which extends from about 1.5 mm below the weld toe, has elongated grains (10–50 μm). Finally the coarse-grained HAZ (which extends to 0.3–0.5 mm in depth) prior austenite grain size is also about 10–50 μm.

Uniform through-cracks have been systematically observed, even at very small crack depths and low nominal stresses. This is mainly due to weld toe unformity in the Z direction, but notch macro-plasticity also bears some responsibility. The overall result is that nearly 90 per cent of the crack propagation life is consumed within the first millimeter of crack growth, and so short crack growth represents a large portion of the fatigue life for those automatic welds. This led us to develop an efficient crack monitoring system.

Crack length and opening level monitoring system

The main features of the system are schematically presented in Fig. 2. When the crack initiates, the response of a strain gauge installed on the main plate near the weld toe is such that there is a progressive deflection of the upper part of the recording. This is due to the fact that the stress flow lines bypass the tip of the propagating crack when it opens.

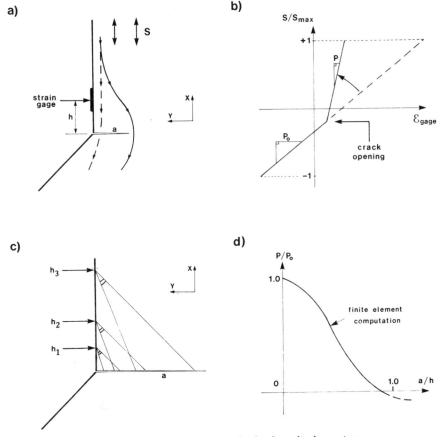

Fig 2   The crack length and opening level monitoring system

The gauge response ratio, P/P_o is calibrated against crack length a by finite element computation (Fig. 2(d)) which depends on the height, h, of the gauge centre from the fracture plane. The response saturates when $a \simeq h$. The first gauge is progressively replaced by others (Fig. 2(c)), which are less sensitive but which cover longer crack lengths.

This system has many advantages. For example it is simple, very sensitive, and accurate. The first gauge, which is installed as close as possible to the fracture plane, easily detects crack lengths of 10–20 μm, and the theoretical calibration, which includes an integration over the 350 μm gauge length, has been verified by 14 marks, obtained via overload or china ink techniques, which were distributed among seven specimens submitted to different loading conditions. The relative error is within 5 per cent over the calibrated crack lengths which ranged from 70 μm to 5 mm. Secondly the ratio P/P_o is independent of many parameters, such as strain gauge calibration, Young's modulus, Poisson's ratio, and the angle θ at the weld toe. Thirdly the system allows monitoring of the crack opening level from a $\simeq 10$–20 μm with an absolute error within ± 6 MPa.

From the results obtained in this investigation, the system, which can be used in many other situations, would appear to be suitable for studying the propagation and closure behaviour of short fatigue cracks.

Elasto-plastic finite element analysis

Some theoretical features are briefly mentioned here because they allow a better physical interpretation of the experimental results. Further details are presented in references (5) and (13).

Except for a very small zone, 6 μm in depth, which is influenced by the 50 μm radius at the weld toe apex, the stress distribution is linear on a log–log plot (Fig. 3), and the slope is identical to that of a V-notch with the same θ angle (15). In other words, the stress field is singular, and the singularity, α (defined as the absolute value of the slope of Fig. 3), varies from 0.5 in the case of a crack ($\theta = 180$ degrees), to 0 in the case of a smooth surface ($\theta = 0$ degrees). However, α decreases only slightly when θ decreases from 180 degrees: $\alpha = 0.407$ for $\theta = 90$ degrees (5), and $\alpha = 0.333$ for $\theta = 45$ degrees (Fig. 3). As a consequence, there is a quasi-similitude between the stress field ahead of a V-notch and the one ahead of a crack tip, i.e., ahead of the most severe notch.

Figure 4 shows the finite element mesh (straight lines) and the contours of the notch cyclic plastic zones which are obtained under fully reversed loading for different nominal stresses, ranging, approximately, from the endurance limit stress to general plasticity; the nominal (cyclic) yield stress chosen for the computations is well representative of the experimental behaviour. Such a directional macro-plasticity, which is more pronounced for $\theta = 90$ degrees is not surprising when considering the quasi-similitude with a crack (in plane strain). This indicates that the severity, plus the triaxiality ahead of the notch, can be sufficient to initially force a newborn crack to propagate along the plane

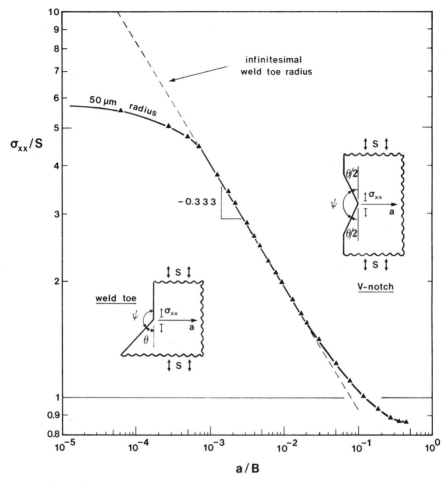

Fig 3   The elastic stress distribution at the weld toe ($\theta = 45$ degrees; $\alpha = 0.333$; $B = 12.7$ mm = plate thickness)

of maximum tensile stress. Hence one can expect there is neither an initiation stage nor a microstructurally short crack propagation stage. This was experimentally verified since the crack initiation life (as detected by the first gauge) was always short even at the endurance limit level and striations were found very close to the edge of the fracture surface.

The elasto-plastic finite element analysis determines the notch plastic zone extent a_{pz} (Fig. 4) as a function of the nominal stress. The computations show that one can expect an important effect of notch plasticity on short crack propagation behaviour since the newborn crack is completely surrounded by a plastic zone which is still about 100 μm in depth and 500 μm in width at the endurance limit level. This is due to the fact that, even in plane strain, the small root radius is insufficient to moderate the V-notch severity.

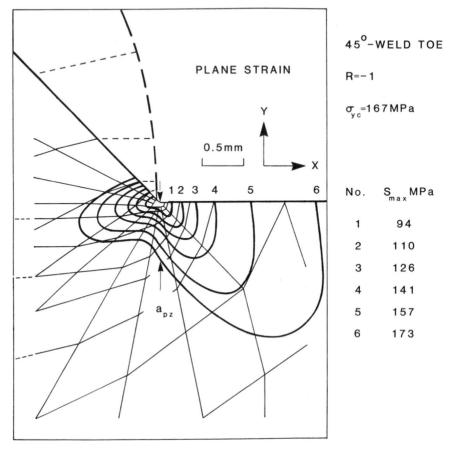

Fig 4   Cyclic plastic zones at the weld toe (kinematic hardening)

Experimental results

Figure 5(a) represents a typical evolution of crack length vs cycle life recorded under fully reversed loading below the nominal cyclic yield stress (σ_{yc} = 167 MPa). The fatigue fracture can be divided into three successive periods.

(1)   The first gauge does not detect any variation greater than 20 μm, but life at this detection level never exceeds 10 per cent of total life.

(2)   A rapid crack growth is recorded (2a), then the crack growth rate decreases by one order of magnitude and remains nearly constant (2b); LEFM cannot account for this short crack behaviour since the factor K increases with crack length.

(3)   The crack growth rate invariably increases until final fracture and is now compatible with LEFM.

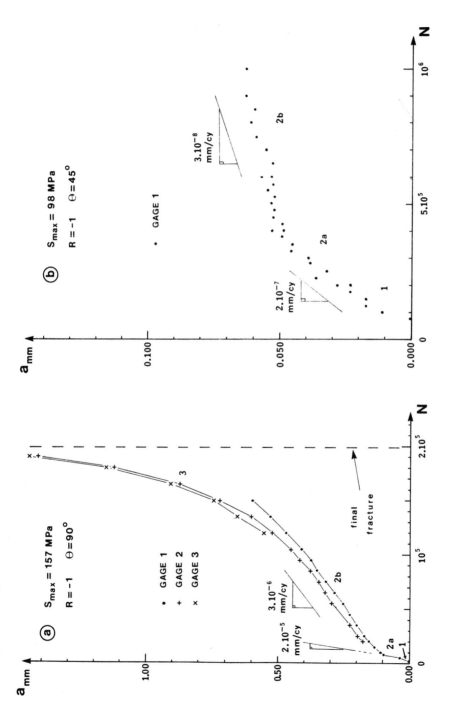

Fig 5   Crack length versus cycle life: (a) below nominal yield stress; (b) at endurance limit level

A similar behaviour is still observed at the endurance limit level (Fig. 5(b)), though crack growth rates differ from Fig. 5(a) by two orders of magnitude. Period 3 is not apparent here because of a power failure after 10^6 cycles. The rate of change of crack length in sub-period 2(b) is very small and may be due to a drift in the electronics with time. Hence, this crack may become non-propagating.

Crack growth data from five specimens subjected to different nominal stress levels are reported in Fig. 6 on a conventional da/dN versus K_{max} plot. As expected, crack growth rates corresponding to part 3 (open symbols in Fig. 6) are well correlated with K_{max} (dashed curve), while short crack growth rates (solid symbols) exhibit important discrepancies from this long crack trend. Behaviour above the nominal yield stress (square symbols of Fig. 6) is not considered here. The major discrepancy occurs at the endurance limit level where cracks grow below the conventional long crack threshold, and can become non-propagating ($S_{max} = 98\,\mathrm{MPa}$; Fig. 6).

Two factors concerning the effect of notch plasticity on short crack behaviour are: (i) crack length of transition to LEFM behaviour increases with nominal stress and is comparable in size with the extent of the notch plastic zone computed by finite element analysis. This length cannot be correlated with any microstructural parameter such as grain size; (ii) the short crack growth rate systematically decreases except above the nominal yield stress. Indeed the 'initial crack driving force' produced by the V-notch is likely to decrease rapidly because of a severe plastic strain gradient. the fact that the same short crack behaviour is still observed at the endurance limit level can be attributed to the V-notch severity. However, in this case, the notch plasticity effect can vanish before the long crack threshold is reached (Fig. 6), i.e., before the crack can grow by its own plasticity in the LEFM regime.

Generally speaking, crack growth data show an important influence of notch plasticity on short crack behaviour whatever the nominal stress level. Notch plasticity not only promotes immediate crack propagation along the plane of maximum tensile stress, but also has an effect over a crack length up to 500 μm at high nominal stresses (Fig. 5(a)). However, comparison with theory can only be semi-quantitative. For example, the crack length at transition to LEFM behaviour cannot be clearly determined because the crack growth rate is nearly constant over an important distance (Fig. 5(a)). In other words crack growth data alone do not allow strict quantitative evaluation of the effect of notch plasticity and other parameters related to short crack behaviour.

The results relative to the variation of the crack opening level with crack length (Fig. 7) are in good agreement with computations from Ohji and Newman (6)(7). Moreover, they show that the initial transient variation is related to notch plasticity. Note that in this figure $2U = 1 - S_{op}/S_{max}$ and that even at the endurance limit, the slope da/dU of the initial transient variation is proportional to the computed plastic zone extent. Further, that the crack length at which the opening level stabilizes is approximately equal to this extent. This was confirmed by two other tests with a weld toe angle θ of 90 degrees.

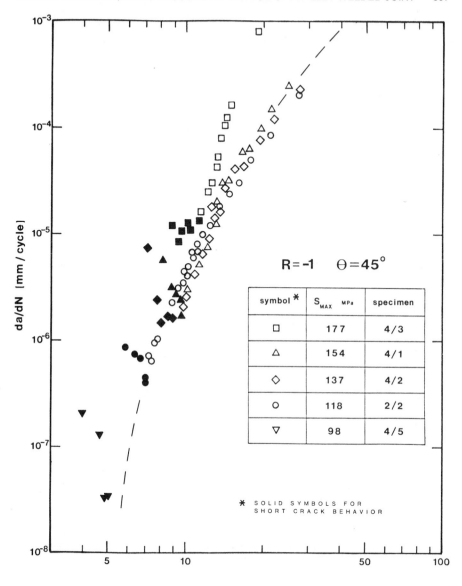

$$R=-1 \quad \Theta=45°$$

symbol *	S_{MAX} MPa	specimen
□	177	4/3
△	154	4/1
◇	137	4/2
○	118	2/2
▽	98	4/5

* SOLID SYMBOLS FOR SHORT CRACK BEHAVIOR

Fig 6   Crack growth rate versus K_{max}

As may be expected, correction of raw crack growth data with corresponding opening levels (Fig. 7) leads to a much stronger relation as witnessed on a da/dN versus ΔK_{eff} plot (Fig. 8). More specifically, the short crack propagation behaviour is quantitatively rationalized by opening level transient variations inside the notch plastic zone. One will note that the knee at low growth rates seen on Fig. 6 disappears on the da/dN vs ΔK_{eff} plot.

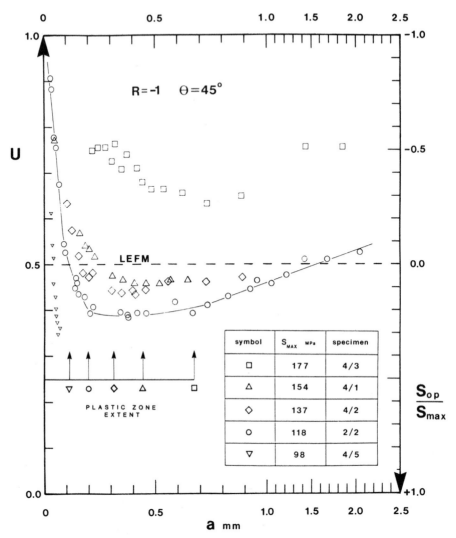

Fig 7 Crack opening level versus crack length. (Note that the crack length scale is linear, but is reduced above 1 mm)

Discussion

The fact that short and long crack propagation data are both correlated by ΔK_{eff} is a further confirmation that the short crack behaviour observed in this investigation is purely 'mechanical'. In addition, the systematic correspondence between the crack length at which the opening level becomes stabilized and the computed plastic zone extent (Fig. 7) shows that notch plasticity is totally responsible for the short crack (closure and propagation) behaviour reported here.

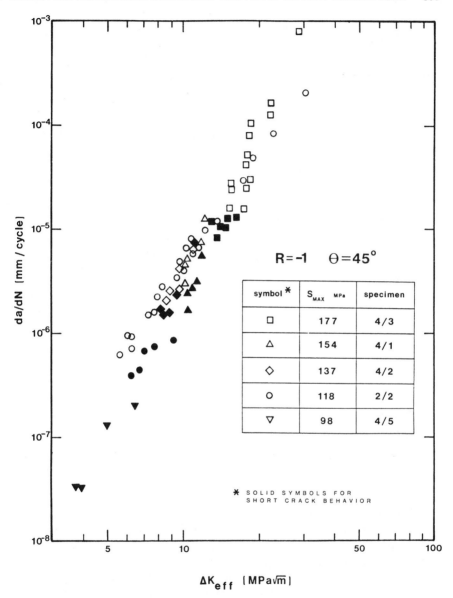

Fig 8   Crack growth rate versus ΔK_{eff}

The absence of closure ($S_{op} < 0$; Fig. 7) is a direct consequence of the notch plasticity which is ahead of the crack tip; note that an initial absence of Elber's mechanism in the crack wake cannot explain negative S_{op} values. The total absence of closure when the crack is starting (Fig. 7) is in agreement with the computations of Ohji and Newman. Although the decrease of U is slightly convex, it is sufficiently large and abrupt for ΔK_{eff}, and thus the crack growth

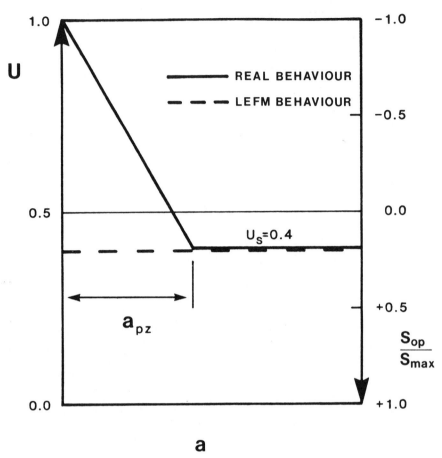

a

Fig 9   Idealization of the crack opening level variation at $R = -1$

rate, to decrease, and the possibility of a non-propagating crack at low nominal stresses is explained if the existence of an effective threshold $\Delta K_{\text{eff,th}}$ is assumed.

Variations of S_{op} are much smaller in the 'quasi-LEFM' regime outside the notch plastic zone (Fig. 7). The fact that the stabilized value of $S_{\text{op}}/S_{\text{max}}$ slightly increases when S_{max} decreases is in agreement with other computations by Newman (16), which show that Elber's mechanism in the crack wake has a maximum effect at low nominal stresses. This increase corresponds to a less effective stress intensity factor range, which explains the disappearance of the knee in lower part of the da/dN versus ΔK_{eff} plot. Reappearance of negative S_{op} values when approaching general plasticity (Fig. 7) is also in agreement with Newman's computations: plasticity ahead of the crack tip overcomes plasticity behind the crack tip, and removes closure. Note that general plasticity has already an important effect on short crack behaviour when nominal stress

Fig 10 **Straightness of the almost non-propagating crack front ($da/dN = 3 \cdot 10^{-8}$ mm/cycle; Fig. 5(b))**

exceeds yield stress (square symbols; Fig. 7), and in this particular case, although validity of the ΔK_{eff} parameter is somewhat doubtful, the much slower transient variation of the opening level inside the notch plastic zone explains why the short crack growth rate does not decrease (Fig. 6).

From a practical point of view, the variations of the crack opening level in the 'quasi-LEFM' regime outside the notch plastic zone and below the nominal yield stress are either not very significant with respect to scatter in the da/dN versus ΔK_{eff} plot or they concern a very small portion of the total fatigue life; note that crack length scale is reduced above 1 mm in Fig. 7. As a consequence, one can idealize the dependence of U on the crack length by a linear decrease inside the notch plastic zone, followed by a single stabilized value outside it (Fig. 9). This confirms our assumptions (**5**) proposed to estimate the resulting errors in fatigue life prediction when the notch plasticity effect is not taken into account. Consequently a computerized life prediction model using the ΔK_{eff} parameter has been developed on the basis of such an idealization of real behaviour (**14**).

The fundamental reasons which allow such an idealization whatever the nominal stress (except for general plasticity) are believed to be as follows.

(1)   Newman's computations (**16**) show that the opening level of long cracks strongly depends on the plastic constraint factor, especially at low R ratios. Absence of triaxiality (e.g., plane stress) amplifies the competition between plasticity behind the crack tip (involving closure by Elber's mechanism) and plasticity ahead of the crack tip (involving absence of closure). As a consequence U rapidly increases with S_{max}. However, in plane strain, which is the present situation, triaxiality is sufficient to induce a quasi-stabilized value of U whatever the nominal stress (Fig. 7).

(2)   Near the endurance limit level, it is well known that growing short cracks have a natural tendency to present a semi-elliptical crack front shape (more rigorously, they should be called 'small cracks' (**9**)). This was not the case in the present investigation where, even at the endurance limit level, uniform through-cracks have been systematically observed all along the weld bead. Figure 10 shows the fracture surface of the specimen where the crack was non-propagating (zone A): the crack is 100 μm in depth while it is 38.1 mm in width. This extreme situation must be attributed to the V-notch severity. As previously shown, notch plasticity is at a macroscopic scale, and the high degree of triaxiality tends to activate numerous slip systems everywhere along the crack width; this also contributes to the regulation of the crack front, and probably explains why short crack (propagation and closure) behaviour is similar to the one observed at higher stresses.

Another practical consequence follows from this investigation. The V-notch macro-geometry is such that crack propagation occurs very early even at the endurance limit level (Fig. 5). As a consequence, small pre-existing crack-like defects at the weld toe (which are seldom observed here) would not represent a critical situation. Their influence on total fatigue life should almost equate to the cycle life necessary for producing an initial crack of the same length. Crack growth curves in Fig. 5 show that this life is negligible up to $a = 100$–$150\ \mu$m at a high nominal stress (Fig. 5(a)), and at least $a = 50\ \mu$m at the endurance limit level (Fig. 5(b)). In this latter case, the presence of an initial defect of 50 μm should not influence the endurance limit since a crack of the same length becomes non-propagating (Fig. 5(b)).

Finally, other finite element computations and fatigue tests have been performed under zero-to-tension loading ($R = 0$) and these can be summarized as follows. (a) For the same maximum nominal stress, S_{max}, the notch monotonic plastic zone has the same size as under fully reversed loading; however the cyclic zone is smaller by a factor of from 5 to 8 depending on the weld toe angle θ. (b) Although reversed yielding is less pronounced, notch severity is such that crack initiation life is still short when compared to total life; however, anomalous (non-LEFM) crack growth rates are barely observed. (c) The variations of crack opening level with crack length are smaller than under fully reversed loading; this confirms the above observations on crack propagation behaviour; S_{op}/S_{max} varies from 0 to nearly 0.35 at a crack length corresponding to the cyclic plastic zone extent. It then decreases to nearly 0.2 at a crack length corresponding to the monotonic plastic zone extent. Subsequent crack growth occurs with an opening level stabilized at the last value, except when approaching general plasticity. (d) From a practical point of view, one can assume a LEFM behaviour with S_{op}/S_{max} equal to 0.2 whatever the crack length.

Conclusions

Fatigue fractures of V-notched members in plane strain have the following characteristics under fully reversed loading, even at the endurance limit level.

(1)   The crack initiation stage is very short and can be neglected and the pre-existence of small crack like defects is not a critical condition.
(2)   The microstructurally short crack propagation stage appears to be non-existent.
(3)   Short crack behaviour is exclusively controlled by notch plasticity (excepting general plasticity).
(4)   Short crack growth rates are well correlated with long crack rates when using the effective stress intensity factor range.
(5)   Short crack closure behaviour is characterized by an initial transient variation within the notch plastic zone.
(6)   A high degree of notch severity and triaxiality allows a simple idealization for fatigue life prediction.

Under zero-to-tension loading, the same conclusions hold; however, the short crack effect is barely pronounced because of a smaller crack closure variation.

Acknowledgements

The authors are grateful to the Natural Science and Engineering Research Council for its partial financial support. They would also like to thank the reviewers for their useful comments.

References

(1)   SMITH, R. A. and MILLER, K. J. (1978) Prediction of fatigue regimes in notched components, *Int. J. Mech. Sci.*, **20**, 201–206.
(2)   EL HADDAD, M. H., SMITH, K. N., and TOPPER, T. H. (1979) A strain-based intensity factor solution for short fatigue cracks initiating from notches, *ASTM STP 677*, pp. 274–289.
(3)   SCHIJVE, J. (1984) The practical and theoretical significance of small cracks. An evaluation, *Proceedings of the Second International Conference on Fatigue and Fatigue Thresholds*, Vol. II, pp. 751–771.
(4)   LEIS, B. N. (1982) Fatigue crack propagation through inelastic gradient fields, *Int. J. Pressure Vessels Piping*, **10**, 141–158.
(5)   VERREMAN, Y., BAÏLON, J. P., and MASOUNAVE, J. (1986) Fatigue life prediction of welded joints – a reassessment, To be published.

(6) OHJI, K. *et al.* (1975) Cyclic analysis of a propagating crack and its correlation with fatigue crack growth, *Engng. Fracture Mech.*, **7**, 457–464.

(7) NEWMAN, J. C., Jr. (1982) A nonlinear fracture mechanics approach to the growth of short cracks, *Proceedings of AGARD Specialists Meeting on Behaviour of Short Cracks*, Toronto, Canada.

(8) SURESH, S. and RITCHIE, R. O. (1984) Propagation of short fatigue cracks, *Int. Met. Rev.*, **29**, 445–476.

(9) LANKFORD, J. (1985) The influence of microstructure on the growth of small fatigue cracks, *Fatigue Fracture Engng Mater. Structures*, **8**, 161–175.

(10) LANKFORD, J. (1982) The growth of small fatigue cracks in 7075–T6 aluminum, *Fatigue Engng Mater. Structures*, **5**, 233–248.

(11) ALLEN, R. J. and SINCLAIR, J. C. (1982) The behaviour of short cracks, *Fatigue Engng Mater. Structures*, **5**, 343–347.

(12) LeMAY, I. and CHEUNG, S. K. P. (1984) Crack closure effects for short cracks in notched aluminum and steel plates, *Proceedings of Fatigue 84*, Vol. II, p. 677.

(13) VERREMAN, Y. (1985) *Comportement en fatigue des joints soudés automatiques*, PhD thesis, Ecole Polytechnique de Montréal, Montréal, Canada.

(14) VERREMAN, Y., BAÏLON, J. P., and MASOUNAVE, J. (1985) Fatigue life prediction of welded joints using the effective stress intensity factor range. To be presented at ASM Conference on Fatigue, Corrosion Cracking, Fracture Mechanics and Failure Analysis, Salt Lake City, Utah, USA.

(15) USAMI, S. *et al.* (1978) Cyclic strain and fatigue strength at the toes of heavy welded joints, *Trans Jap. Weld. Soc.*, **9**, 118–127.

(16) NEWMAN, J. C. Jr (1981) A crack closure model for predicting fatigue crack growth under aircraft spectrum loading, *ASTM STP 748*, pp. 53–84.

FRACTURE MECHANICS

F. Guiu* and R. N. Stevens*

Thermodynamic Considerations of Fatigue Crack Nucleation and Propagation

REFERENCE Guiu, F. and Stevens, R. N., **Thermodynamic Considerations of Fatigue Crack Nucleation and Propagation**, *The Behaviour of Short Fatigue Cracks*, EGF Pub. 1 (Edited by K. J. Miller and E. R. de los Rios) 1986, Mechanical Engineering Publications, London, pp. 407–421.

ABSTRACT Attention is drawn to two paradoxes in the problem of nucleation and propagation of fatigue cracks. One is the fact that an external stress system alone cannot provide the driving force for the nucleation of a crack on a well polished specimen surface since there is a considerable energy barrier to the growth of a crack of virtually zero length. This results in the inability of fracture mechanics to describe such events. Fatigue cracks can only nucleate at the expense of strain energy stored in the material during stress–strain cycling and they are, therefore, formed at regions where, due to material inhomogeneities, large internal strain fields are set up. The existence of non-propagating fatigue cracks is the consequence of this thermodynamic requirement and an excellent example of this is provided by the nucleation of cracks at persistent slip bands.

Another paradox appears when fracture mechanics concepts are applied to the problem of fatigue crack growth, and it refers to the observation that fatigue cracks grow under values of K_{max} lower than the critical value K_{IC} measured in monotonic stressing. This 'thermodynamic impossibility' is difficult to understand by invoking plasticity effects at the crack tip and by modifications to LEFM because, in the context of this theory, localized plastic deformation would be expected to make propagation more difficult, not easier.

It is suggested that the paradox can be resolved by noting that the instability criterion of 'maximum free energy change' for crack propagation is a *necessary* condition only and not a *sufficient* one, since the plastically relaxed stress levels at the crack tip must reach values capable of producing atomic bond rupture. With this consideration in mind the low values of K_I measured in the growth of fatigue cracks can be explained by the existence of processes which can produce atomic bond rupture under stresses lower than those needed to break bonds in tension.

Introduction

The mechanism of nucleation and growth of small cracks on a smooth surface is a central issue in the problem of fatigue failure. Many cracks may be formed on the surface of a material during its fatigue life, but most of these stop growing and only a few become the propagating cracks which will ultimately cause fatigue fracture (1)(2). It is, therefore, of great practical importance to be able to predict the growth behaviour of short cracks, but efforts towards this goal have been fraught with difficulties because it is claimed that linear elastic fracture mechanics (LEFM) methods are not applicable to this problem (3). The conventional way of representing data on fatigue crack growth is illustrated schematically in Fig. 1 where the crack length increment per cycle, da/dN, is plotted against the alternating stress intensity factor, ΔK ($= K_{max} - K_{min}$) in double logarithmic coordinates. The linear region in this plot corresponds to

* Department of Materials, Queen Mary College, Mile End Road, London E1 4NS.

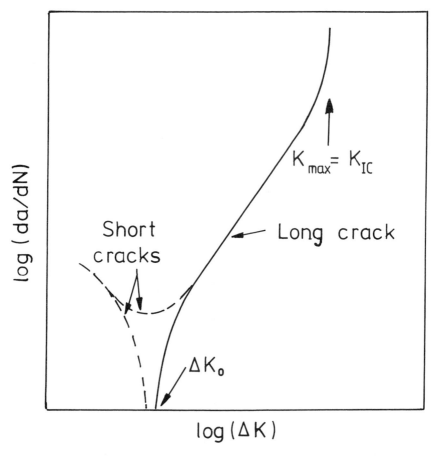

Fig 1 Double logarithmic plot of crack growth rate, da/dN against ΔK showing the normal behaviour of 'long' cracks (full line) and the anomalous behaviour of 'short' cracks (broken lines)

the stable growth rate of fatigue cracks as described by the Paris–Erdogan relation (**4**),

$$\frac{\mathrm{d}a}{\mathrm{d}N} = C \Delta K^n \tag{1}$$

where C and n are material constants. The unstable crack propagation regime is approached as K_{max} tends to K_{IC}, the critical value of the stress intensity factor. Below the threshold value, ΔK_0, fatigue cracks grow at undetectably small rates and appear to remain dormant. The anomalous behaviour of 'short' cracks is represented by the broken lines of Fig. 1 and attempts have been made to explain and quantify this anomalous behaviour in terms of local plasticity effects, microstructural and environmental factors, and by using empirical elastic–plastic constitutive laws (3). It seems that most investigations into the problem of nucleation and growth of physically short cracks fail to recognize the real reason for the limitations of conventional LEFM methods and continuum mechanical approaches.

If crack growth is to occur at all, in either static or dynamic loading conditions, there must be a positive driving force, i.e., growth of the crack must reduce the energy of the system. Since crack growth creates new surfaces, thereby increasing the energy of the system, energy must be extracted from the external loading system or from the general strain energy stored in the body if a net energy decrease is to be achieved. The essential idea of the classical Griffith theory (5) is that this results in a critical length of crack below which the driving force for growth is negative. It is, therefore, paradoxical that cracks can grow from virtually zero length in fatigue conditions. The use of both fracture mechanics and continuum mechanics formulated in terms of the stress intensity factor, ΔK, and critical strain or displacement criteria tends to obscure this difficulty, but if the physics of fatigue crack growth are to be understood the problem must be addressed.

Another paradox arises when we consider the stable growth of 'long' fatigue cracks within a range of K_{max} values smaller than K_{IC}. This paradox becomes more evident when we consider the modifications to the Griffith theory proposed by Orowan (6) and Irwin (7). In this modification crack growth is supposed to increase the energy of the system by an amount greater than the energy of the new surfaces because of plastic deformation. A greater critical length of crack is predicted by this modification than by the original theory. But the slow, cyclic growth of fatigue cracks occurs below this modified limiting length (i.e., at values of $K_{max} < K_{IC}$) and the accumulated evidence is that plastic deformation is essential to this growth process in spite of the fact that plastic deformation is supposed to impose an additional energy penalty on crack growth.

This paper aims to consider these two paradoxes and point to the way in which they can be resolved. In the process it is hoped that some progress will be made towards a better understanding of the physics of fatigue fracture. Clearly, a theory implying that crack growth violates the laws of thermodynamics is unlikely to provide much gain in our understanding of the processes involved, and classical fracture mechanics comes into this category if improperly applied to the initiation and growth of fatigue cracks.

The driving force for crack growth

The classical theory of fracture originated by Griffith (5) is a thermodynamic theory. Unlike most of the applications of thermodynamics, the equilibrium is unstable rather than stable, and the appropriate thermodynamic potential has a maximum rather than a minimum, the maximum giving the instability criterion.

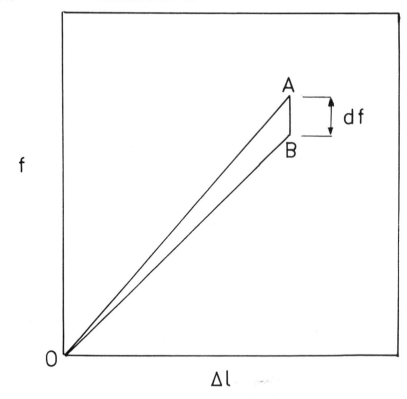

Fig 2   Load elongation curves for a plate with a crack of length a (OA) and for a plate with a crack of length $a + da$ (OB), when no plastic deformation has occurred

Consider a system consisting of a plate of length l and width w under a longitudinal tension and having an edge crack, length a, normal to the tension direction. If crack growth occurs with the system held at fixed length (the fixed grips condition) the Helmholtz energy F of the system is at a maximum value at the critical crack length. If, on the other hand, the tensile force is held constant (the constant load condition) then the thermodynamic potential whose maximum indicates the unstable condition is a modified Gibbs free energy, G' ($= F - fl$, where f is the tensile force). The strain energy of the plate and the surface energy of the crack are both Helmholtz free energies. For simplicity we shall use the fixed grip condition throughout, but, of course, the same results are obtained irrespective of the external constraints.

The load–elongation curve for the plate with a crack of length a is shown as the line OA in Fig. 2. If the crack increases in length by da then the force at A drops by df and the load–elongation curve is now OB. The area OAB is the decrease in strain energy of the system. The rate of change, $\mathscr{G}$, of elastic energy, F_e, with crack length per unit thickness of plate for a specimen in the fixed grip condition is given by (**8**)

$$\mathcal{G} = -\frac{1}{t}\left(\frac{\partial F_e}{\partial a}\right)_{T,l} = \frac{\alpha\pi(1 - \nu^2)\sigma^2 a}{E} \qquad (2)$$

where t is the thickness of the plate, T is thermodynamic temperature, ν is Poisson's ratio, σ the stress remote from the crack, α a constant ~1.25 for large l and w, and E is Young's modulus. The quantity $\mathcal{G}$ is called the energy release rate, or the crack extension force. The rate of change in the total Helmholtz free energy of the system per unit thickness of plate with crack length is now found by adding the term due to the increase in surface area of the crack. The result is

$$\frac{1}{t}\left(\frac{\partial F}{\partial a}\right)_{T,l} = -\mathcal{G} + 2\gamma = -\frac{\alpha\pi(1 - \nu^2)\sigma^2 a}{E} + 2\gamma \qquad (3)$$

where γ is the surface energy. We shall call the negative of the left-hand side of equation (3), i.e., $-(1/t)(\partial F_e/\partial a)_{T,l}$, the driving force for crack growth. Clearly a positive force as defined means that crack growth will reduce the Helmholtz free energy of the system. Setting the right-hand side of equation (3) to zero yields the well known result for the critical crack length, a_0

$$a_0 = \frac{2\gamma E}{\alpha\pi(1 - \nu^2)\sigma^2} \qquad (4)$$

It is worth observing that the strain energy of the system falls because the average stress in the plate diminishes as the crack grows. The energy motivating crack growth therefore comes from the whole system and not from the region around the crack tip as is sometimes supposed.

Real materials do not appear to obey equation (4). Although the stress for rapid crack propagation is inversely proportional to the square root of the crack length (providing the cracks are 'sharp'), the values of the surface energy, γ, calculated from experiment are 1 to 3 orders of magnitude higher than the thermodynamic surface energy. Orowan (6) and Irwin (7) proposed that crack growth involved not only the creation of new surfaces but also plastic deformation even in apparently brittle materials and that a plastic work term had to be added to the surface energy. This is a view widely held today and has been emphasized more recently by Weertman (9). The surface energy, γ, in equation (4) has therefore to be replaced by a quantity, γ', which is much larger and includes the plastic work done per unit area of crack surface.

If this view is accepted then there arises the problem of explaining how a crack which is sub-critical with respect to equation (4) with γ equal to γ' (the surface energy plus the plastic work) can propagate in fatigue, even if this propagation is slow and progressive rather than catastrophic. The problem is compounded by the fact that it is undoubtedly plastic deformation which allows such cracks to grow in the fatigue situation, whereas according to the Orowan/Irwin modification of the Griffith theory it is plastic deformation which renders them sub-critical.

Since the controlling factor in the propagation of a fatigue crack seems to be the localized plasticity at the crack tip, critical strain or displacement criteria are used rather than stress intensity criteria in the fracture mechanics approach to fatigue crack growth, and this makes the paradox less evident.

The role of plastic deformation

To resolve the problem raised above we reconsider the effect of plastic deformation on the energetics of crack growth. We can, to a first approximation, employ the methods already used for the classical theory. Although plastic deformation is irreversible we may still be able to evaluate the change in Helmholtz free energy. We assume that the plate is not undergoing general plastic yield and that the plastic deformation occurring is *necessarily* linked with crack growth. We shall also assume, for the purpose of envisaging the energy changes taking place, that crack growth and the accompanying plastic deformation take place sequentially. It is noted that this last assumption is not strictly necessary and does not affect the conclusions.

The energy changes are illustrated in Fig. 3. The crack length increases by d*a*

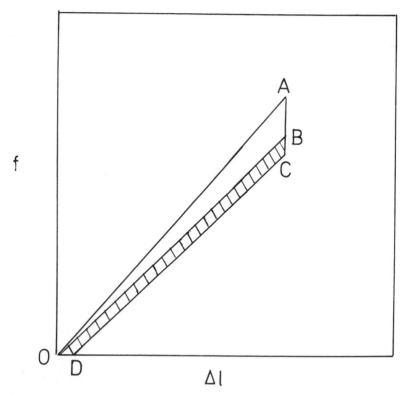

Fig 3   Load elongation curves for a plate with a crack of length *a* (OA) and for a plate with a crack of length *a* + d*a* (CO) in which plastic deformation has occurred during crack growth. The shaded area OBCD is the plastic work

under fixed grips. The change in Helmholtz free energy of the system is given by the area OAB as before. Plastic deformation causes the load to fall further from B to C. It is clear that an additional fall in load has to occur since the plastic deformation can only take place by the conversion of elastic strain into plastic strain and the degradation of some of the strain energy into heat. The unloading curve will not now pass through the origin. The additional strain energy lost is equal to the area OBCD.

The implications of this are considerable. It is clear that the sign of the energy change brought about by plastic deformation is the same as the sign of the elastic strain energy change calculated by Griffith. In other words, plastic work, if it has any effect at all on the energetics of crack growth, far from opposing it, will actually provide an increased driving force. We have, however, omitted one small complication. Not all of the energy OBCD is degraded into heat; some of it is stored as the energy of the dislocations created during the plastic deformation which is necessarily associated with crack growth. This gives a term in the energy balance equation having the same sign as the surface energy term. The net effect is to reduce the absolute value of the energy changes associated with plastic deformation somewhat below that represented by the area OBCD.

It will be noted that we have drawn the area OBCD as a rather small fraction of the total OACD. That this is the case can be shown by using an approximate argument based on the Dugdale–Barenblatt (10)(11) model for the plastic zone length at the crack tip and its effect on the compliance of the system. The plastic zone length, w_p, is given by

$$w_p = \frac{\alpha \pi^2 \sigma^2 a}{8 \sigma_y^2} \qquad (5)$$

where σ_y is the (tensile) yield stress. The effect of the plastic zone is to reduce the compliance of the specimen by an amount equivalent to increasing the crack length by a fraction β ($\sim \frac{1}{2}$) of the plastic zone length, which incidentally illustrates that plastic deformation does decrease the strain energy of the system as argued above. If we put a virtual crack length of $a(1 + \alpha\beta\pi^2\sigma^2/8\sigma_y^2)$ into the Griffith equation for the strain energy of the cracked plate we can find the rate of change of strain energy and separate out the classical term, dF_e (given by multiplying equation (2) throughout by $t\,da$), and the term due to plastic deformation. The total change, dF_e', is given by

$$dF_e' = -\frac{\alpha\pi(1 - \nu^2)\sigma^2 at}{E} \, da\left(1 + \frac{\beta\alpha\pi^2\sigma^2}{8\sigma_y^2}\right)^2 \qquad (6)$$

Evidently, when multiplied out, the first term on the right-hand side will be the classical strain energy term, dF_p and the remainder will be the changes due to plastic deformation. Hence the strain energy change dF_p due to plastic deformation at the crack tip is given by

$$\mathrm{d}F_\mathrm{p} = \mathrm{d}F_\mathrm{e}\left(\frac{\alpha\beta\pi^2\sigma^2}{4\sigma_\mathrm{y}^2} + \frac{\alpha^2\beta^2\pi^4\sigma^4}{64\sigma_\mathrm{y}^4}\right) \tag{7}$$

For small scale yielding, $\sigma_\mathrm{y} > \sigma$, and $\mathrm{d}F_\mathrm{p} < \mathrm{d}F_\mathrm{e}$. If $\sigma_\mathrm{y} = 3\sigma$ then $\mathrm{d}F_\mathrm{p} \sim 0.1$ $\mathrm{d}F_\mathrm{e}$. It is, therefore, clear that the plastic work does not oppose crack growth but, since it is small, it will not greatly affect the thermodynamic instability criterion. The term due to the dislocations introduced will be of opposite sign and necessarily be of smaller absolute value than the plastic work term since most of the work done by plastic deformation is degraded into heat. It will, therefore, have even less effect.

If this is correct, it is clear that cracks in real materials are not in unstable equilibrium when they begin to propagate rapidly. They are well past the unstable equilibrium condition and must be in a metastable state when their length lies between a_0 given by equation (3) and the experimentally observed length, a', at which rapid propagation occurs. We conclude that there is no energy balance at the point of rapid propagation; an excess of energy is available for crack growth in this circumstance.

The condition that the driving force (the negative of equation (3), slightly modified to take account of the small effects of plastic deformation) must be equal to, or greater than, zero is certainly a necessary one. Unless the energy of the system can be reduced thereby, there will be no crack growth, slow or rapid. However, it may be that the condition is not sufficient, and as a possible explanation for the non-propagation of thermodynamically unstable cracks it is proposed that the stresses at the crack tip must exceed the theoretical cohesive stress as well as the crack satisfying the energy criterion (**12**). If the theoretical cohesive stress is not exceeded at the crack tip, separation of the surfaces will not occur even if the Helmholtz free energy of the system were to be reduced in the process. The stresses at the crack tip depend on the curvature at this point and it is suggested that the role of plastic deformation is to blunt the crack, thus reducing the local stresses and rendering it metastable. This view explains why crack-tip geometry is so important in practice, whereas it has negligible influence on the magnitude of the crack extension force.

It perhaps should be mentioned that all this has no great consequence for fracture mechanics, which can be legitimately regarded as a phenomenological theory based on the well-founded observation that crack propagation takes place at a critical value of the elastic plus mechanical energy release rate for a given crack geometry. What is affected is our understanding of the physics of fracture and this is important when considering the growth of small cracks.

Slow propagation of metastable cracks in fatigue

If this interpretation is accepted, one of the paradoxes of fatigue crack growth can be easily resolved. There is plenty of energy available to drive crack growth, by *any suitable mechanism*, for crack lengths between the Griffith critical length a_0 and the length, a' corresponding to the value of K_{IC} (see Fig. 1). Hence, in fatigue conditions, stable crack growth is possible within a range of stress intensity factors smaller than K_{IC}, and the plastic deformation which is intimately associated with the crack tip provides the mechanism necessary for it.

All the existing models of fatigue crack growth, which are well supported by experimental evidence, envisage that the crack grows by a 'shear decohesion' mode rather than by 'tensile decohesion' (13)(16). This can be achieved by the generation, or annihilation, of dislocations at the crack tip (i.e., plastic deformation), but this shear mode of crack growth will also need to satisfy both an energy criterion and a stress criterion. In the shear mode, however, the stress criterion is easier to satisfy because the theoretical shear stress for creation of dislocations is lower than the theoretical cohesive stress by a factor of 2 to 30, depending on the material (17), and crack growth by a shear mode can proceed under a lower value of stress than for tensile cracking (i.e., at a value of K_{max} less than K_{IC}). In addition, the alternating stresses and the plastic deformation associated with them can re-sharpen the crack, keeping the crack tip stresses sufficiently near the theoretical shear stress to allow an increment of growth each cycle. A limited amount of propagation per cycle is a natural consequence of these models since the dislocations produced to allow shear crack growth will themselves reduce the shear stresses at the tip and work hardening will limit the numbers created and the distances they move. Re-sharpening then takes place in the compressive part of the stress cycle.

Nucleation and growth of sub-critical cracks in fatigue

We now turn to the problem of short fatigue cracks, or cracks which are so short that they are sub-critical with respect to the classical Griffith equation. It is a fact that the growth of such cracks from virtually zero length occurs in fatigue (1)(2) and it is certain that this cannot violate energy conservation. The energy deficit must be made up by processes reducing the energy and which are necessarily coupled to the crack growth process.

In fatigue deformation, and in particular during the stage which precedes the nucleation of a crack, the applied stress does work on the fatigued sample each cycle and a great deal of energy is continually extracted from the mechanical system. Most of this energy is irreversibly degraded into heat, as is evidenced by the stress–strain hysteresis curves observed in fatigue. The main dissipative mechanism is plastic deformation and it is a characteristic of plastic deformation that not all the work done is converted into heat. A small fraction of the plastic work is stored in the material as the energy of the strain fields of the dislocations created. These internal strain fields depend not only on the number of dislocations but also on their spatial arrangements, which can be highly non-uniform and can lead to large energy densities in local regions. In low amplitude fatigue conditions these regions of high energy density tend to

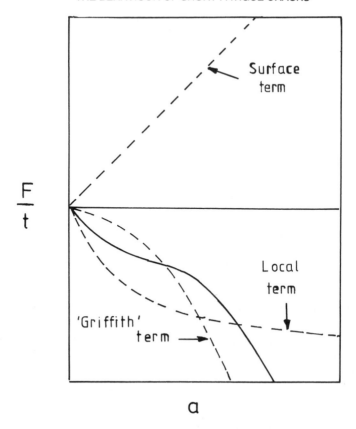

Fig 4 Helmholtz free energy per unit thickness of plate as a function of crack length, *a*. The
various terms contributing to the free energy are shown as broken lines. The total free energy is
shown as the full line

develop over many cycles in places where, due to flaws, defects, or
inhomogeneities, the plastic strain is concentrated. It is then possible for a
crack to nucleate and grow in these regions if the elastic energy stored in them
decreases sufficiently when the crack grows.

The energetic conditions for this process are schematically illustrated in
Fig. 4, which shows clearly how the strain released from the local regions of
internal strain provides the driving force for the nucleation of the crack and part
of that required for its initial growth.

The various contributions to the free energy of a stressed plate system when
a surface crack nucleates and grows are represented in Fig. 4. The surface
energy term is positive and proportional to *a*; the free energy term arising from
the external loading, or from strain energy in regions remote from the crack
(the classical Griffith term), is negative and proportional to a^2. The negative of
the slope of the curve representing this term is, of course, the crack extension

force, $\mathcal{G}$. The term F_L/t, representing the local strain energy released is also negative but the absolute value of its slope decreases with a as the crack grows and the local strain energy is consumed. It is convenient to call the negative of the slope of this term the *local* crack extension force, $\mathcal{G}_L$. The sum of all these terms gives the total Helmholtz free energy per unit thickness of plate as a function of crack length, a, and is shown as a full line in Fig. 4.

In order for a crack to nucleate from zero length the resultant curve in Fig. 4 must have a negative slope at the origin. Since $\mathcal{G}$ is zero at $a = 0$, the condition that the nucleation of a crack reduces the energy of the system is

$$\mathcal{G}_L(a = 0) > 2\gamma \tag{8}$$

and this is a necessary condition. Provided that it is satisfied, then the three typical situations illustrated in the upper diagrams in Fig. 5 arise when the three free energy terms are added up. If the local crack extension force is very high then the total free energy curve may have no maximum, and there will be no energy barrier for the nucleation and propagation of the crack to any length. This situation is shown in Fig. 5(a).

If $\mathcal{G}_L$ decreases more rapidly than shown in Fig. 5(a) then the total free energy curve has a minimum and a maximum, as in Fig. 5(c). In this case a crack which has nucleated can grow to a stable size, a_0, corresponding to the minimum of the free energy curve, where it will remain a non-propagating crack. There is no driving force for any further crack growth.

The Helmholtz free energy curve in Fig. 5(b) corresponds to the critical situation where $\mathcal{G}_L$ remains just large enough to give a positive driving force for any length of crack.

It is to be noted that since the size of the region of high strain energy density cannot be very large, the local crack extension force, $\mathcal{G}_L$, is expected to drop to zero quite rapidly as crack length increases. Hence, providing that a crack nucleated and propagated by a local region of high strain energy does not get trapped in a thermodynamic equilibrium state, as is the case Fig. 5(c), it will sooner or later begin to show normal 'long' crack behaviour.

In order to show that this energetic argument can qualitatively reproduce well the observed behaviour of short fatigue cracks, the conventional logarithmic plots of growth rate, da/dN, against ΔK are also shown schematically in the lower diagrams in Fig. 5. These have been drawn on the assumption that the crack growth rate is some increasing function of the driving force, i.e., the negative of the slope of the full curves in the upper part of Fig. 5. It can be seen that there are short cracks with anomolously high growth rates relative to the 'nominal' values of ΔK calculated by classical LEFM and that these growth rates tend to *decrease* with increasing ΔK. This is clearly a direct consequence of the fact that the actual driving force is higher than that apparent because of the high value of $\mathcal{G}_L$. The decreasing growth rate with nominal ΔK is the result of the shape of the curve of the local strain energy released by crack growth. Thus cracks apparently nucleate with great ease on the surface of a material

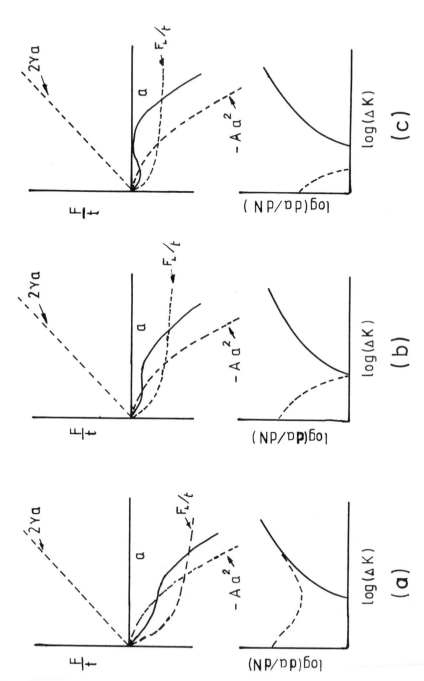

Fig 5 The upper diagrams represent change in Helmholtz free energy with crack length for three different cases: (a) no energy barrier to crack growth, (c) an energy barrier is met after initial crack growth, and (b) the critical transition between (a) and (c). The lower diagrams represent schematically the corresponding conventional crack growth rate curves assuming that the growth rate increases with the driving force

undergoing fatigue. Most grow to a certain length and then stop, having reached a situation in which the driving force is insufficient for further growth (Fig. 5(c)). A few go on to grow to a length at which they become 'long' fatigue cracks (Fig. 5(a)) and one eventually grows to a length at which rapid propagation and failure occurs.

It should be noted that, as pointed out before, a critical stress criterion needs also to be satisfied for a crack to nucleate and grow. It is, therefore, understandable that fatigue cracks are observed to nucleate at regions of high stress concentration which can arise from the geometrical magnification of the remote applied stress as well as from the local internal stress which can reach in some cases very high values. These local sources of internal strain and stress fields, necessary for the nucleation and initial propagation of a fatigue crack can be easily identified in real fatigue situations. Perhaps the best known is the persistent slip band (PSB) which has been subjected to intensive study both theoretical and experimental (18)(19). Persistent slip bands are the source of large internal stresses, as clearly explained by Brown and Ogin (19), who discussed the formation of non-propagating fatigue cracks from PSBs in a way which is a particular example of the general principle outlined above. The PSB stores a high density of strain energy locally and in an appendix by Eshelby to the paper of Brown and Ogin an expression is given for the local crack extension force, $\mathscr{G}_L$, for a crack of length a originating at the intersection of the PSB boundary with the specimen surface and lying along the boundary of the band. This is given by

$$\mathscr{G}_L = \frac{B}{a} \simeq \frac{2\mu h^2 \varepsilon_t^2}{(1 - \nu)a} \tag{9}$$

where h is the width of the PSB, μ is the shear modulus, ν Poisson's ratio, and ε_t the shear strain in the band. This implies that the local strain energy released varies with crack length according to

$$(F_L - F_{L,0})/t = -B/\ln a \tag{10}$$

in agreement with the general requirement outlined above and illustrated in Fig. 5. The local crack extension force has a singularity at $a = 0$. This means that the condition in equation (8) would always be satisfied, but the singularity is physically impossible and caution is required in using equation (9) and (10) to discuss the conditions pertaining to the nucleation of a crack.

The expression can be used, however, to illustrate the behaviour of a very short crack with a concrete example. After adding all the free energy terms the necessary condition for the nucleation and propagation of a crack is

$$\frac{1}{t}\left(\frac{\partial F}{\partial a}\right)_{T,l} = 2\gamma - 2Aa - B/a = 0 \tag{11}$$

where $2Aa = \mathcal{G}$ is the classical crack extension force given by equation (2). The quadratic equation (11) has solutions

$$a_0 = \frac{\gamma \pm (\gamma^2 - 2AB)^{1/2}}{2A} \qquad (12)$$

corresponding to the unstable and stable equilibrium situations illustrated in Fig. 5. When $\gamma^2 < 2AB$ there are no real roots and there are, therefore, no positions of stable or unstable equilibrium and no energy barrier for the nucleation and growth of a crack. This is the situation illustrated in Fig. 5(a). When $\gamma^2 > 2AB$ then the equation has two roots, one corresponding to the minimum and one to the maximum of the curve in Fig. 5(c), and a short crack can nucleate and grow at first, but then becomes a stable non-propagating crack, remaining stuck at the minimum energy position. The transition between the two cases, represented by Fig. 5(b), occurs when $\gamma^2 = 2AB$. The condition which must be satisfied for a short crack to continue growing and eventually become a long crack is $\gamma^2 \leqslant 2AB$. Using the expressions defined by equations (2) and (9) for the constants A and B this can also be written as $\gamma/\sigma \leqslant 2.6h\varepsilon_t$.

It is noted that if B is zero the solution of equation (11) is the Griffith critical length, $a_0 = \gamma/A$. If the condition $\gamma^2 = 2AB$ is applied then the solution of equation (11) yields $a = \gamma/(2A)$. Thus if the local strain energy release rate is sufficient to allow a crack to grow to a length greater than half the Griffith critical length there will be sufficient energy to allow the crack to become a long crack.

If we use values which are typical for copper then we can put $\sigma \simeq 60$ MPa (the saturation stress for Cu single crystals) and $\gamma \simeq 1$ J/m^2. With these values the Griffith crack length, $a_0 \simeq 20$ μm and the short/long transition length is half this, $\simeq 10$ μm. The value of $h\varepsilon_t$ is then 6 nm and if we take a typical observed value for h of 2 μm the strain in the band must be 3×10^{-3} which is certainly a reasonable value (19).

The nucleation and initial growth of a fatigue crack at a PSB has been used here as an illustrative example because it is the only case in which an expression for the rate of release of locally stored energy is available. There are, of course, other sources of internal strain which can develop during fatigue and can provide the driving force for crack nucleation. Examples of these are the incompatible deformation of surface grains which promotes the initiation of cracks at grain boundaries (20)(21), the formation of soft regions and dislocation free channels in a hard matrix with corresponding extrusions at the surface (22)(23), and the incompatibilities of deformation developed around second phase particles (23).

We believe that detailed knowledge of the structure of these regions of high strain energy density and the manner in which the strain energy is released by crack growth is essential if the initiation and growth of short cracks by fatigue is to be properly understood.

References

(1)   MILLER, K. J. (1985) Initiation and growth rates of short fatigue cracks, *Fundamentals of deformation and fracture* (Cambridge University Press, Cambridge), pp. 477–500.

(2)   BASINSKI, Z. S. and BASINSKI, S. J. (1985) Surface geometry in fatigued copper crystals, *Fundamentals of deformation and fracture* (Cambridge Univerity Press, Cambridge), pp. 583–594.

(3)   SURESH, S. and RITCHIE, R. O. (1984) Propagation of short fatigue cracks, *Int. Metals Revs*, **29**, 445–476.

(4)   PARIS, P. C. and ERDOGAN, F. (1963) A critical analysis of crack propagation laws, *Trans AIME, J. Basic. Engng*, **85**, 528–534.

(5)   GRIFFITH, A. A. (1920) The phenomena of rupture and flow in solids, *Trans R. Soc. (Lond.)*, **A221**, 163–198.

(6)   OROWAN, E. (1952) Fundamentals of brittle behaviour in metals, *Fatigue fracture of metals* (John Wiley, New York), p. 154.

(7)   IRWIN, G. R. (1948) Fracture dynamics, *Fracture of metals* (American Society of Metals), p. 152.

(8)   KNOTT, J. F. (1973) *Fundamentals of fracture mechanics* (Halstead Press (John Wiley), New York).

(9)   WEERTMAN, J. (1978) Fracture mechanics: a unified view for Griffith–Irwin–Orowan cracks, *Acta Met.* **26**, 1731–1738.

(10) DUGDALE, D. S. (1960) Yielding of steel sheets containing slots, *J. Mech. Phy. Solids*, **8**, 100–104.

(11) BARENBLATT, G. I. (1962) The mathematical theory of equilibrium cracks. *Advances in Applied Mechanics* (Academic Press, New York), Vol. 7, p. 55.

(12) PETCH, N. J. (1968) Metallographic aspects of fracture, *Fracture* (Academic Press, New York), Vol. 1, p. 351.

(13) LAIRD, C. and SMITH, G. C. (1962) Crack propagation in high stress fatigue, *Phil. Mag.*, **7**, 847–857.

(14) NEUMANN, P. (1974) New experiments concerning the slip processes at propagating fatigue cracks – I, *Acta Met.*, **22**, 1155–1166.

(15) NEUMANN, P. (1974) The geometry of slip processes at a propagating fatigue crack – II, *Acta Met.*, **22**, 1167–1178.

(16) WEERTMAN, J. (1984) Crack growth for the double-slip-plane and the modified DSP crack model, *Acta Met.*, **32**, 575–584.

(17) KELLY, A. (1966) *Strong solids* (Clarendon Press, Oxford).

(18) MUGHRABI, H., WANG, R., DIFFERT, K., and ESSMAN, U. (1983) Fatigue crack initiation by cyclic slip irreversibilities in high-cycle fatigue, *Fatigue mechanisms: advances in quantitative measurement of physical damage, ASTM, STP 811* (American Society for Testing and Materials), p. 5.

(19) BROWN, L. M. and OGIN, S. L. (1985) Role of internal stresses in the nucleation of fatigue cracks, *Fundamentals of deformation and fracture* (Cambridge University Press, Cambridge), pp. 501–528.

(20) MUGHRABI, H. (1975) Wechselverformung und Erdmüdungsbruch von α-Eisenrielkristallen, *Zeitschrift Metalkunde*, **66**, 719–724.

(21) GUIU, F., DULNIAK, R., and EDWARDS, B. C. (1982) On the nucleation of fatigue cracks in pure polycrystalline α-iron. *Fatigue Engng Mater. Structures*, **5**, 311–321.

(22) MUGHRABI, H., ACKERMANN, F., AND HERZ, K. (1979) Persistent slip bands in fatigued face-centered and body-centered cubic metals, *Fatigue mechanisms, ASTM-STP 675* (American Society for Testing and Materials), p. 69.

(23) MUGHRABI, H. (1983) Cyclic deformation and fatigue of multi-phase materials, Proceedings of the 4th RISØ International Symposium on Metallurgy and Materials Science: Deformation of Multi-phase and Particle Containing Materials, Roskilde, Denmark, p. 65.

*M. W. Brown**

Interfaces Between Short, Long, and Non-Propagating Cracks

REFERENCE Brown, M. W., **Interfaces Between Short, Long, and Non-Propagating Cracks**, *The Behaviour of Short Fatigue Cracks*, EGF Pub. 1 (Edited by K. J. Miller and E. R. de los Rios) 1986, Mechanical Engineering Publications, London, pp. **423–439**.

ABSTRACT A wide range of experimental techniques is used to study fatigue crack behaviour, covering the regimes of (a) long cracks (LEFM), (b) high strain fatigue, (c) microstructurally short cracks, and (d) non-propagating cracks. A full range of stress levels and crack lengths have been considered to show where each mechanism for crack extension is valid, what modes of crack tip opening operate, and what model is appropriate for describing crack growth rate. A fatigue diagram is developed to illustrate six regimes of fatigue crack behaviour, and this is contrasted with the Kitagawa–Takahashi diagram for finite/infinite life fatigue.

Notation

A	Constant, $A > 2$, for Mode I/III transition
a	Crack length
d	Spacing between microstructural barriers
E	Young's modulus
K	Stress intensity factor
k	Cyclic strength coefficient
N	Number of cycles
n	Cyclic strain hardening exponent
x	Percentage deviation of strain from LEFM formulation
Y	Geometry factor
Δ	Range of stress or strain
ε	Strain
σ	Stress

Subscripts

fl	Fatigue limit
M	Microstructural/continuum transition
PSB	Threshold stress for PSB nucleation
th	Threshold
u	Ultimate strength
y	Yield or proof stress

Introduction

The failure of metals by fatigue has received extensive industrial attention and academic research activity for many years, with a variety of methods being

* Department of Mechanical Engineering, University of Sheffield, Mappin Street, Sheffield, UK.

developed to observe the behaviour of fatigue cracks. Different disciplines from mathematics to metallurgy have come together to tackle this problem, each bringing their own areas of insight, but the unification of results from different laboratories has been largely overlooked, due primarily to the vast amount of literature published. However the overriding importance of the mechanisms of cyclic crack extension to the deterioration of metals in fatigue has been firmly established; therefore this paper attempts firstly to list the different types of propagation experimentally observed, and secondly to quantify the regimes within which each mechanism is dominant. Such mechanisms are rarely contrasted because they are usually studied in different types of laboratory test, namely those concerned with (a) high strain fatigue, (b) linear elastic fracture mechanics, (c) the behaviour of notches, (d) threshold experiments, and (e) fractographic analyses employing the surface replication technique.

All but one of these areas of investigation fall under the heading of 'short' cracks, even though the crack lengths involved differ by orders of magnitude. The exception is the linear elastic fracture mechanics (LEFM) case, for which propagating cracks at low levels of stress are termed 'long' cracks. Notwithstanding the normal usage of the word 'short', a long crack can become short simply by raising the level of applied stress in order to violate the allowable limits of LEFM. So in this instance the term 'short' has no relationship to the actual size of the crack, but rather it describes the intensity of load or degree of plasticity experienced.

This paper is limited to cracks growing in plain specimens. However the concepts may be extended to cover notched specimens if the effect of the stress gradient due to the notch can be taken into account. The behaviour of microstructurally short cracks, high strain or elasto-plastic fracture mechanics cracks, and non-propagating cracks are discussed briefly, but more detailed reviews have been published elsewhere (1)–(4).

The Kitagawa–Takahashi diagram

A significant advance in the understanding of short crack behaviour was made by Kitagawa and Takahashi (5), who proposed a diagram to show the effect of defect size on the fatigue limit stress, see, for example, Fig. 1. For large defects, the allowable stress for infinite life must be low, within the linear elastic fracture mechanics regime, and, therefore, the limiting condition is given by a straight line of slope minus one half corresponding to the threshold stress intensity factor ΔK_{th}, such that

$$\Delta K_{\text{th}} = Y \cdot \Delta \sigma \sqrt{(\pi a)} \tag{1}$$

At the other end of the spectrum, for vanishingly small defects the allowable stress level must relate to the plain specimen fatigue limit. The Kitagawa–Takahashi diagram for the medium carbon steel in Fig. 1 shows the observed behaviour between these two extremes.

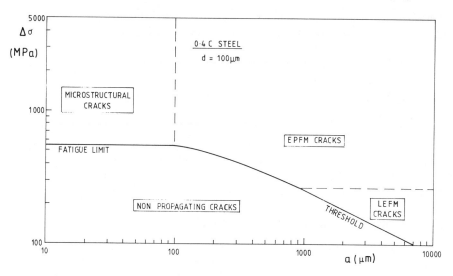

Fig 1   A Kitagawa–Takahashi diagram

Deviation from the LEFM straight line can usually be observed for peak stress levels above one third of the yield stress. Throughout this paper, we will be concerned with fully reversed cycling only, or zero mean stress. Thus the limiting stress range, $\Delta\sigma$, for LEFM behaviour will be two thirds of the yield stress. This requirement arises from the necessity for small scale yielding only at the crack tip if the LEFM characterization of crack tip stress conditions is to be valid. The basis of the description of fatigue crack propagation by the stress intensity factor range, ΔK, is that crack extension is governed by an elastic stress field at the crack tip, which determines various parameters such as plastic zone size and crack opening displacement. For small scale yielding, the plastic zone size is generally taken to be less than one fiftieth of the crack length (6). Note that this corresponds to a 1.5 per cent error in the elastic stress field for a centre-cracked panel, and as much as 7 per cent for a compact tension specimen. Since the plane strain plastic zone size is approximately one third of the plane stress value (6), one may write, for LEFM validity

$$(K_\mathrm{I}/\sigma_\mathrm{y})^2/6\pi \leqslant 0.02a \tag{2}$$

But for a crack with a geometry factor, Y, equal to unity

$$K_\mathrm{I} = \sigma\sqrt{(\pi a)}$$

which, on substitution into equation (2), gives approximately

$$\sigma/\sigma_\mathrm{y} \leqslant 1/3 \tag{3}$$

Once this stress limit is exceeded, the errors in stress values derived from the stress intensity factor will build up rapidly, and therefore one should strictly use elasto-plastic fracture mechanics (EPFM), for which a variety of parameters have been proposed to correlate fatigue crack growth behaviour (2). Even in the EPFM regime, a threshold condition can be observed (5), below which cracks are unable to grow, giving rise to the non-propagating crack regime in Fig. 1.

A fourth regime for crack growth can be identified in Fig. 1, where the cracks are sufficiently small to interact markedly with the microstructure. This strong dependence on microstructure may be associated with the fatigue limit (7), and the insensitivity of fatigue strength to small defect sizes shown by the Kitagawa–Takahashi diagram can be related to the ease of fatigue crack growth in the first grain where the crack initiates (8). Thus the use of the continuum based discipline of fracture mechanics is inappropriate for this regime.

Continuum and microstructural crack growth

The behaviour of propagating short cracks is typically presented in the literature using graphs of crack length versus crack growth rate, as shown in Fig. 2. An initial period of decelerating crack growth is frequently observed (1), where the minimum growth rate corresponds to the crack tip meeting some barrier to growth. However, if the crack has sufficient driving force to pass through that barrier, then a period of accelerating growth ensues until fracture takes place. The most common representation of this behaviour employs logarithmic axes (1)(7)(9)(10), as in Fig. 2(a), since this choice for the abscissa facilitates comparison of the accelerating, or EPFM, phase with the LEFM parameter, ΔK. However, a better understanding of the nature of microstructural crack growth (MCG) is gained from Fig. 2(b), which uses linear scales.

Figure 2 has been constructed from the equations for EPFM and MCG cracks derived empirically by Hobson (11)(12). Working on a medium carbon steel with uniaxial low cycle fatigue specimens, Hobson was able to derive the short crack growth characteristics from surface replication studies, which provided a detailed history of each crack. The MCG phase was described by the equation

$$da/dN = 153620(\Delta\varepsilon)^{3.51}(d - a) \tag{4}$$

where d, a and da/dN each have the same units for length. Here a is the surface crack length determined by replication, and d, the distance between the microstructural barriers for each end of the individual crack monitored, is found by fitting equation (4) to the data. Details of the material, test procedures, the relationship between d and the microstructure, and derivation of equation (4) are given elsewhere (11)(12).

For cracks spanning a number of grains, the effect of individual barriers to growth is small (13), and a continuum mechanics representation may be employed. Hobson derived the simple equation

$$da/dN = 4.102(\Delta\varepsilon)^{2.06}a - 4.237 \tag{5}$$

where da/dN is in nm/cycle, and a is in nm. This equation describes both LEFM

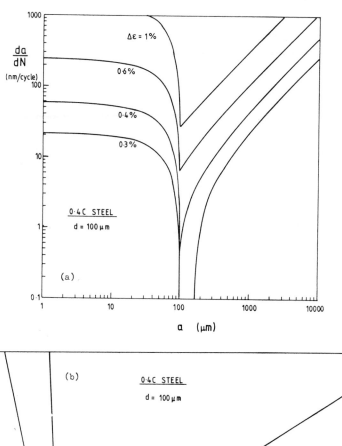

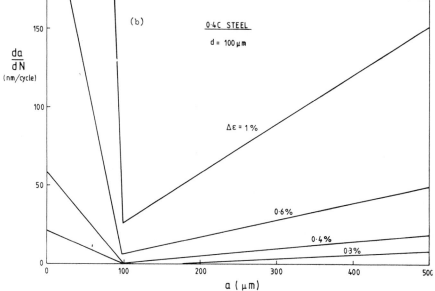

Fig 2   Growth of short cracks on (a) logarithmic axes, and (b) linear axes

behaviour and EPFM crack advance by adopting a form widely used in EPFM studies (2), but with the addition of a small threshold term.

Equations (4) and (5) have been used to construct Fig. 2, taking a nominal value for d in a medium carbon steel of 100 μm. This compares favourably with values measured by Hobson (11). The growth rate was chosen as the greatest of those determined from equations (4) and (5), respectively, for any given value of strain range, following the crack propagation criterion of Pineau *et al.* (14) that a crack will always adopt the mechanism that gives the fastest growth rate (15). The change in mechanism on reaching 100 μm crack length should correspond to the transition from Stage I to Stage II growth, depicted in Fig. 3 (15).

Stage I or MCG is typically a shear mode of cracking strongly associated with crystallographic planes. If a Stage I crack traverses two grains, a slight change of plane will be noticed at the first grain boundary. Similarly, MCG can cover two grains where a low angle boundary is involved, since a low angle boundary fails to provide a significant barrier to propagation (13)(16). Thus equation (4) appears to relate to the Stage I crack growth process.

However, Stage II cracking, or striation crack growth, is well known as a Mode I controlled mechanism, with a fracture surface normal to the maximum principal stress (15)(17). This clearly relates to a different plane to the MCG phase, so that a sharp transition in growth rate also, as predicted in Fig. 2, can be expected to occur in practice. Note that the early portion of Stage II propagation may have, faceted appearance indicating a crystallographic dependence, before the clear striation mechanism is able to operate (18). Nevertheless, this slow Mode I growth in the near threshold regime is normally described

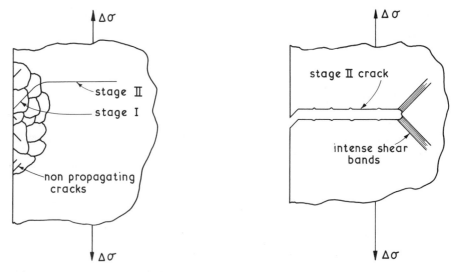

Fig 3   Two stages of fatigue crack growth

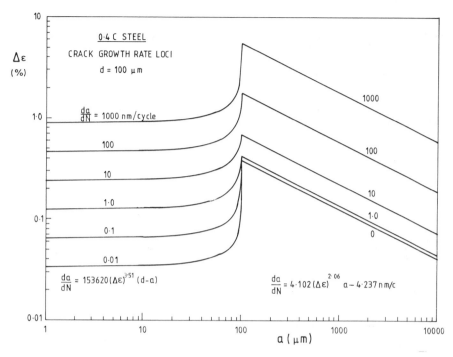

Fig 4   Crack propagation rates for strain controlled fatigue

by a fracture mechanics formulation. Thus equation (5) relates to the EPFM or Stage II growth regime.

Having derived the two crack growth equations above, an alternative method of presenting crack growth rate loci may be developed, as in Fig. 4. For any strain range applied to a crack of length a, the relevant mechanism for fatigue may be derived from Pineau's criterion, see above, and the growth rate determined. Contours or loci of constant growth rate have been plotted in Fig. 4, to show a clear demarcation between the two regimes of short crack growth, namely those of MCG and EPFM.

The fatigue diagram in Fig. 4 may be used to predict the history of a crack in a constant amplitude low cycle fatigue test. For example, for a strain range of 0.6 per cent, a crack growth rate of about 250 nm/cycle will be observed at the start of a test for a specimen with small initial defects ($< 1 \ \mu$m). This growth rate will be essentially constant up to 20 μm crack length, as can be seen by sketching a horizontal line on Fig. 4 at $\Delta\varepsilon = 0.6$ per cent, since the loci plotted are essentially flat and parallel to this line. Thus in only 80 cycles, the crack will attain 20 μm length, followed by a period of deceleration to just 2.5 per cent of its initial speed. However, the change in mechanism to Mode I growth enables the crack to accelerate, and it will continue to accelerate up to failure. Figure 5 depicts the crack history for various strain ranges.

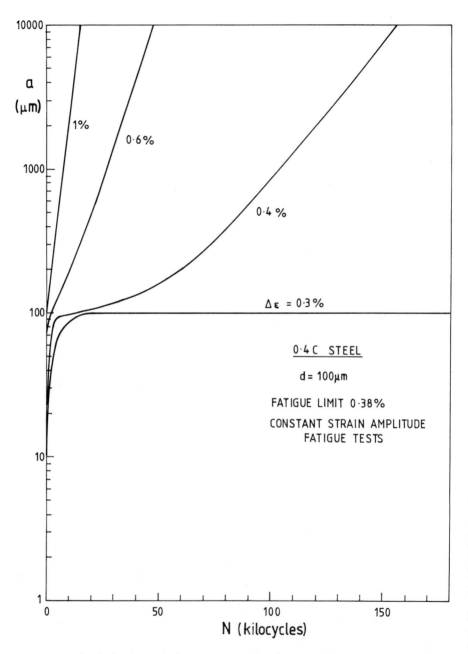

Fig 5   Predicted crack histories for constant amplitude fatigue tests

The case of $\Delta\varepsilon = 0.3$ per cent is of interest since it falls below the fatigue limit, and no transition to EPFM growth can be obtained (**19**). Such a specimen will contain non-propagating cracks of length 99 μm after only 21 500 cycles. The fatigue limit condition is clearly governed by the threshold value, together with critical crack length, d, of 100 μm. From equation (5), for zero crack growth rate, putting a equal to d

$$0 = 4.102(\Delta\varepsilon)^{2.06} \times 10^5 - 4.237 \tag{6}$$

which gives 0.38 per cent for the fatigue limit strain range (550 MPa stress range).

The fatigue diagram in Fig. 4 could also be used to derive variable strain amplitude crack histories, or even to deal with stress gradients due to notches or thermal stresses. Here the horizontal line used for the constant strain amplitude case should be replaced by a curve to depict the actual load history, or the stress gradient in a component. Such an approach, although a simplification of real behaviour, has been applied to EPFM cracks under cyclic thermal stress (**20**).

The fatigue and Kitagawa–Takahashi diagrams

Figure 4, a graph of strain range versus crack length, has a clear affinity to the Kitagawa–Takahashi diagram in terms of stress range. In order to provide a direct comparison, the cyclic stress–strain curve was used to replot Fig. 4 in terms of stress, as shown in Fig. 6.

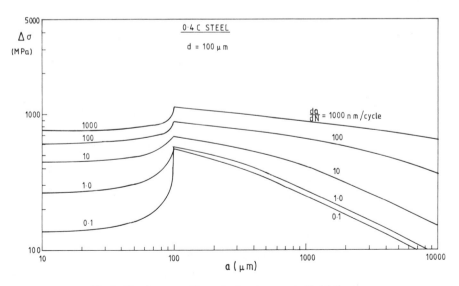

Fig 6   Crack propagation rates for stress controlled fatigue

It was assumed that the Ramberg–Osgood formulation of cyclic behaviour was applicable (11), such that

$$\Delta\varepsilon = \Delta\sigma/203\,000 + (\Delta\sigma/2013)^{1/0.190} \tag{7}$$

where the stress is given in MPa, and 0.19 is the strain hardening coefficient. Equation (7) gave reasonable fit to the stable half-life stress range measured in strain controlled experiments, and also to a multiple step test (12).

Comparison of Figs 1 and 6 shows that the Kitagawa–Takahashi line between non-propagating and fracture mechanics cracks corresponds closely with the 0.1 nm/cycle contour, i.e., the threshold condition. (Note 0.1 nm/cycle is widely used as a practical definition in the determination of threshold values.) Indeed the contours for 0 and 0.1 nm/cycle were so close that they could not be shown separately with the scales used in Fig. 6.

The fatigue limit from equation (6) gives a stress range of 550 MPa, as plotted in Fig. 1. This compares favourably with an experimentally measured value of 487 ± 15 MPa for a very similar batch of 0.4C steel (21). However, the discrepancy between these values is due to the arbitrary choice of d as 100 μm in Fig. 6, so that this figure represents the behaviour of a particular crack that initiates in a given grain where $d = 100$ μm. Clearly a representative value for d must be derived if the actual fatigue limit of a material is to be obtained. Now Hobson (12) found a mean value of d from 35 measurements, being 116 μm with a standard deviation of 51 μm. Thus a reasonably large well-orientated grain in a polycrystalline metal would probably correspond to a value for d of the mean plus one standard deviation, i.e., 167 μm, and a grain of about this size should initiate the failure crack that governs the material fatigue limit. A value of 167 μm for d predicts a fatigue strength, $\Delta\sigma$, of 486 MPa from equations (6) and (7).

Note that the deviation from linearity in Fig. 6 for the EPFM lines on leaving the LEFM regime derives purely from the cyclic stress–strain curve, if the formulation of the crack growth law in equation (5) is taken to be accurate. If the cyclic yield is taken as a 0.02 per cent proof stress, equation (7) gives a value of 228 MPa, compared to the monotonic yield of 392 MPa which was used in Fig. 1. The 0.02 per cent proof stress, if substituted in equation (3), implies that the allowable stress range for LEFM calculation is only 152 MPa, at which there is just 0.16 per cent deviation from the LEFM line of slope minus one half on Fig. 6. Thus, for this material, equation (3) provides a very stiff restriction on the use of LEFM. However, in general, for a material following the Ramberg–Osgood formulation, the percentage deviation of the stress range from the elastic line at 2/3 of the 0.02 per cent proof stress is given by x, where

$$x = (E/k)(3)^{1-1/n}(0.0002)^{1-n} \times 100 \text{ per cent} \tag{8}$$

and E, k, and n are the elastic modulus, cyclic strength coefficient, and cyclic strain hardening exponent, respectively. This formula is particularly sensitive to the value of n; for example a strain hardening exponent of 0.3 gives a

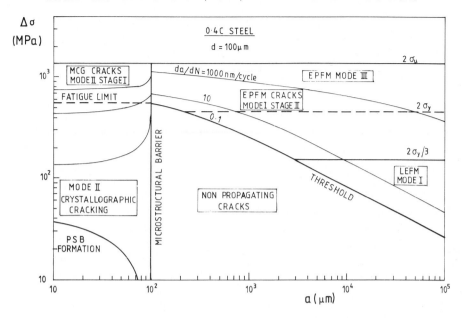

Fig 7   Fatigue diagram for 0.4C steel, based on equations (4) and (5)

deviation of 3.5 per cent, justifying the choice of the LEFM boundary at $2\sigma_y/3$ as reasonable.

Interfaces between crack growth regimes

The fatigue diagram of Fig. 6 has been replotted in Fig. 7 with six distinct regimes of crack behaviour defined. As in the Kitagawa–Takahashi diagram, the LEFM regime is limited by the horizontal line at $2\sigma_y/3$, with lower bound crack lengths given by the fatigue threshold condition. Below-threshold non-propagating cracks exist, but the microstructural barrier, characterized by the dimension d, gives a lower bound crack length to this regime.

The upper bound to EPFM behaviour must correspond with the tensile strength $\sigma_u = 683$ MPa, i.e., for fully reversed cycling, $2\sigma_u$. This line indicates the special case of infinite crack growth rate. However, the EPFM regime can be subdivided into two regimes corresponding to Mode I behaviour and, at higher stresses, Mode III cracking. It has not yet been shown conclusively what constitutes the critical condition for a transition to Mode III, but such transitions have been observed regularly (**22**), and the preference for Mode I at low stress and Mode III at high stress has been elegantly demonstrated by Ritchie *et al.* (**23**). It is possible that the transition to Mode III relates principally to yield stress, i.e., the line $2\sigma_y$ in Fig. 7, but it is more likely that a higher stress range is required for the tearing mechanism to dominate.

Considering microstructurally short cracks, above the fatigue limit these correspond to Stage I crack growth, as discussed previously. However, the fatigue limit stress is not necessarily a limit on the MCG mechanism, and Mode II cracks are free to develop on crystallographic slip systems well below the fatigue limit stress. Although equation (4) provides no lower limit of strain or stress for MCG, it is reasonable to assume that microscopic plasticity is an essential feature of crack extension, if only to provide the necessary irreversibility of deformation. Since many such cracks propagate along persistent slip bands (PSB), a very tentative lower bound to the crystallographic cracking region has been drawn in Fig. 7 by using a resolved shear stress range for PSB formation of 98 MPa, observed in α iron (**14**). In addition a stress concentration factor of 5 for the defect of length a acting on a net section stress, taken over the ligament between a and d, was assumed to determine the resolved shear stress on the PSB that must emanate from the crack tip. In spite of the lack of rigour in these three assumptions, it is nevertheless apparent that the lower bound to the Mode II cracking regime is an order of magnitude below the fatigue limit. Non-propagating fatigue cracks have been observed both in short crack studies (**24**) below the fatigue limit and in mixed mode LEFM experiments well below the Mode I threshold (**25**), and in both cases the cracks arrested on reaching defined microstructural features.

The fatigue diagram in Fig. 7 shows the six regimes of crack growth behaviour, together with lines denoting the interfaces between each regime, labelled with the principal physical causes for each interface. It is also apparent that the three fundamental modes of cracking, traditionally defined in fracture mechanics texts, each have a distinctive role to play in the process of fatigue failure. Although specific laws, equations (4) and (5) for 0.4C steel, have been used to derive Fig. 7, it appears that the general form of this diagram can have a much wider applicability to most, and probably all, ductile metals. The

Table 1   Regimes of crack propagation

Type	Mode	Stage	Equation	Mechanism	Crack length limits	Stress limits
LEFM	I	—	5	Striation	$a > a_{th}$	$\Delta\sigma < 2\sigma_y/3$
EPFM	I	II	5	Striation	$a > a_{th}$	$\Delta\sigma < A\sigma_y$
					$a > a_M$	
EPFM	III	III	*	Tearing	$a > a_{th}$	$A\sigma_y < \Delta\sigma < 2\sigma_u$
					$a > a_M$	
MCG	II	I	4	Crystallographic shear	$a < a_M$	$\Delta\sigma_{fl} < \Delta\sigma < 2\sigma_u$
CC†	II	—	4	Crystallographic shear (PSB)	$a < d$	$\Delta\sigma_{PSB} < \Delta\sigma < \Delta\sigma_{fl}$
NPC‡	I/II	—	1		$d < a < a_{th}$	$\Delta\sigma < \Delta\sigma_{fl}$

* See (**22**).
† CC: crystallographic cracking.
‡ NPC: non-propagating cracks

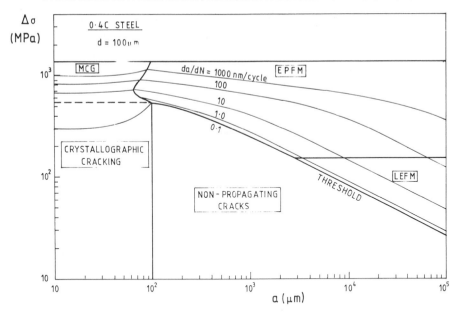

Fig 8 **Fatigue diagram for 0.4C steel with reduced microstructural crack growth rates (equation (9))**

interface lines separating each regime may move considerably with a change of properties, such as yield stress, fracture ductility, microstructure, etc., but these six regimes will remain, with the possible addition of other mechanisms for propagation. The basic types of propagation are listed in Table 1.

As an example, Fig. 8 has been drawn using equation (5) for EPFM cracks, but with a different MCG equation

$$da/dN = 1069(\Delta\varepsilon)^{3.18}(d - a) \qquad (9)$$

derived for the same material (**26**). Whereas the right-hand side of the figure is unchanged, the MCG regime is reduced in extent, illustrating a significant feature of Pineau's criterion in that the transition to Mode I can occur before the microstructural barrier is reached. This is because the MCG rates from equation (9) are considerably lower than those of equation (4). Although the transition to fracture mechanics control generally occurs at grain boundaries, or other microstructural features (**18**), this is not always the case, as in, for example, single crystals (**27**).

Discussion

The fatigue diagram, plotted from crack growth rate considerations, provides a logical basis for the frequently observed shape of the Kitagawa–Takahashi diagram. While Fig. 7 has been developed for the specific case of 0.4C steel, with just one microstructural morphology, expressions of the type given by equations (4) and (5) in this paper only provide a convenient vehicle with which to derive the fatigue diagram. It's basic form, showing six different regions of crack growth, is of general applicability, since it is sufficient for the actual crack propagation laws to have (1) a region of MCG where growth rate decreases as crack length increases, (2) a region of fracture mechanics controlled fatigue where growth rate increases with crack length, and (3) a well defined fatigue threshold for long cracks.

The diagram in Fig. 7 has been based on stress range in order to relate to the well known Kitagawa–Takahashi diagram. This may be the most convenient form while designers continue to use S–N data in infinite life fatigue assessments. However, a strain based diagram, e.g., Fig. 4, presents a simpler picture, relates to strain controlled fatigue design methods used for finite life assessments, and, insofar as the simple EPFM equation (5) remains applicable as originally proposed by Boettner et al. (28)–(30)(2), it obviates the need to distinguish the region of validity for LEFM. Furthermore, for Mode III crack growth it has been shown that the use of a strain intensity parameter enables crack growth rates to be represented by simple equations which should facilitate the inclusion of a Mode I/Mode III transition on the diagram (22)(23).

The effects of both mean stress and multiaxial loading should be assessed from the point of view of design utility. This may prove difficult while maintaining a two dimensional plot, but the concepts embodied in the Goodman diagram show how at least the low stress fatigue behaviour might be represented in three dimensions, the third axis relating to mean stress. Work on multiaxial effects for both short cracks (26), LEFM cracks (31), and EPFM cracks (32)(33) indicates how equations (4) and (5) can be extended to describe Mode II and Mode I behaviour, respectively, but considerable work is required on Mode III (34) before a multiaxial stress fatigue diagram can be completed.

The interface lines have been discussed above, but it is clear that so far as MCG is concerned, the choice of d, the characteristic microstructural dimension for an individual crack, is crucial. Many papers in this volume show that a variety of microstructural barriers can arise, and for any metal or alloy, it is important from a design viewpoint not to underestimate the value of d. If the crack growth obstacles are not strong, and cracks penetrate and keep on propagating from one obstacle to the next, albeit very slowly, then no definite fatigue limit will be obtained. This is the case for many aluminium alloys, for example (35). Obstacles may take a variety of forms, such as grain boundaries, triple points, second phases of high strength, pearlite, large inclusions, or precipitates. A good example of grain boundaries acting as highly effective barriers to slip is given by Kompek et al. (36) for low carbon steels, and correspondingly, they may arrest microstructural fatigue cracks. In the particular case of 0.4C steel studied by Hobson, the dominant barrier was pearlite, being of high strength compared to the ferrite in which cracks nucleated. Thus d could also be estimated from a statistical analysis of the microstructure to find the greatest linear ferrite path between pearlite colonies. The concept of pearlite as a crack growth barrier is not new, and for textured steels the relationship between crack growth direction (or mode) and the local texture has an important bearing on the fatigue strength (37), for both short and long cracks.

The limitations to LEFM applicability have received much attention in the literature (8)(2)(7), and equation (3) may appear to be unduly restrictive for some materials. However, simple rules for predicting l_2, the crack length above which LEFM can be used (8), are liable to give non-conservative estimates for metals with low yield stress and high strain hardening exponents, if the prediction of equation (8) is correct.

The development of MCG is frequently cited in the literature as being associated with the PSB. However, this is not always the case, and Mode II cracks have been observed to form on grain boundaries due to incompatibility of deformation in adjacent grains (38). Other mechanisms for crack nucleation could be postulated, such as corrosive attack; therefore, the lower bound to the crystallographic cracking regime should be estimated according to the relevant mechanism for the material and environment concerned.

The fatigue diagram provides a visual method of selecting the correct mechanism, together with the corresponding characteristic equation and crack growth mode, for any crack size and stress level. This is an important choice for the scientist assessing fatigue crack behaviour, and also for the designer seeking to use a defect tolerant approach to finite life prediction or remnant life assessment.

Conclusions

Various distinct regimes of short and long crack growth in fatigue have been identified, each with separate mechanisms for crack extension. Since it is inappropriate to use a model for one mechanism when examining data related to another mode of cracking, a fatigue diagram has been proposed to show which is the dominant mode of growth for any given stress range and crack length.

Such diagrams may be used to interpret the Kitagawa–Takahashi diagram, to predict the history of a crack under general loading, and to estimate the fatigue limit for a known microstructural morphology.

Acknowledgements

The author is indebted to the Central Electricity Generating Board for a Research Fellowship. The concepts developed in this paper are the direct result of detailed investigations by numerous research students and visitors in the fracture research group at the Mechanical Engineering Department, Sheffield University.

References

(1) LANKFORD, J. (1985) The influence of microstructure on the growth of small fatigue cracks, *Fatigue Fracture Engng Mater. Structures*, **8**, 161–175.
(2) SKELTON, R. P. (1982) Growth of short cracks during high strain fatigue and thermal cycling, *ASTM STP 770* (American Society for Testing and Materials, Philadelphia), pp. 337–381.
(3) TAYLOR, D. and KNOTT, J. F. (1981) Fatigue crack propagation behaviour of short cracks; the effect of microstructure, *Fatigue Engng Mater. Structures*, **4**, 147–155.
(4) MUGHRABI, H. (1980) Microscopic mechanisms of metal fatigue, *Strength of metals and alloys* (Edited by P. Haasen, V. Gerold, and G. Kostorz), Vol. III (Pergamon, Oxford), pp. 1615–1638.
(5) KITAGAWA, H. and TAKAHASHI, S. (1976) Applicability of fracture mechanics to very small cracks or the cracks in the early stage, *Proc. 2nd Int. Conf. Mech. Behaviour of Materials*, Boston, pp. 627–631.
(6) KNOTT, J. F. (1973) *Fundamentals of fracture mechanics* (Butterworths, London), p. 134.
(7) MILLER, K. J. (1984) Initiation and growth rates of short fatigue cracks, *Fundamentals of deformation and fracture* (Edited by B. A. Bilby, K. J. Miller, and J. R. Willis) (Cambridge University Press), pp. 477–500.
(8) TAYLOR, D. (1984) The effect of crack length on fatigue threshold, *Fatigue Engng Mater. Structures*, **7**, 267–277.
(9) DE LOS RIOS, E. R., TANG, Z., and MILLER, K. J. (1984) Short crack fatigue behaviour in a medium carbon steel, *Fatigue Engng Mater. Structures*, **7**, 97–108.
(10) NAM SOON CHANG and HAWORTH, W. L. (1985) Fatigue crack initiation and early growth in an austenitic stainless steel, *Proc. 7th Int. Conf. Strength of Metals and Alloys*, Vol. II (Pergamon Press, Oxford), pp. 1225–1230.
(11) HOBSON, P. D. (1985) The growth of short fatigue cracks in a medium carbon steel, PhD thesis, University of Sheffield.
(12) HOBSON, P. D., BROWN, M. W., and DE LOS RIOS, E. R. (1986) Two phases of short crack growth in a carbon steel, *The behaviour of short fatigue cracks* (Edited by K. J. Miller and E. R. de los Rios), (Mechanical Engineering Publications, London), pp. 441–459 (This volume).
(13) LANKFORD, J. (1982) The growth of small fatigue cracks in 7075-T6 aluminium, *Fatigue Engng Mater. Structures*, **5**, 233–248.
(14) HOURLIER, F., d'HONDT, H., TRUCHON, M., and PINEAU, A. (1985) Fatigue crack path behaviour under polymodal fatigue, *ASTM STP 853* (American Society for Testing and Materials, Philadelphia), pp. 228–248.
(15) BROWN, M. W. and MILLER, K. J. (1979) Initiation and growth of cracks in biaxial fatigue, *Fatigue Engng Mater. Structures*, **1**, 231–246.
(16) DE LOS RIOS, E. R., MOHAMED, H. J., and MILLER, K. J. (1985) A micromechanics analysis for short fatigue crack growth, *Ibid*, **8**, 49–63.
(17) MILLER, K. J. and BROWN, M. W. (1984) Multiaxial fatigue: an introductory review, *Subcritical crack growth due to fatigue, stress corrosion and creep* (Edited by L. H. Larsson) (Elsevier Applied Science, London), pp. 215–238.
(18) PLUMBRIDGE, W. J. (1972) Review: fatigue-crack propagation in metallic and polymeric materials, *J. Mater. Sci.*, **7**, 939–962.
(19) HOBSON, P. D. (1982) The formulation of a crack growth equation for short cracks, *Fatigue Engng Mater. Structures*, **5**, 323–327.
(20) MARSH, D. J. (1981) A thermal shock fatigue study of type 304 and 316 stainless steels, *Fatigue Engng Mater. Structures*, **4**, 179–195.
(21) IBRAHIM, M. F. E. (1981) *Early damage accumulation in metal fatigue*, PhD thesis, University of Sheffield.
(22) HAY, E. and BROWN, M. W. (1986) Initiation and early growth of fatigue cracks from a circumferential notch loaded in torsion, *The behaviour of short fatigue cracks* (Edited by K. J. Miller and E. R. de los Rios), (Mechanical Engineering Publications, London), pp. 309–321 (This volume).
(23) RITCHIE, R. O., McCLINTOCK, F. A., NAYEB-HASHEMI, H., and RITTER, M. A. (1982) Mode III fatigue crack propagation in low alloy steel, *Met. Trans*, **13A**, 101–110.

(24) MILLER, K. J., MOHAMED, H. J. and DE LOS RIOS, E. R. (1986) Fatigue damage accumulation above and below the fatigue limit, *The behaviour of short fatigue cracks* (Edited by K. J. Miller and E. R. de los Rios), (Mechanical Engineering Publications, London), pp. 491–511 (This volume).

(25) GAO, H., BROWN, M. W., and MILLER, K. J. (1982) Mixed-mode fatigue thresholds, *Fatigue Engng Mater. Structures*, **5**, 1–17.

(26) PEREZ CARBONELL, E. and BROWN, M. W. (1986) A study of short crack growth in torsional low cycle fatigue for a medium carbon steel, *Fatigue Engng Mater. Structures*, **9**, 15–33.

(27) KLESNIL, M. and LUKAS, P. (1969) Dislocation substructure associated with propagating fatigue crack, *2nd Int. Conf. Fracture*, Brighton, pp. 725–730.

(28) BOETTNER, R. C., LAIRD, C., and McEVILY, A. J. (1965) Crack nucleation and growth in high strain low cycle fatigue, *Trans Met. Soc. AIME*, **233**, 379–387.

(29) HAIGH, J. R. and SKELTON, R. P. (1978) A strain intensity approach to high temperature fatigue crack growth and failure, *Mater. Sci. Engng*, **36**, 133–137.

(30) TOMKINS, B. (1975) The development of fatigue crack propagation models for engineering applications at elevated temperatures, *J. Engng Mater. Tech.*, **97H**, 289–297.

(31) KITAGAWA, H., YUUKI, R., TOHGO, K., and TANABE, M. (1985) ΔK-dependency of fatigue growth of single and mixed mode cracks under biaxial stress, *ASTM STP 853* (American Society for Testing and Materials, Philadelphia), pp. 164–183.

(32) BROWN, M. W. and MILLER, K. J. (1985) Mode I fatigue crack growth under biaxial stress at room and elevated temperature, *ASTM STP 853*, pp. 135–152.

(33) BROWN, M. W., DE LOS RIOS, E. R., and MILLER, K. J. (1984) A critical comparison of proposed parameters for high strain fatigue crack growth, *ASTM Symp. on Fundamental Questions and Critical Experiments in Fatigue*, Dallas.

(34) MILLER, K. J. and BROWN, M. W. (Editors) (1985) *Multiaxial fatigue, ASTM STP 853* (American Society for Testing and Materials, Philadelphia), pp. 203–377.

(35) BLOM, A. F., FATHULLA, A., HEDLUND, A., STICKLER, R., and WEISS, B. (1986) Short fatigue crack growth behaviour in Al-2024 and Al-7475 alloys, *The behaviour of short fatigue cracks* (Edited by K. J. Miller and E. R. de los Rios), (Mechanical Engineering Publications, London), pp. 37–66 (This volume).

(36) KOMPEK, G., MATZER, F. E., and MAURER, K. L. (1982) Crack initiation and crack propagation by torsional fatigue in low carbon steels, *Fracture and the role of microstructure* (Edited by K. L. Mauer and F. E. Matzer), (EMAS, Warley), Vol. 2, pp. 398–406.

(37) OHJI, K., OGURA, K., and HARADA, S. (1975) Observation of low cycle fatigue crack initiation and propagation in anisotropic rolled steel under biaxial stressing, *Bull. Japan Soc. Mech. Engrs*, **18**, 17–24.

(38) GUIU, F., DULNIAK, R., and EDWARDS, B. C. (1982) On the nucleation of fatigue cracks in pure polycrystalline α-iron, *Fatigue Engng Mater. Structures*, **5**, 311–321.

P. D. Hobson,* M. W. Brown,* and E. R. de los Rios*

Two Phases of Short Crack Growth in a Medium Carbon Steel

REFERENCE Hobson, P. D., Brown, M. W., and de los Rios, E. R. **Two Phases of Short Crack Growth in a Medium Carbon Steel**, *The Behaviour of Short Fatigue Cracks*, EGF Pub. 1 (Edited by K. J. Miller and E. R. de los Rios) 1986, Mechanical Engineering Publications, London, pp. 441–459.

ABSTRACT Many studies of cracks growing under high strain fatigue conditions have revealed growth rates in excess of those predicted by LEFM analyses. In addition, anomalously high growth rates have been measured for very small cracks, of length less than typical microstructural dimensions. Both types of fatigue crack have been classified as 'short'.

An experimental fatigue study using medium carbon steel specimens has shown that both types of crack growth occur. Surface replication enables individual small cracks to be monitored throughout a constant load amplitude test, and their growth rates to be determined. Two equations have been derived, corresponding to each type of short crack growth. These equations calculate the relative proportions of each phase of the fatigue life and therefore they may be used to predict the *S–N* curve for medium carbon steel. The relationship of crack development to microstructural features is also discussed.

Introduction

Although anomalies in the growth of short cracks have been noted for several years (**1**)(**2**), attempts to model such behaviour have only been made recently. The heightened interest in this area of crack growth arises not only from the practical need for increased levels of service stress but also from the inability of established methods of fatigue analysis, such as fracture mechanics, to provide adequate mathematical models. One of the fundamental assumptions in fracture mechanics, that a material is an isotropic continuum, has been shown not to be valid in the case of short crack growth for several materials (e.g., (**3**)–(**5**)) in which microstructural features have been observed to play a dominant role. Using a study of an aluminium alloy (**6**), a crack growth equation has been suggested which incorporates a microstructural dimension to describe the growth of a crack in a single grain (**7**). Combining a general form of this equation together with a second equation to represent crack growth beyond the first grain, it is shown here that the complete history of crack growth for a medium carbon steel may be described adequately.

Experimental work

All fatigue tests were performed in fully reversed tension–compression loading on a servo-hydraulic fatigue testing machine with a load capacity of ±250 kN.

* Department of Mechanical Engineering, University of Sheffield, Mappin Street, Sheffield, UK.

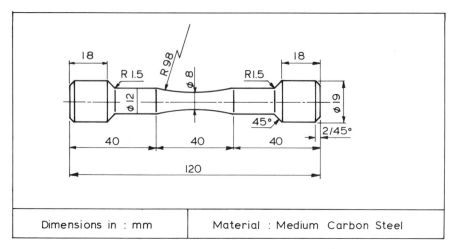

| Dimensions in : mm | Material : Medium Carbon Steel |

Fig 1   Specimen geometry

The material tested was a medium carbon steel of composition (wt%) 0.4C, 1.0Mn, 0.10Si, 0.001S, 0.005P, with yield and tensile stengths of 392 MPa and 683 MPa, respectively, an elongation to fracture of 44 per cent, and a reduction in area of 36 per cent.

Specimens having a mild hour glass profile were used, as shown in Fig. 1, because this shape limited the area of cracking to the centre of the specimen, thus restricting the region requiring replication to locate cracks. To replicate the surface of the specimen, acetate sheet was softened with the solvent acetone and pressed lightly onto the specimen. For each test, replicas were taken for at least seven stages in life. As most interest was in the growth of short cracks, replicas were taken more frequently during the initial period. Cycling was stopped at maximum tensile load so that the cracks were held open during replication.

Tests with surface replication were performed in load control at stress levels of 998, 816, and 638 MPa, corresponding to lifetimes of approximately 1000, 6000 and 30 000 cycles, respectively. Several tests were carried out at each load range to enable a number of cracks to be studied. A triangular waveform was used at frequencies in the range 0.016–0.3 Hz.

Strain controlled tests were also performed with a diametral extensometer monitoring extension at the minimum diameter of the specimen. Results from all the tests are shown in Figs 2 and 3. In Fig. 2 a least-squares analysis was used for data where $\Delta\sigma > 450$ MPa, in order to model the plastic portion of the cyclic stress–strain curve. Also shown in this figure is the line representing the linear elastic response. The stable cyclic stress–strain curve in Fig. 2 shows both elastic and power law strain-hardening regions in the classical manner, and therefore this behaviour can be adequately represented by the Ramberg–Osgood formulation

$$\Delta\varepsilon = \Delta\sigma/E + (\Delta\sigma/k)^{1/n}$$
$$= \Delta\sigma/203\,000 + (\Delta\sigma/2013)^{1/0.19} \tag{1}$$

Fatigue failure was defined as fracture in load controlled tests, but for strain control a more suitable definition was a 5 per cent decay in the measured load range from the stable level. The diametral extensometer was not used for the load controlled tests as it would have been necessary to remove the extensometer to enable replication of the surface cracks. However, one load controlled experiment was conducted while monitoring the strain to compare with the strain controlled fatigue strength. Figure 3(a) shows that slightly longer life was found for load control, and that the strain controlled life may be found from the Coffin–Manson equation

$$\Delta\varepsilon_p \cdot N_f^{0.673} = 2.23 \tag{2}$$

The S–N curve is shown with a stress scale in Fig. 3(b), giving for strain controlled tests the Basquin equation

$$\Delta\sigma \cdot N_f^{0.137} = 2553 \tag{3}$$

The difference in lifetime observed between load and strain controlled tests may be attributed to the choice of failure definition. Assuming that a 5 per cent load decay would correspond to a total crack area of 5 per cent for the low cycle strain controlled tests, and that there are a number of independent fast growing

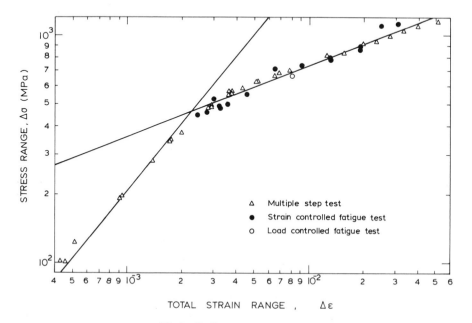

Fig 2 Cyclic stress–strain curve

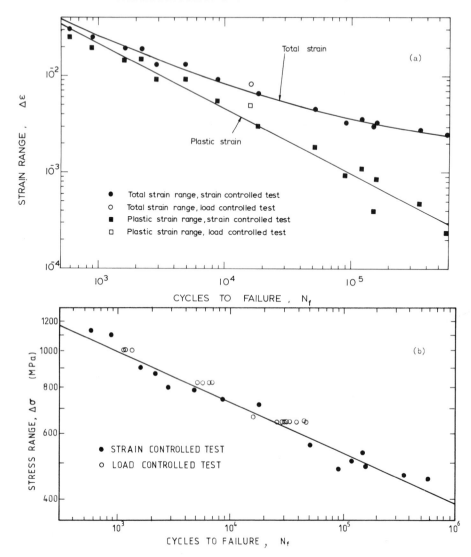

Fig 3   **(a) Total and plastic strain life relationships.**
            (b) The S–N curve

cracks, say five, of semi-circular shape, then five such cracks correspond to a crack area of $5(\pi R^2/2)$ for cracks of radius R. If 5 per cent of the cross-section is cracked, this must be equal to $(0.05\pi R_1^2)$ where R_1 is the radius of the specimen cross-section, which implies for $R_1 = 4$ mm a value for $2R$ of 1.1 mm for the surface initiation length. This failure crack length will clearly give a reduced endurance compared to the load controlled case where specimen fracture occurred, which gives a surface crack length of 4 mm required to raise the net section stress to the tensile strength.

Full characterization of the microstructure of this material is given elsewhere (5). Briefly it is a typical normalized structure of a 0.4% C steel where the ferrite phase, in the form of long ferrite plates, has formed along the previous austenite grain boundaries. The ferrite plates have an average thickness of 8 μm and an average length of 116 μm. The volume fraction of ferrite is approximately 5 per cent; the rest is pearlite.

Plastic replicas taken from polished and etched specimens showed that cracks initiated in, and grew along, the ferrite plates. Short cracks contained within a single plate exhibited faster growth rates than those that would be predicted by elasto-plastic fracture mechanics. As a crack approached the edge of the ferrite plate that contained the initiation site, the crack growth rate decreased. However, once the crack had extended beyond one ferrite plate it showed a steadily increasing growth rate with subsequent microstructural variations apparently having little effect on it.

This type of growth behaviour is shown in Fig. 4, the plotted data points are surface crack lengths, a_s, obtained during life at the lowest stress level tests, namely 638 MPa. Here the minimum crack growth rate occurred at a crack length of 85 μm after only 6000 cycles out of a life of 29 300 cycles. The growth curve beyond 10 000 cycles conforms with established continuum fracture mechanics predictions, i.e.

$$da/dN = f(\Delta\varepsilon) \cdot a \tag{4}$$

which gives on integration

$$\ln(a) = f(\Delta\varepsilon) \cdot N + A \tag{5}$$

This is a straight line for a constant strain amplitude test, as in Fig. 4. However, for microstructurally short cracks, classical fracture mechanics cannot be employed and a new model for crack propagation must be used.

Theoretical model for microstructural short crack growth

In order to formulate a crack growth equation to model the behaviour of microstructurally short cracks, it is essential to account for microstructural influences. By considering a list of parameters which might be expected to affect the propagation of a crack embedded within the grain containing the nucleation site, a general expression for crack growth may be postulated (8), using the method of dimensional analysis

$$da/dN = f(\Delta\sigma, \Delta\varepsilon, E, k, n, \sigma_y, a, d) \tag{6}$$

where f is an unknown function, a is the crack length, σ_y is the yield stress and d is a characteristic dimension representative of the distance between adjacent microstructural obstacles to crack propagation. For stresses greater than yield

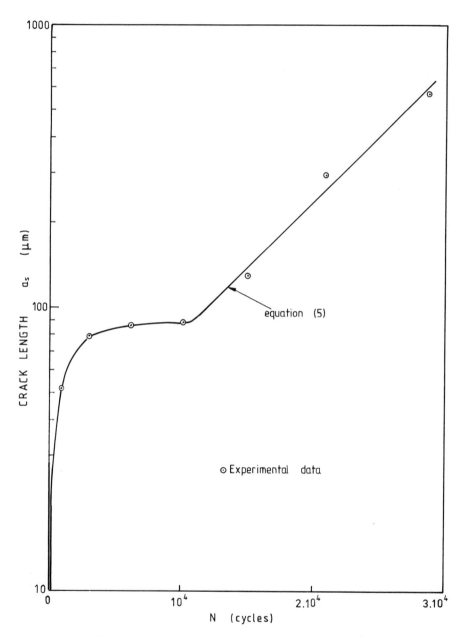

Fig 4   Crack growth curve for one crack with $\Delta\sigma = 638$ MPa, $N_f = 29\,300$

$\Delta\sigma = k\,\Delta\varepsilon_p^n$ as in equation (1), and hence equation (6) includes sufficient parameters to describe the cyclic deformation behaviour of a given material in addition to some microstructural influences. Thus $\Delta\varepsilon$ is a dependent variable and may be excluded from equation (6). The grain size or any other single dominant feature may be represented by the parameter d.

Dividing the left-hand side of equation (6) by a (with $\Delta\varepsilon$ eliminated), and collecting all the quantities into dimensionless groups, one obtains

$$\frac{1}{a}\frac{da}{dN} = f\left(\frac{\Delta\sigma}{k}, \frac{E}{k}, n, \frac{\sigma_y}{k}, \frac{d}{a}\right) \tag{7}$$

Replacing d by $(d-a)$ for convenience on the right-hand side, and adopting a series form leads to

$$\frac{1}{a}\frac{da}{dN} = \sum_i C_i\left(\frac{d-a}{a}\right)^{1-\alpha_i} \tag{8}$$

where

$$C_i = C_i\left(\frac{\Delta\sigma}{k}, \frac{E}{k}, n, \frac{\sigma_y}{k}\right)$$

However, if only the first term of the series in equation (8) is employed, the solution

$$\frac{da}{dN} = Ca^\alpha(d-a)^{1-\alpha} \tag{9}$$

is obtained. This crack growth equation has previously been postulated to describe the propagation of short cracks in an aluminium alloy (7). It can also describe the growth of cracks under elastic–plastic conditions when $\alpha = 1$, as characterized by equation (4).

Application of the short crack model

Comparing the experimental observations with equation (9), a value of d equal to the ferrite plate length seems to provide a reasonable choice, such that the propagation rate decreases on approaching the edge of the ferrite. However, in practice it was difficult to measure this dimension for each crack, so a value for d was calculated for each crack using the following method. Firstly, the values of da_s/dN were plotted (where a_s is surface crack length), as shown schematically in Fig. 5. The growth rate was calculated from the secant formula for adjacent pairs of (a, N) data, and plotted against the mean value of a over the intervening cycles (9). A least squares analysis was performed for those data points which showed a decrease in growth rate with increasing crack length. The point of intersection of the regression line with the abscissa in Fig. 5 gives the value of d, as equation (9) predicts zero crack growth for $a_s = d$. This

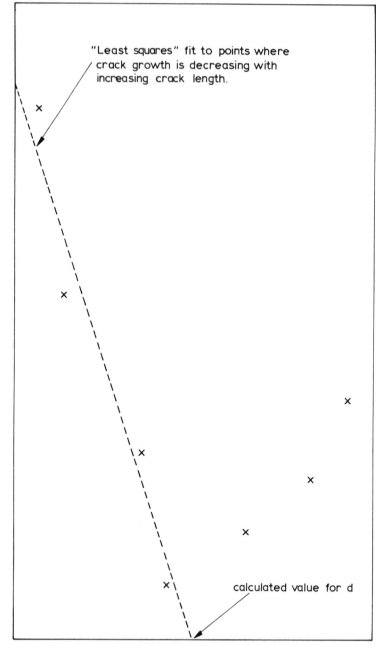

Fig 5   Schematic of calculation procedure to determine *d*

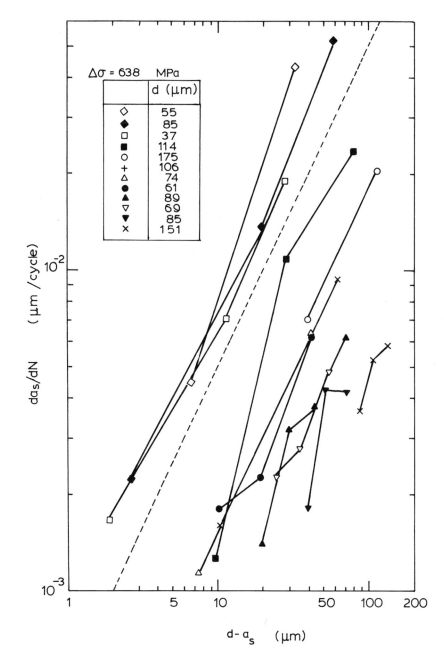

Fig 6   Observed growth rates for microstructurally short cracks

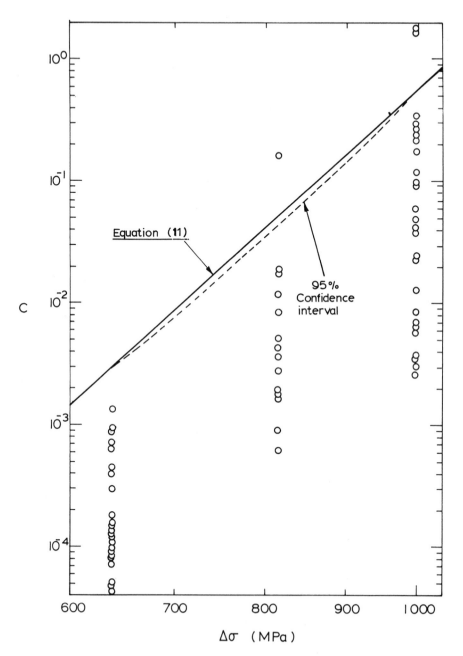

Fig 7   Dependence of the crack growth rate constant C on stress range. Dashed line shows the upper 95 per cent confidence interval

method covers the possibility that the propagation rate minimum may not correspond exactly to the instant where the crack is at the end of a ferrite plate.

In order to be of practical use, equation (9) must predict a decreasing crack growth rate as a_s approaches d. To satisfy this condition, it can be shown (8) that α must lie in the range $0 \leq \alpha < 1$. With a value for α of zero, equation (9) becomes, for a surface crack

$$\frac{da_s}{dN} = C(d - a_s) \tag{10}$$

Using the above calculated crack growth rate, da_s/dN, a plot of the short crack growth data for $\Delta\sigma = 638$ MPa is shown in Fig. 6, for twelve separate fatigue cracks. The dotted line represents an example of equation (10), which defines a slope of unity for lines connecting successive data points assuming $\alpha = 0$. It can be seen that this value of α defines a slope that gives a reasonable representation of microstructural short crack behaviour.

From equation (10), a value of C was calculated for every successive pair of replica data points where $a < d$. Then all these values of C were plotted against the relevant stress range, as shown in Fig. 7. This produces scatter in the values of C, but such dispersion is to be expected due to some cracks being able to propagate much more quickly than others, because of variations in the crystallographic orientations of individual grains.

It can be reasonably assumed that fast growing cracks are more likely to be the cause of final failure than those cracks which are growing slowly, so any equation expressing C as a function of stress, should take this spread of data into consideration. Therefore a 95 per cent confidence interval was taken, shown by the dotted line in Fig. 7, as being a reasonable statistical representation of the fastest observed cracks. Approximately twenty cracks were studied for each stress level, so a 95 per cent confidence interval is representative of the fastest growing crack that was measured on the replicas. Taking values for a 95 per cent confidence interval at both the lower stress of 638 MPa and the upper stress of 998 MPa, a straight line between these points is given by the equation

$$C = 1.64 \times 10^{-34}(\Delta\sigma)^{11.14} \tag{11}$$

Substituting for C into equation (10) gives

$$\frac{da_s}{dN} = 1.64 \times 10^{-34}(\Delta\sigma)^{11.14}(d - a_s) \tag{12}$$

where $\Delta\sigma$ is measured in MPa.

Continuum fracture mechanics crack growth

Crack growth data were obtained from replicas where the surface crack length (a_s) was greater than d for each of the three stress levels. An equation of the form

$$\frac{da_s}{dN} = Ga_s - D \tag{13}$$

was assumed to describe the crack growth rate where G is dependent on the stress level. This corresponds with equation (4), but with an added constant D to represent the crack growth threshold. Since equation (13) cannot accommodate microstructural influences on crack growth, a regression line was fitted to the data points obtained from replicas where the crack length was greater than 400 microns. By expressing G as a function of the total strain range, equation (13) became (8)

$$\frac{da_s}{dN} = 4.10(\Delta\varepsilon_t)^{2.06}a_s - D \tag{14}$$

where a_s is measured in microns.

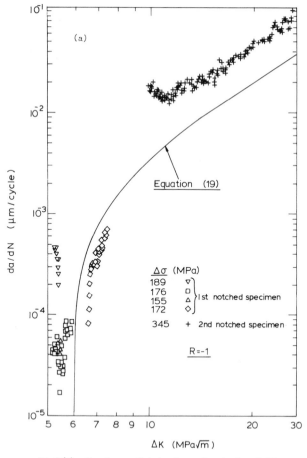

Fig 8(a)   Crack growth behaviour close to threshold

To determine D, use was made of threshold data obtained on a notched specimen. These data, shown in Fig. 8(a), have an average value for ΔK_{th} of 6.0 MPa$\sqrt{\text{m}}$, where threshold is taken at a propagation rate of 10^{-10} m/cycle following the normal convention. The specimen used was of the same geometry employed in the fatigue tests, loaded in fully reversed tension–compression at a frequency of 16 Hz with a triangular waveform. Rearranging the stress intensity equation, namely

$$\Delta K = Y \, \Delta\sigma\sqrt{(\pi a)} \tag{15}$$

at threshold conditions

$$a_s = \frac{2 \, \Delta K_{th}^2}{\pi Y^2 \, \Delta\sigma^2} \tag{16}$$

when $a_s = 2a$, assuming a semicircular crack shape. Noting that for linear

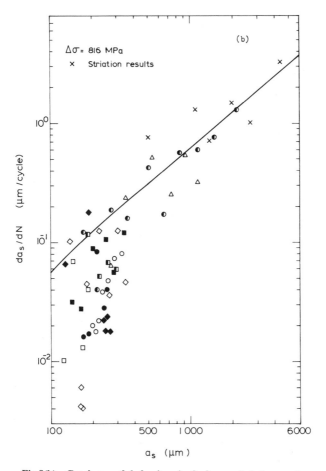

Fig 8(b)   Crack growth behaviour in the low-cycle fatigue regime

elastic conditions $\Delta\sigma = E\Delta\varepsilon$, and taking $\Delta K_{th} = 6.0$ MPa$\sqrt{m}$, $Y = 2/\pi$, $E = 203$ GPa, equations (14) and (16) give for threshold conditions of zero crack growth rate

$$D = 5.63 \times 10^{-3}(\Delta\varepsilon_t)^{0.06} \qquad (17)$$

Although D appears to vary with strain range, the lowest strain level used in the fatigue tests gave a value for $(\Delta\varepsilon_t)^{0.06}$ of 0.70 and the highest strain level gave $(\Delta\varepsilon_t)^{0.06}$ a value of 0.81. Thus D did not vary significantly with strain level, and an average value of $(\Delta\varepsilon_t)^{0.06}$ equal to 0.75 may be used to calculate D, giving, in equation (14)

$$\frac{da_s}{dN} = 4.10(\Delta\varepsilon_t)^{2.06}a_s - 4.24 \times 10^{-3} \qquad (18)$$

which was assumed to describe crack growth for $a_s > d$ at all strain levels.

This equation is given by the line in Fig. 8(b) showing for $a_s > 400\ \mu$m a fair fit to the data, and further justification for equation (18) is provided by the striation spacing measurements in the figure, which reflect crack growth rates on the fracture surface. Each individual crack is represented by a different symbol in this figure. By substituting for a_s in equation (18) using the relationship between a_s and ΔK in equation (15) gives

$$\frac{da}{dN} = \frac{1}{2}(1.177 \times 10^{-4}\Delta K^2 - 4.237 \times 10^{-3}) \qquad (19)$$

which is also plotted in Fig. 8(a). It can be seen that the equation is an approximate fit to this low stress data, especially when considering that it is an extrapolation from tests conducted at higher strains.

Fatigue lifetime calculations

The preceding sections have shown the development of two equations, (12) and (18), to describe the rates of the respective phases of crack growth. Next it is necessary to determine the range of valid crack lengths for the application of each equation.

Equation (12) predicts crack arrest when $a = d$, which implies that the crack cannot continue propagating for $a > d$, unless equation (18) begins to contribute to crack advance. So to apply the two crack growth equations three regions of integration may be defined (7), as follows.

(i)　Microstructural crack zone
This is defined to be the region where the crack grows from its initial length (assumed here to be the peak to trough surface roughness measurement R_t of 0.4 μm) to the crack length corresponding to the threshold stress intensity, a_{th}, which was calculated for each strain range by taking the value of a_s which makes da_s/dN zero in equation (18).

Table 1 Fatigue lifetime calculations

Cyclic ranges			Threshold crack lengths		Calculated lifetimes (Number of cycles)			Calculated lifetime	Actual lifetime
	Strain		Short cracks	Long cracks	Zone 1	Zone 2	Zone 3	Using equations (12) and (18)	Using equation (3)
Stress	Plastic	Total	$d\,(\mu m)$	$a_{th}\,(\mu m)$	Eqn (12)	Eqn (12) + (18)	Eqn (18)		
MPa								Number of cycles	
998	0.0250	0.0261	116.37	1.89	0	12	1586	1598	1254
816	0.0086	0.0138	116.37	7.08	1	97	6014	6112	6214
700	0.0039	0.0085	116.37	19.29	22	455	16907	17384	12643
638	0.0024	0.0063	116.37	35.20	122	1276	32308	33706	35778
550	0.0011	0.0039	116.37	93.41	2912	4907	113248	121067	73508

(ii) The interactive zone

This is the zone of crack growth where both the microstructural and fracture mechanics 'short' growth mechanisms, equations (12) and (18) respectively, may operate. Here the crack extends from a length a_{th} to length d, which for the purposes of integration was taken to be the average ferrite plate length observed, i.e., 116 μm.

(iii) Continuum fracture mechanics zone

This is defined to be the final region of accelerating crack growth from the end of the first ferrite plate ($a = d$) to the crack length at failure, a_f. A value of 4.0 mm was chosen for a_f representing half the diameter of the specimen, since for load controlled tests final fracture corresponds to a crack of such magnitude.

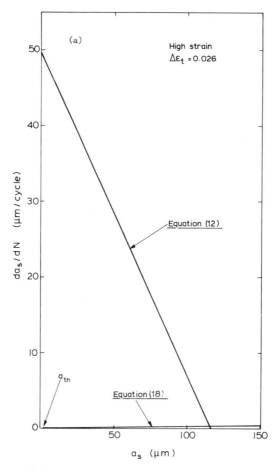

Fig 9(a)   Crack growth characteristics at high strain range

By summing the three areas of crack growth the lifetime was calculated. Results of the integrations for some typical strain levels are shown in Table 1, along with the actual lifetimes obtained from equation (3). Graphs showing equations (12) and (18) for the highest and lowest strain levels are shown in Fig. 9.

It can be seen from the results of Table 1 that the calculated lifetimes are in good agreement with actual fatigue lifetimes in load controlled tests. Even for the lowest stress range, less than 2.5 per cent of the calculated lifetime is spent in region 1, suggesting that the number of cycles spent in initiating a crack may be taken as zero.

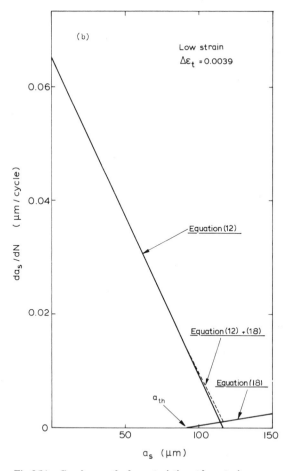

Fig 9(b)   Crack growth characteristics at low strain range

Discussion

Two equations have been derived to describe the growth of cracks that are (a) short compared to dominant microstructural features, and (b) larger than those features. Equation (12) for short cracks, that is, short with respect to microstructure, shows much faster crack growth than the subsequent continuum fracture mechanics stage, described by equation (18), implying that the major proportion of lifetime is spent in propagating longer cracks, as shown in Table 1. Indeed the contribution of the initial microstructural short crack period to overall endurance is negligible, which concurs with the widely held viewpoint that low cycle fatigue can be described in terms of integrated crack propagation laws (10).

Having derived two equations for crack growth, and noting that no period of nucleation is required for fatigue cracks, one is in a position to attempt predictions of fatigue behaviour under variable amplitude loading. Conventionally damage is defined in terms of cycle ratio (N/N_f) where N cycles are applied at a strain range corresponding to N_f cycles to failure, but the concept of damage can be replaced by a more satisfactory physical measure of damage, i.e., crack length. Indeed it has previously been demonstrated that (N/N_f) and the logarithm of crack length are directly related (11)(12). It remains, however, to include the effect of mean stress in equations (12) and (18) to obtain greater generality in the predictive capability for variable amplitude behaviour.

It is recognized that equation (12) incorporates the effect of microstructure on the simplest level through the term d. Other factors that influence crack growth should include the orientation of both the slip plane and the slip direction to the loading axis. Secondly, the representation of detailed microstructure by one dimension reflecting just a single barrier to growth is, for many microstructures, a gross simplification of actual behaviour. However, an engineering approach has been adopted in deriving equation (12), seeking to obtain an expression that is easy to apply while retaining the most important features of the growth of those cracks that eventually determine fatigue endurance.

Metallographic studies of the etched carbon steel samples showed that the values of d derived mathematically for a number of short cracks corresponded closely to the point where short cracks ran into a pearlite region. Thus, in this instance, the dominant or strong barrier to short crack growth in ferrite is provided by the ferrite/pearlite boundaries at the edges of the ferrite plates in which the cracks initiate. This conforms with the behaviour of microstructurally short cracks observed at the fatigue limit in torsional tests for the same material (5).

Conclusions

(1) Two equations have been derived to describe the propagation of fatigue cracks: one for microstructurally dominated short crack growth and one for elasto-plastic fracture mechanics controlled growth.

(2)   The period for crack nucleation in this medium carbon steel has been shown to be negligibly small.

(3)   Fatigue lifetime may be predicted by integration of the two crack growth laws, giving a good representation of the $S-N$ curve.

(4)   Pearlite provides a barrier to continued short crack propagation in a medium carbon steel where cracks form in ferrite plates.

Acknowledgements

The authors would like to thank the Central Electricity Generating Board for their sponsorship of the project, through an SERC Case Award (PDH), and a research fellowship (MWB). On of the authors (ERdlR) is also indebted to the Rio Tinto Zinc Corporation for the award of a research fellowship.

References

(1)   HUNTER, M. S. and FRICK, W. M. G., Jr. (1956) Fatigue crack propagation in aluminium alloy, *Proc. ASTM*, **56**, 1038–1046.

(2)   DE LANGE, R. G. (1964) Plastic replica methods applied to a study of fatigue crack propagation in steel 35CD4 and 26 SE aluminium alloy, *Trans AIME*, **230**, 644–648.

(3)   LANKFORD, J. (1977) Initiation and early crack growth of fatigue cracks in high strength steels, *Engng Fracture Mech.*, **9**, 617–624.

(4)   BROWN, C. W. and HICKS, M. A. (1983) A study of short fatigue crack growth behaviour in titanium alloy IMI 685, *Fatigue Engng Mater. Structures*, **6**, 67–75.

(5)   DE LOS RIOS, E. R., TANG, Z., and MILLER, K. J. (1984) Short crack fatigue behaviour in a medium carbon steel, *Fatigue Engng Mater. Structures*, **7**, 97–108.

(6)   LANKFORD, J. (1982) The growth of small fatigue cracks in T6-7075 aluminium alloy, *Fatigue Engng Mater. Structures*, **5**, 233–248.

(7)   HOBSON, P. D. (1982) The formulation of a crack growth equation for short cracks, *Fatigue Engng Mater. Structures*, **5**, 323–327.

(8)   HOBSON, P. D. (1985) *The growth of short fatigue cracks in a medium carbon steel*, PhD thesis, University of Sheffield.

(9)   ASTM Standard E647-83 (1983) *Constant load amplitude fatigue crack growth rates above 10^{-8} m/cycle*, Appendix XI.

(10) WAREING, J. (1983) Mechanisms of high temperature fatigue and creep-fatigue failure in engineering materials, *Fatigue at high temperature* (Edited by R. P. Skelton) (Applied Science Publishers, London), pp. 135–185.

(11) MILLER, K. J. (1984) The propagation behaviour of short fatigue cracks, *Subcritical crack growth due to fatigue, stress corrosion and creep* (Edited by L. H. Larsson), (Elsevier Applied Science Publishers, London), pp. 151–166.

(12) MILLER, K. J. and IBRAHIM, M. F. E. (1981) Damage accumulation during initiation and short crack growth regimes, *Fatigue Engng Mater. Structures*, **4**, 263–277.

*H. Nisitani** and M. Goto†

A Small-Crack Growth Law and its Application to the Evaluation of Fatigue Life

REFERENCE Nisitani, H. and Goto, M., **A Small-Crack Growth Law and its Application to the Evaluation of Fatigue Life**, *The Behaviour of Short Fatigue Cracks*, EGF Pub. 1 (Edited by K. J. Miller and E. R. de los Rios) 1986, Mechanical Engineering Publications, London, pp. 461–478.

ABSTRACT In fatigue tests on carbon steels, the life of a plain specimen is occupied mainly by the life in which a crack propagates from an initial size up to about 1 mm. This means that the growth law of a small crack must be known in order to evaluate the fatigue life of plain members.

The growth rate of a small crack cannot be predicted usually by linear long-crack fracture mechanics, but is determined uniquely by the term $\sigma_a^n l$ where σ_a is the stress amplitude, l is the crack length, and n is a constant. However, if the condition of small scale yielding is satisfied during the propagation of such a small crack, it is possible to quantify the behaviour by linear fracture mechanics.

In this paper, a unifying treatment for small-crack growth is presented, and a method for determining fatigue life is suggested.

Introduction

The fatigue life of a plain specimen is controlled mainly by the life in which a crack propagates from a certain initial size up to about 1 mm. This means that we must know the growth law of small cracks if we want to predict the fatigue life of the specimen. Linear fracture mechanics cannot be applied usually to such a small crack. The fatigue life of a structure may also be closely related to the growth of small cracks.

Many investigations of fatigue crack propagation have been carried out and many growth laws have been reported (**1**). All of these growth laws have succeeded in describing crack growth behaviour over a limited experimental programme, though some equations formally contradict each other.

It seems natural to use the crack tip opening displacement (CTOD) as a measure of control in fatigue crack propagation studies (**2**)(**3**). However, the definition of CTOD is not clear, and is somewhat obscure and difficult to measure in polycrystalline metals; therefore, using an assumption that crack growth rate dl/dN is proportional to the reversible plastic zone size (r_{pr}) (**4**), since r_{pr} seems to be substantially similar to CTOD, a unified explanation of fatigue crack growth laws was made by Nisitani (**5**). He suggested two crack growth laws, that is, $dl/dN = C\,\Delta K^m$ for low nominal stresses and $dl/dN = B\sigma_a^n l$ (**6**)–(**10**) for high nominal stresses.

* Faculty of Engineering, Kyushu University, Fukuoka, 812 Japan.
† Faculty of Engineering, Oita University, Oita, 870–11 Japan.

In this paper, fatigue tests under axial loading on plain steel specimens are reported, and the unifying treatment for fatigue crack growth under low and high nominal stresses is shown from data obtained on small cracks on the specimen surface. Moreover, a method of fatigue life prediction based on the fatigue crack growth laws is suggested.

Material, specimen, and experimental procedures

The material used in a rolled round bar (about 18 mm in diameter) of 0.45% C steel. The specimens were machined from the bar after annealing for 60 minutes at 845°C. The chemical composition (wt%) was 0.45C, 0.25Si, 0.79Mn, 0.01P, 0.01S, 0.09Cu, 0.03Ni, 0.18Cr, remainder Ferrite and mechanical properties are 364 MPa lower yield stress, 631 MPa tensile strength, 1156 MPa true breaking stress, and 45.8 per cent reduction of area. The shape and dimensions of the specimen are shown in Fig. 1(a). Although the specimen has a fine shallow partial notch, see Fig. 1(b) its strength reduction factor is only about 1.1 and can be considered as a plain specimen. Before testing, all the specimens were re-annealed in vacuum at 600°C for 60 minutes in order to remove residual stresses, and were then electro-polished to remove about 20

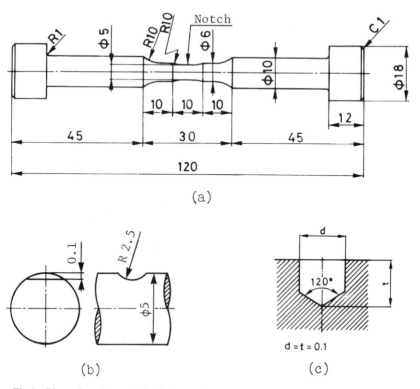

Fig 1   Dimensions (in mm) of: (a) the specimen, (b) the transverse notch, (c) the hole

μm from the surface layer. In some specimens, a small hole was drilled on the central part of the shallow notch after electro-polishing. The shape and dimensions of the hole are shown in Fig. 1(c). Both the diameter and depth of the hole are 0.1 mm.

All the tests were carried out under constant stress using an electrohydraulic control tension–compression fatigue machine at a frequency of 5–32 Hz. The observations of fatigue damage on the specimen surface and the measurement of crack length were made via plastic replicas using an optical microscope ($\times 400$). The crack length means the length along the circumferential direction on the specimen surface and when appropriate includes the 0.1 mm diameter of the hole.

The value of stress in this paper means the nominal stress σ_a at the minimum cross section (5 mm in diameter) calculated by neglecting the existence of the shallow notch and the small hole. The stress range $\Delta\sigma$ is given by $2\sigma_a$.

Experimental results and discussion

Crack initiation and propagation behaviour in plain specimens

Figure 2 shows the change in surface state. Until the initiation of a crack, fatigue damage is accumulated gradually at the same region whose dimension is closely related to the grain size, and then this region turns into a crack. After initiation, the crack propagates by the concentration of stress and strain near the crack tip and then final fracture occurs. It can be concluded that the crack initiation process is essentially different from the crack propagation process (**11**).

Figure 3(a) shows the relation between crack length and the relative number of cycles N/N_f. It is found that a crack initiates at $N/N_f \cong 0.2$ and the fatigue life of a plain specimen is controlled mainly by the life in which a crack propagates from an initial size ($\cong 10 \ \mu$m in this material) to about 1 mm; this life is about $0.7N_f$. This means that we must known the growth law of small cracks if we want to predict the fatigue life of plain members. In Fig. 3(a), the marks A, B, C, D, E, F, and G correspond to the marks in Fig. 2.

Figure 3(b) shows the relation between the crack growth rate and the stress intensity factor range ΔK. Here ΔK is the effective parameter for the propagation of large cracks in which the condition of small scale yielding is satisfied. However, dl/dN of a small crack under a high stress is not determined uniquely by ΔK, as shown in this figure.

Figure 3(c) shows the relation between dl/dN and the term $\sigma_a^n l$. The growth rate of a small crack is determined uniquely by $\sigma_a^n l$ (**6**)–(**10**) and this can be seen more clearly in later figures.

Figure 4(a) shows the dependency of dl/dN on crack length l. For every constant stress range, a straight line can be drawn approximately for crack

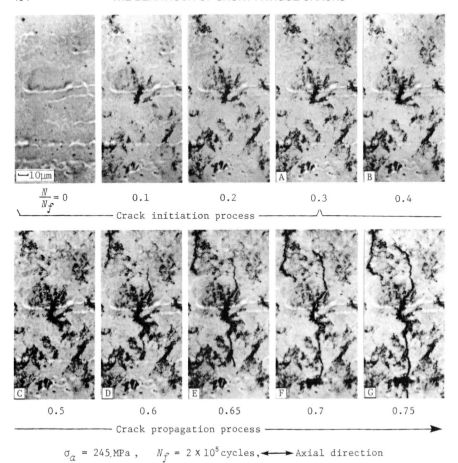

Fig 2   Change in surface states of a plain specimen. (Marks A, B, C, D, E, F, and G correspond to those in Fig. 3(a))

lengths smaller than 1 mm. The slope is about unity. Accordingly, dl/dN is proportional to l for plain specimens.

Figure 4(b) shows the dependency of dl/dN on stress amplitude σ_a. In the high stress levels above the fatigue limit, the dependency is expressed by the relation dl/dN $\propto \sigma_a^n$. The value of n is constant and about 8 in this case. The dotted lines show the results obtained from drilled specimens and will be mentioned in the following section.

Putting the results of Fig. 4(a) and Fig. 4(b) together, we can obtain the growth law of a small crack, i.e.

$$\frac{\mathrm{d}l}{\mathrm{d}N} = B\sigma_a^n l \qquad (1)$$

where n is approximately 8.

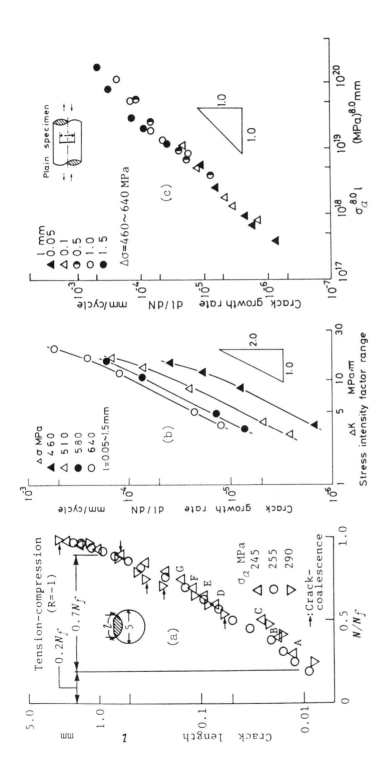

Fig 3 Crack initiation and propagation in a 0.45% C steel plain specimen.
(a) A crack initiates at $N/N_f = 0.2$. The value of l is mainly controlled by N/N_f alone
(b) dl/dN is not determined uniquely by ΔK in the case of $\sigma_a > 0.6\sigma_y$
(c) dl/dN is not determined by $\sigma_a^8 l$ in the case of $\sigma_a > 0.6\sigma_y$

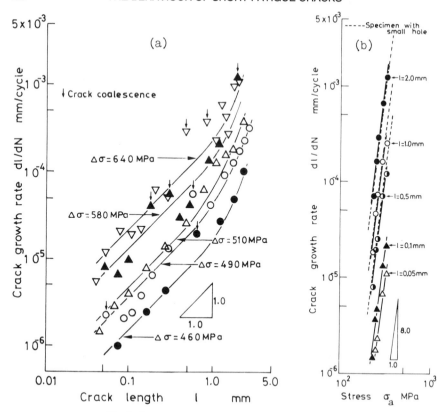

Fig 4   Propagation behaviour of small cracks.
(a) d*l*/d*N* versus *l* relations in plain specimens
(b) d*l*/d*N* versus σ_a relations in plain specimens and specimens with a small hole

Crack propagation behaviour in drilled specimens

The fatigue tests on drilled specimens and 0.5 mm pre-cracked specimens were also carried out to measure the crack growth rate at stress levels below the fatigue limit of plain specimens.

Figure 5 shows the dependency of d*l*/d*N* on *l*. For every constant stress range, a straight line can be drawn approximately for cracks smaller than 1 mm. The slope of the straight lines in the range of $\Delta\sigma > 420$ MPa (approx.) is about unity. The slope increases with decrease in $\Delta\sigma$ and is about 2 in the range of $\Delta\sigma < 380$ MPa (approx.). Thus the dependency of d*l*/d*N* on *l* is influenced by the stress level.

Figure 6 shows the results obtained from all of the specimens (plain specimens, drilled specimens, and pre-cracked specimens). Figure 6(a) shows the d*l*/d*N* versus ΔK relation and if the stress is relatively low ($\sigma_a < 0.5\sigma_y$ (approx.), where σ_y is the yield stress), dl/dN is determined by ΔK alone and the following relation holds

$$\frac{dl}{dN} = C\Delta K^m \qquad (2)$$

with $m = 4$ (approx.).

When the stress is relatively high however, dl/dN is not determined by ΔK. In this case, the parameter $\sigma_a^n l$ is effective. Figure 6(b) shows the dl/dN versus $\sigma_a^n l$ relation at high stress levels ($\sigma_a > 0.6\sigma_y$ (approx.)).

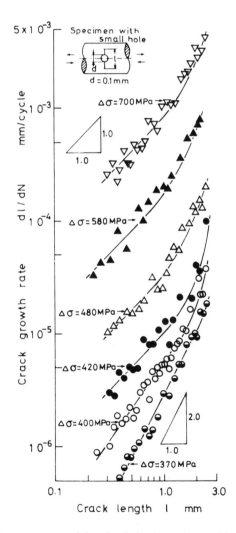

Fig 5   **Crack growth rate versus crack length relation in specimens with a small hole**

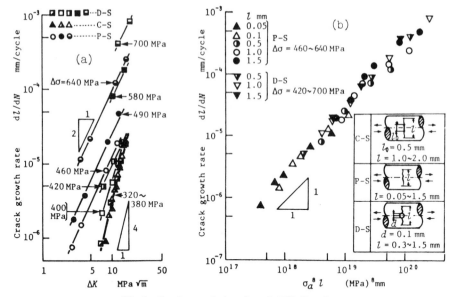

Fig 6   Crack growth data for a 0.45% C steel.
(a) dl/dN is determined by ΔK alone in the case of $\sigma_a \leqslant 0.5\sigma_y$ approximately ($\Delta\sigma \leqslant 380$ MPa approximately). dl/dN is not determined by ΔK in the case of $\sigma_a \geqslant 0.6\sigma_y$ approximately
(b) dl/dN is determined by $\sigma_a^8 l$ alone in the case of $\sigma_a \geqslant 0.6\sigma_y$ approximately ($\Delta\sigma \geqslant 420$ MPa approximately)

A unifying treatment for large-crack growth and small-crack growth

The propagation of a fatigue crack occurs by the accumulation of irreversible deformation at the crack tip. Therefore, the cyclic crack tip opening displacement (CTOD$_r$) seems to be appropriate as a measure of crack propagation. However, it is probably more advantageous to use the cyclic plastic zone size r_{pr} which is closely related to CTOD$_r$, because the measurement of CTOD$_r$ is generally difficult. In this section, using the assumption shown in the following relation, a unifying explanation of fatigue crack growth was made (4)(5)

$$\frac{\mathrm{d}l}{\mathrm{d}N} \propto r_{pr} \tag{3}$$

It has been shown experimentally that the relation dl/dN $\propto r_{pr}$ holds for a Fe–3% Si alloy (12).

(1) A crack growth law for large cracks. In general equation (2) gives good results in the case when a large crack propagates by a small nominal stress. Kikukawa et al. (13) have measured systematically the value of $U(<1)$, the opening ratio equal to $\Delta K_{eff}/\Delta K$ (14), for cracks in many materials. Their results indicate that U is nearly proportional to ΔK in a limited range of ΔK. This indicates that the effective stress intensity factor range ΔK_{eff} is propor-

tional to ΔK^2, and r_{pr} is considered to be proportional to ΔK_{eff}^2. From these considerations, we can obtain the crack growth law given by equation (2), namely $dl/dN = C\Delta K^4$.

(2) *A crack growth law for small cracks.* A small crack does not propagate unless the nominal stress is high enough. Accordingly, when a sufficiently small crack propagates with a finite growth rate (for example, $10^{-6} \sim 10^{-3}$ mm/cycle), the condition of small scale yielding is not satisfied. In this case, equation (1) is valid.

It is natural to assume that U is nearly constant when the nominal stress is high and the condition of small scale yielding does not hold. In this case, the dependency of r_{pr} on σ and l can be estimated from the dependency of r_p, the monotonic plastic zone size, on σ and l under unidirectional loading. In the Dugdale model (15) for unidirectional loading, the relation $r_p/l \propto (\sigma/\sigma_y)^n$ holds. The index n is 2 for $\sigma/\sigma_y \ll 1$ and as σ/σ_y tends to unity, the value of n becomes much larger than 2. The plastic zone size r_p is proportional to crack length l under constant stress. In this case, equation (1) holds and is explained on the assumption based on equation (3).

As mentioned above, although equations (1) and (2) contradict each other, these two growth laws are explained consistently from the same physical background, based on an assumption that the crack growth rate is proportional to r_{pr}. The schematic explanation of the above statements is shown in Fig. 7.

The effect of material properties on the small-crack growth law

From the results of the previous section, it is found that the two crack growth laws given by equations (1) and (2) must be used according to the magnitude of stress amplitude. Until now, only equation (2) has been studied for a wide range of materials. In this section, the relation between equation (1) and material properties is investigated.

Equation (1) can express the growth rate of a small crack in a material, but is not convenient for a comparison between different materials. Therefore, we propose the following crack growth law in which the effect of material properties is partly considered

$$\frac{dl}{dN} = D\left(\frac{\sigma_a}{\sigma_y}\right)^n l \qquad (4)$$

Table 1 compares the constants B and D for various carbon steels. In the comparison of experimental results under the different conditions, it is necessary to know the effects of specimen size and shape of a specimen on the growth law. However, it is found that these effects are small (see Appendix 1).

The table shows that D is a stable constant for materials with similar mechanical properties and is suitable for the comparison of different materials. The constant D represents the resistance for crack propagation in each

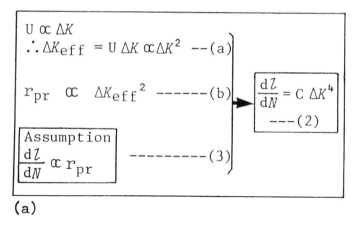

(a)

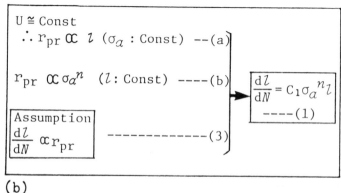

(b)

Fig 7   Schematic explanation of crack growth laws.
(a) Large cracks ($\sigma_a \leqslant 0.5\sigma_y$ approximately)
(b) Small cracks ($\sigma_a \geqslant 0.6\sigma_y$ approximately)

material. The dl/dN versus $(\sigma_a/\sigma_y)^n l$ type relations of Table 1 are shown in Figs. 8(a) and (b), respectively.

Though the experiments No. 1 and No. 2 in Table 1 are of the same material, the values of n are different ($n = 6.5$ for No. 1 and $n = 8.5$ for No. 2). This difference of n is clearly based on a change of material property due to macroscopic yielding (see Appendix 2). However, even in such a case the value of D is stable.

Application of the small-crack growth law in evaluating fatigue life

It is reasonable to predict the fatigue life of plain members based on the small-crack growth law, because the fatigue life of a plain specimen is dominated by the life in which a crack propagates from a certain initial size up to about 1 mm; see Fig. 3(a).

Table 1 Values of B and D and n in equations (1) and (3) (16)

(a) for the case of annealed steels (1 to 11)

No.	Material	Shape of cross section	σ_y (MPa)	σ_B (MPa)	Range of σ_a (MPa)	Range of σ_a/σ_y	n	B ($\times 10^{-23}$)	D ($\times 10^{-4}$)	Range of l (mm)	Type of loads
1	S 10 C	D = 10 mm	203	373	196 ~ 265	0.95 ~ 1.29	6.5	4.4×10^2	4.4×10^{-2}	0.5 ~ 3.0	R–B
2	S 10 C	5 mm	206	373	190 ~ 220	0.87 ~ 1.07	8.5	6.4×10^{-3}	3.0×10^{-2}	0.3 ~ 1.0	
3	S 35 C	6 mm	331	592	196 ~ 314	0.59 ~ 0.95	8.4	1.2×10^{-2}	1.8	0.5 ~ 2.0	
4	S 45 C	5 mm	364	631	235 ~ 300	0.73 ~ 0.93	7.5	1.0	1.6	0.3 ~ 1.5	
5	S 50 C	10 mm	347	674	279 ~ 373	0.86 ~ 1.09	7.5	1.1	1.2	0.5 ~ 3.0	
6	S 20 C	5 mm	276	469	215 ~ 287	0.78 ~ 1.04	7.5	1.2	2.4×10^{-1}	0.05 ~ 1.0	
7	S 35 C	5 mm	331	592	235 ~ 358	0.71 ~ 1.08	7.3	6.7	1.7	0.05 ~ 1.5	
8	S 45 C	5 mm	364	631	235 ~ 372	0.65 ~ 1.02	7.5	1.1	1.8	0.05 ~ 1.5	
9	S 45 C	5 mm	364	631	210 ~ 350	0.58 ~ 0.96	8.0	0.3	8.3	0.05 ~ 1.0	T–C
10	S 45 C	5 mm	364	631	235 ~ 300	0.65 ~ 0.83	8.0	0.2	7.3	0.3 ~ 1.0	
11	S 45 C	8 mm	284	543	294 ~ 543	1.04 ~ 1.87	6.7	5.1×10^2	1.4	0.3 ~ 1.0	

(b) for the case of heat-treated steels (12 to 16)

No.	Material	Shape of cross section	$\sigma_{0.2}$ (MPa)	σ_B (MPa)	Range of σ_a (MPa)	Range of $\sigma_a/\sigma_{0.2}$	n	B ($\times 10^{-13}$)	D ($\times 10^{-4}$)	Range of l (mm)	Type of loads
12	SCM 4-35	D = 5 mm	748	853	400 ~ 470	0.53 ~ 0.63	4.5	2.9×10^{-4}	2.5	0.5 ~ 1.5	R–B
13	S 45 C	5 mm	871	981	400 ~ 470	0.46 ~ 0.54	4.0	6.3×10^{-3}	3.6	0.5 ~ 1.5	
14	S 45 C	5 mm	1376	1510	800 ~ 900	0.58 ~ 0.65	3.5	6.2×10^{-2}	6.0	0.05 ~ 0.5	
15	S 50 C	10 mm	1133	1247	364 ~ 736	0.32 ~ 0.65	3.0	3.5	5.1	0.5 ~ 3.0	
16	S 50 C	10 mm	1133	1247	343 ~ 540	0.30 ~ 0.48	3.5	0.17	8.1	0.5 ~ 2.0	T–C

R–B: Rotating–bending, T–C: Tension–compression
$\sigma_{0.2}$: Proof stress, σ_B: Ultimate tensile strength

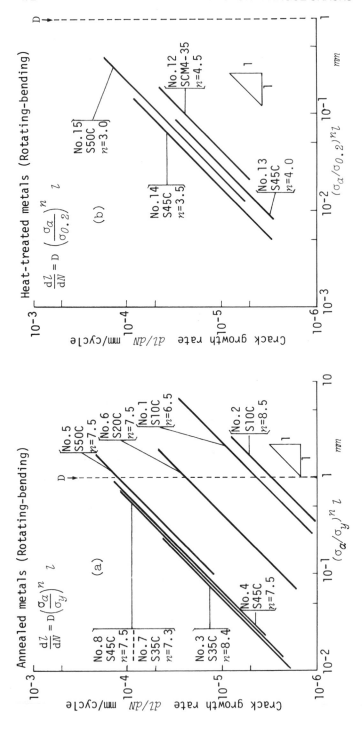

Fig 8   The crack growth relation for (a) annealed carbon steels, and (b) heat treated carbon steels

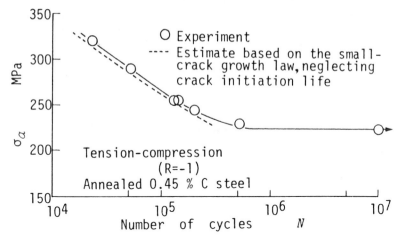

Fig 9   Evaluation of fatigue life based on the growth law of small cracks

Figure 9 shows the comparison of fatigue life between experiment and the estimate based on the behaviour of cracks smaller than 1 mm, neglecting the crack initiation life. The prediction is in good agreement with experiment.

In the cases where we predict the life of large machine members, the prediction based on the behaviour of cracks smaller than 1 mm is also meaningful. For example consider three crack sizes: l_1, an initial size; l_2, an arbitrary size in the propagating period; and l_3, the final size corresponding to fracture. The number of cycles to failure, N_f, is approximately given by

$$N_f \cong \frac{1}{D} \left(\frac{\sigma_y}{\sigma_a}\right)^n \left(\ln \frac{l_2}{l_1} + \ln \frac{l_3}{l_2}\right) \tag{5}$$

If we take $l_1 = 0.01$ mm, $l_2 = 1$ mm, and $l_3 = 100$ mm, the life from l_1 to l_2 is equal to the life from l_2 to l_3. All members having a crack of length 100 mm soon break, and so the real fatigue life in this case is approximately twice the life predicted from the behaviour of a crack smaller than 1 mm. This means that the behaviour of a crack smaller than 1 mm is dominant in the estimation of the fatigue life of machines and structures.

Conclusions

The following conclusions were reached with regard to small crack propagation of plain carbon steel specimens under constant stress amplitudes.

(1) In annealed 0.45% C steel plain specimens subjected to axial loading, the propagation life of a crack smaller than 1 mm occupies about 70 per cent of the total fatigue life.

(2) The crack growth rate of a plain specimen is not determined by the stress intensity factor range ΔK. The crack growth rate increases with increase in the applied stress range $\Delta\sigma$, for a given value of ΔK and the growth rate of a plain specimen is determined uniquely by the term $\sigma_a^n l$ where σ_a is stress amplitude, l is crack length, and n is a constant.

(3) The following crack growth laws are obtained from experiments on specimens with a small hole:

(a) When a crack is propagating under high nominal stress ($\sigma_a \gtrsim 0.6\sigma_y$, σ_y: the lower yield stress),

$$\frac{dl}{dN} = B\sigma_a^n l$$

(b) When a crack is propagating under low nominal stress ($\sigma_a \lesssim 0.5\sigma_y$).

$$\frac{dl}{dN} = C\Delta K^m$$

Both these crack growth laws can be explained consistently from the same physical background, based on the assumption that crack growth rate is proportional to the reversible plastic zone size r_{pr}.

(4) At high stress levels the growth rate of a small crack in one material cannot compare with that of a different material. Therefore the following small crack growth law is proposed

$$\frac{dl}{dN} = D\left(\frac{\sigma_a}{\sigma_y}\right)^n l$$

The effect of material properties is therefore partly considered. The constant D represents the resistance to crack propagation in each material and is a stable constant for those materials with similar mechanical properties.

(5) The fatigue life of 0.45% C steel plain specimens can be evaluated based on the behaviour of cracks smaller than 1 mm. The prediction is in good agreement with experiments.

Appendix 1

Fatigue tests on geometrically similar specimens and specimens having various cross sections were carried out to examine the effectiveness of the relation $dl/dN \propto \sigma_a^n l$ when evaluating fatigue life. Figure 10(a) shows this relation for the geometrically similar specimens having square sections. It is found that the difference in specimen size does not affect the small-crack growth law.

Similar experiments were carried out on plain specimens having various cross sections, i.e., the square section, the circular section and the circular section having a fine shallow notch. Again the difference in propagation behaviour of cracks is hardly observable, see Fig. 10(b).

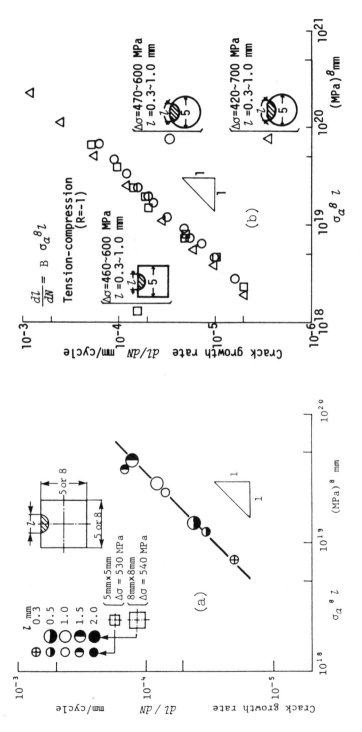

Fig 10 Relation between propagation behaviour of small surface cracks and specimen shape.

(a) dl/dN versus $\sigma_a^8 l$ relation in geometrically similar specimens

(b) Crack growth behaviour in specimens having various cross sections

From these results on the same material, it is found that the propagation behaviour of surface cracks is not affected by the specimen shape. This suggests that the behaviour of small surface cracks in actual machines and structures can be predicted from specimens such as that shown in Fig. 1.

Appendix 2

The behaviour of small cracks in annealed and pre-strained 0.45% carbon steel specimens was investigated in order to clarify the effect of macroscopic yielding on the growth of small cracks (17).

Figure 11 shows the loading cycle for both materials. In the case of annealed material Fig. 11(a) the growth law is shown in Fig. 6. That is, the growth rate of small cracks at high nominal stress is given by $\sigma_a^8 l$. In the case of pre-strained material, see Fig. 11(b), the experimental results are shown in Fig. 12. From Fig. 12(a) and (b), it is clear that the dl/dN versus $\sigma_a^{6.4} l$ relation holds for high nominal stress ($\Delta\sigma \gtrsim 630$ MPa) and the dl/dN versus ΔK relation holds for low nominal stress ($\Delta\sigma \lesssim 570$ MPa).

The value of n is a material property. That is, n has a value 7.5–8.5 for annealed carbon steels and 4–5 for heat-treated carbon steels. The difference of n between annealed specimen ($n \cong 8$) and pre-strained specimen ($n \cong 6.4$) is explained by material property changes caused by macroscopic yielding from pre-straining. This means that the value of n depends on the magnitude of stress (above or below yield stress) despite the material being the same; see No. 1 and No. 2 in Table 1.

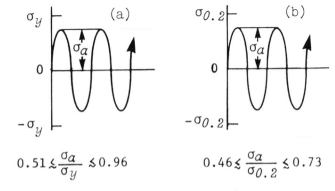

$$0.51 \lesssim \frac{\sigma_a}{\sigma_y} \lesssim 0.96 \qquad\qquad 0.46 \lesssim \frac{\sigma_a}{\sigma_{0.2}} \lesssim 0.73$$

Fig 11   Condition of loads in (a) annealed, and (b) pre-strained specimens

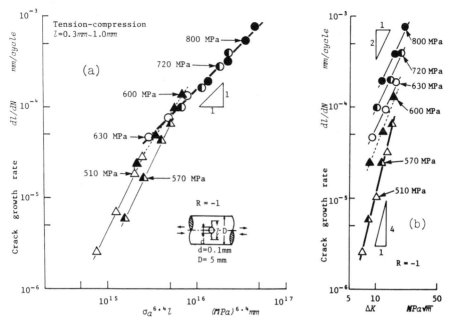

Fig 12   Crack growth data for a pre-strained 0.45% carbon steel

References

(1)   KOCANDA, S. (1978) Fatigue failure of metals (Sijthoff & Noordhoff, Amsterdam).

(2)   LAIRD, C. and SMITH, G. C. (1962) Crack propagation in high stress fatigue, *Phil. Mag.*, **7**, 847–857.

(3)   NEUMANN, P. (1974) New experiments concerning the slip processes at propagating cracks, *Acta Met.*, **22**, 1155–1165.

(4)   LIU, H. W. (1963) Fatigue crack propagation and applied stress range – an energy approach, *J. Basic Engng*, **85**, 116–122.

(5)   NISITANI, H. (1981) Unifying treatment of fatigue crack growth laws in small, large and non-propagating cracks, *Mechanics of Fatigue-AMD* (Edited by Mura, T.) (ASME), Vol. 43, pp. 151–166.

(6)   NISITANI, H. and MORIMITSU, T. (1976) Estimation of crack propagating property by the rotating bending fatigue tests of specimens with a small hole, *Bull. JSME*, **19**, 1091–1099.

(7)   NISITANI, H. and KAWAGOISHI, N. (1983) Fatigue crack growth law in small cracks and its application to the evaluation of fatigue life, *Trans JSME.*, **49**, 431–440.

(8)   NISITANI, H. and GOTO, M. (1983) Investigation of Miner's law and modified Miner's law based on the behavior of micro-cracks, *Trans JSME.*, **49**, 779–787.

(9)   NISITANI, H. GOTO, M., and KAWAGOISHI, N. (1984) Comparison of fatigue crack growth laws under a high nominal stress and a low nominal stress, *Trans JSME.*, **50**, 23–32.

(10) NISITANI, H. and GOTO, M. (1984) Effect of stress ratio on the propagation of small crack of plain specimens under high and low stress amplitudes, *Trans JSME.*, **50**, 1090–1096.

(11) NISITANI, H. and TAKAO, K. (1981) Significance of initiation, propagation and closure of microcracks in high cycle fatigue of ductile metals, *Engng Fracture Mech.*, **15**, 445–456.

(12) NISITANI, H. and KAWAGOISHI, N. (1984) Relation between fatigue crack growth law and reversible plastic zone size in Fe–3% Si alloy, *Trans JSME.*, **50**, 277–282.

(13) KIKUKAWA, M., JOHNO, M., TANAKA, K., and TAKATANI, M. (1976) Measurement of fatigue crack propagation and crack closure at low stress intensity level by unloading elastic compliance method, *J. Soc. Mat. Sci. Japan.*, **25**, 899–903.
(14) ELBER, W. (1971) The significance of fatigue crack closure, *ASTM. STP. 486*, pp. 230–242.
(15) DUGDALE, D. S. (1960) Yielding of steel sheets containing slits, *J. Mech. Phys. Solids,* **8**, 100–108.
(16) NISITANI, H. and GOTO, M. (1985) Relation between small-crack growth law and fatigue life of machines and structures, *Trans JSME.*, **51**, 332–341.
(17) NISITANI, H. and GOTO, M. (1984) Effect of macroscopic yielding on the small-crack growth law (investigation based on the fatigue tests under axial loading of annealed and pre-strained 0.45%C steels), *Trans JSME,* **51**, 440–444.

D. Taylor*

Fatigue of Short Cracks: the Limitations of Fracture Mechanics

REFERENCE Taylor, D., **Fatigue of Short Cracks: the Limitations of Fracture Mechanics**, *The Behaviour of Short Fatigue Cracks*, EGF Pub. 1 (Edited by K. J. Miller and E. R. de los Rios) 1986, Mechanical Engineering Publications, London, pp. 479–490.

ABSTRACT Cracks less than some critical length, defined as l_2, show anomalous growth behaviour which cannot be quantified by LEFM. At present there are no reliable methods for predicting the growth rates and thresholds of these cracks. This paper presents a simple approach which has had some success in predicting the value of l_2 for various materials. This parameter is shown to depend on the material properties yield strength and grain size. The prediction of l_2 for a given material enables the designer to make conservative predictions of the fatigue life of components in situations where short crack behaviour dominates.

Introduction

In recent years there have been many investigations into the behaviour of short fatigue cracks; this body of work is well summarized by three recent reviews(1)– (3). Possibly the most important point to emerge is that, where short cracks are concerned, present design procedures, whether based on S/N type data or on fatigue crack propagation data, are in danger of being non-conservative. This can be illustrated using the now-common method of displaying short crack data, i.e., the plot of threshold stress range as a function of crack length. Figure 1 shows schematically the type of results obtained; the open-circle data points represent the region in which use of either the fatigue limit or the long-crack threshold value gives too high a value for the threshold stress and is thus non-conservative.

In a paper concerned with the fatigue behaviour of an aluminium bronze alloy (4) the author defined the terms l_1 and l_2 as the values of crack length at the limits of this non-conservative region. The parameter l_1 is possibly only of academic interest, since there are few practical situations for which defect sizes as small as l_1 are important. However, the parameter l_2 has considerable importance in design and failure analysis work; l_2 values may be greater than 1 mm in some materials, so in components for which there is good control of surface condition and defect size, the initial defect sizes may be below l_2.

Given the practical difficulties of measuring l_2 values, some method of estimating l_2 from other material parameters is needed. The author himself has had to deal with failure analysis problems which have required the estimation of an l_2 value, for example, the failure of cast ships' propellers (5). If the practical situation is such that inherent cracks larger than l_2 exist, than a normal

* Department of Mechanical Engineering, Trinity College, Dublin, Ireland.

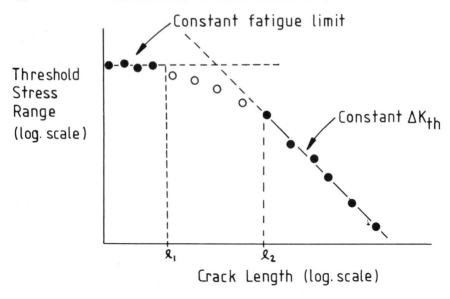

Fig 1 Typical short crack fatigue results; the open circles indicate the region of non-conservative
behaviour

conservative approach has to be adopted; the simplest conservative approach in such cases is to take the defect size as l_2.

This paper outlines a simple method which aims for an approximate and conservative estimate of l_2 for any given metallic material. After outlining the form of the prediction and showing that it agrees with known data on short cracks, a detailed justification of the approach follows.

The prediction of l_2

In two previous publications (**4**)(**6**) a microstructural argument was used to suggest that l_2 could be approximated by

$$l_2 \approx 10d \tag{1}$$

Here d is a 'characteristic microstructural dimension' which is usually equal to the grain size but becomes carbide lath width or precipitate spacing in microstructures dominated by fine precipitates. Thus d is similar to the 'effective grain size', $\bar{l}$, defined by Yoder *et al.* (**7**) in their work on long crack thresholds.

It was shown (**4**)(**6**) that quite good correlation existed between l_2 and $10d$ for available data on a range of materials. However, the microstructural argument is clearly not a complete one. A number of other hypotheses have been advanced to account for short crack behaviour, not all of which are concerned with microstructural details; these include the effects of crack closure (**1**), crack-tip stress distribution (8), crack deflection (9), and constraint (1). These various models will be discussed in more detail below, where it will be shown that they can be divided into two types: those based on microstructure, for which the parameter d is important, and those based on plasticity for which the appropriate parameter is the plastic zone size. The extent of the reversed plastic zone r_p in fatigue can be estimated as $0.04(\Delta K/\sigma_{y,c})^2$, where $\sigma_{y,c}$ is the cyclic yield strength.

In what follows it will be shown that the plasticity based arguments can be reduced to the condition that, for LEFM to apply, the crack length must be greater than $10r_p$; hence a second estimate of l_2 is

$$l_2 \approx 10r_p \tag{2}$$

Equations (1) and (2) represents two independent conditions on the value of l_2; thus, for any given material, logic dictates that l_2 will be given by whichever is the larger of $10d$ and $10r_p$.

There is a fundamental point which should be borne in mind which makes this simple analysis of l_2 possible; there are many different factors which affect the behaviour of a short crack, including the five listed above, and this makes the prediction of the growth rate and threshold of a short crack very difficult. However, each of these different effects can be thought of as having an l_2 value associated with it, so the observed value of l_2 for the material will be whichever is the largest of these, assuming only that the various effects act independently. For instance, if microstructural considerations lead to a prediction of, say, $l_2 = 10\ \mu m$, but a closure argument predicts $l_2 = 100\ \mu m$ for the same material, then $100\ \mu m$ will be the relevant value and microstructural effects can be discounted as far as l_2 is concerned.

Comparison of predictions and results

Figure 2 shows measured values of l_2 for various materials (3)(4)(10)–(17) (19)(34) plotted against $10d$ and $10r_p$. The materials covered include mild steels, high strength steels, titanium, copper, and aluminium alloys; the open circles are the $10d$ values and the solid circles are the $10r_p$ values. The letters alongside the points refer to the original publications, as shown in Table 1.

The accuracy of the determined values of l_2 is estimated to be within 20 per cent unless the error bars on Fig. 2 indicate to the contrary, likewise the accuracy of the $10d$ and $10r_p$ determinations is estimated at 20 per cent unless otherwise indicated.

Figure 2 shows that there is good predictive capacity with this model over the whole range of l_2 values and different materials. Taking the largest of $10d$ and $10r_p$ (when both are available) then of the 17 different results presented, 15 of the predictions are either correct or slightly conservative. For the 11 cases where both d and r_p values were available, there are two cases in which $d > r_p$, three cases in which $r_p > d$ and six cases where d and r_p are equal within the errors of estimation.

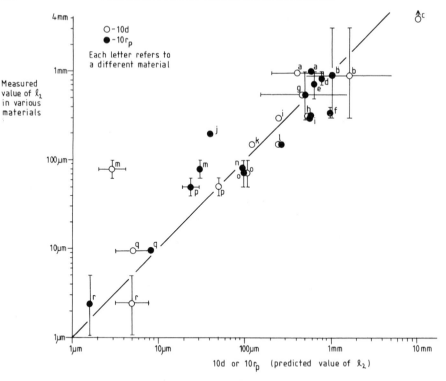

Fig 2 Comparison of measured and predicted values of l_2. See Table 1 for source references

Table 1   Sources of data on Figure 2

Reference letter on Fig. 2	Publication
a	(11)
b	(13)
c	(37)
d	(16)
e	(16)
f	(16)
g	(14)
h	(4)(10)
i	(13)
j	(17)
k	(15)
l	(13)
m	(3)
n	(19)
o	(3)
p	(3)
q	(12)
r	(13)

Another important point to note is that the values of d and r_p are always similar in magnitude here; in only one case (3) do they differ by more than a factor of three, though the data range over three orders of magnitude. This arises because we are considering near-threshold behaviour; it has been shown (7)(20)–(22) that d and r_p are approximately equal near the long-crack threshold, ΔK_{th}. The implication of these results is that, if the value of l_2 is controlled by r_p rather than by d, then the value of l_2 might be expected to increase with increasing applied load. In the present paper, only near-threshold results are used, and l_2 is defined only at the long-crack threshold; few results exist at present to show how anomalous short-crack behaviour varies with applied load, though clearly this point should be pursued in the future.

Hypotheses relating to short crack behaviour

The above section has shown that the simple hypothesis described by equations (1) and (2) is capable of giving good predictions for l_2 values. The following section examines various proposed models of short crack behaviour in order to justify this hypothesis in the light of the various mechanisms advanced by other workers. Five distinct reasons for anomalous short-crack behaviour have been advanced:

(1) microstructure (**4**);
(2) closure (**1**);
(3) K-estimation errors (**8**);
(4) crack deflection (**9**);
(5) constraint (**1**).

These have already received much attention and discussion elsewhere (**1**)–(**3**). Here it will be shown that the microstructure and deflection arguments can be related to the $10d$ prediction and that the other three arguments can be related to the $10r_p$ prediction.

(1) Microstructure

Observations by the present author and others (**4**)(**5**)(**15**)(**17**) showed that anomalous growth rates in short cracks frequently occurred when cracks were growing in a single grain or a small number of grains. It was noted that linear elastic fracture mechanics demands a homogeneous continuum, which is unlikely to be the case until the crack front length becomes considerably greater than the grain size. Considering a typical surface crack, semi-circular in shape, a crack length of $10d$ would give a crack front length of about $31d$, which would satisfy a homogeneity requirement such as the one outlined in section (4) below. Thus equation (1); i.e., $l_2 \approx 10d$ can be thought of as a sufficient condition on the value of l_2 in this case, and therefore a conservative prediction. It must be remarked, however, that d is the mean of a grain-size distribution; thus, some cracks of length $10d$ will pass through fewer than 31 grains along their fronts. It can be shown (23) that there is only a very small probability of the crack front passing through less than 10 grains at this length, given the normal form of the grain-size distribution, and assuming no grain-shape texture effects, such as are common in wrought alloys.

(2) Crack closure

Since crack closure is caused by residual stresses in the crack wake, it has been proposed (e.g., (19)) that a short crack will experience less closure because there is insufficient length behind the crack tip for the wake field to fully develop. This seems to be a very powerful argument, and it has certainly been shown that the closure characteristics of long and short cracks differ considerably. The picture is confused by the difficulty of accurately measuring short-crack closure; only very small changes in compliance are detected, and by the effect of microstructural features such as grain boundaries (15)(24).

It has been proposed (1) that for a homogeneous continuum the crack length should be greater than the reversed plastic zone size, r_p, in order to establish the full wake field and therefore ensure normal closure behaviour. This seems a rather short length; as yet there is no reliable evidence to settle the question, but in that case the condition of equation (2), i.e., $l_2 \approx 10r_p$ will express a sufficient condition on l_2.

(3) K estimation errors

Perhaps not enough attention has been given to the fact that the normally-used equation for stress distribution at a sharp crack

$$\sigma = \sigma_o \sqrt{(a/2r)} \qquad (3)$$

where

σ = stress at a distance, r, from the crack tip

and

σ_o = applied nominal stress

is valid only for r very much less than a and should be replaced in other cases by the more complete form of the Westergaard equation

$$\sigma = \frac{\sigma_o\{1 + (r/a)\}}{\{2(r/a) + (r/a)^2\}^{1/2}} \qquad (4)$$

Sinclair and Allen (6) have shown that for short cracks which show anomalous behaviour, the difference between equations (3) and (4) is significant. However it is difficult to define a K value from equation (4) because of the lack of a simple r singularity. Using a method based on comparison of plastic zone size, Sinclair and Allen define an effective K value, K_{eff}, as

$$K_{eff} = K(1 + r_p/a)^{1/2} \qquad (5)$$

Equation (5) implies that for any value of r_p, K_{eff} will be greater than K. However, if r_p/a is small, K_{eff} will approach K in magnitude. If K_{eff}/K is arbitrarily set to 1.05 (i.e., taking a 5 per cent difference between K_{eff} and K), this will lead to a value for R_p/a of approximately 0.1, which would amount to a restatement of the condition in equation (2). The choice of the figure of 5 per cent is arbitrary, but represents the amount of scatter usually observed on threshold data measurement. The use of 10 or 2 per cent would lead to estimates of l_2 which would be of the same order of magnitude.

(4) Crack deflection

Figure 3 shows schematically the difference between an idealized straight crack and a real crack. The crooked, deflected crack path tends to lower the effective K value, as has been considered very thoroughly by Suresh (9) and Kitagawa et al. (25). Suresh also showed qualitatively that a short crack, because it has only a few deflections in it, would be expected to have a K value which varied from one crack to another, whereas in a long crack these effects would average out, giving a consistent value for K. Suresh was unable to use this deflection argument to explain the increased growth rates of short cracks, and indeed this does not seem to be possible unless one postulates a very strong dependence on the K_{II} value.

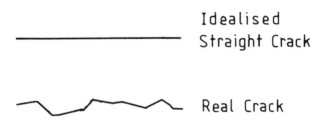

Fig 3   Schematic of idealized and real cracks

Considering the definition of l_2, this deflection condition amounts to another microstructural effect, since it has been shown that for cracks growing under near-threshold conditions a faceted or 'structure sensitive' growth mode occurs (21) in which the facets are equal to the grain size.

Two effects can be considered here: the effect of crack length and the effect of crack front length. Considering crack front length, a deflected crack will be modelled here as shown in Fig. 4, i.e., a crack with a straight section of length a and an end portion of length d deflected through an angle θ. Using (27) it is possible to estimate the reduction in effective K value resulting from deflecting the end portion. For example, if $\theta = 45$ degrees the k value is reduced from its nominal value by about 20 per cent. This figure is only slightly dependent on the values of a and d, because an increase in the ratio d/a tends to decrease K_I but ⁺o increase K_{II} for a given θ.

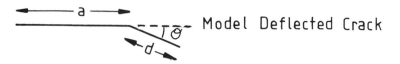

Fig 4   Model deflected crack to be used in analysis

Taking a crack front only one grain in length, and allowing θ to vary at random from $+45$ degrees to -45 degrees gives a distribution of possible K values; expressed as the ratio $K/K_{nominal}$ these vary from 0.99 to 0.87 if one takes two standard deviations from the mean of the distribution. Thus a very small crack front such as this would be expected to show about 13 per cent variation in its K value if a large number of cracks were examined. If the crack front length is now increased, using a model crack front, as shown in Fig. 5, then the distribution of possible K values tends to narrow, as the effects from separate grains tend to cancel each other out. Figure 5 shows the results of a simple computer analysis of this model; here $K/K_{nominal}$ is plotted as a function of crack front length, expressed as a number of grain diameters. It can be seen that for a crack front longer than $10d$ the distribution becomes narrow and roughly constant at about 2 per cent of the mean.

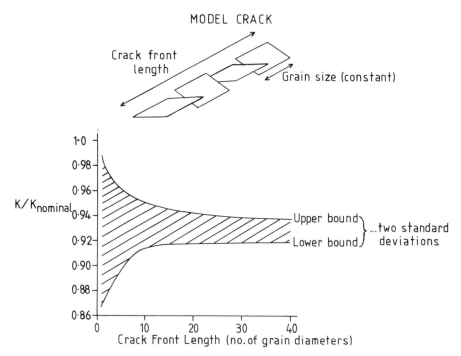

Fig 5   Results of analysis on the effect of crack front length on K. The scatter bands are placed at two standard deviations from the mean of the data

This analysis implies that, as far as this condition is concerned, a crack with a short front may grow faster or slower than a crack with a long front. Taking again the case of a semi-circular surface crack, if the crack length is $10d$ then the crack front length will be $31d$, so the condition $l_2 \approx 10d$ will again express a sufficient condition for l_2.

The effect of deflections along the crack length is more difficult to analyse. Suresh (9) considered the singly and doubly deflected cracks and developed a method to deal with a long crack containing many deflections, but as yet the critical range of lengths around $a = 10d$ cannot be treated.

The other postulated effect of crack deflection, as advanced by Beevers and others, is that it induces premature closure, the roughened crack surfaces coming into mutual contact at a higher applied K value in the cycle than expected. If we note again that the degree of surface roughness is proportional to the grain size at near-threshold growth, a similar microstructural condition, i.e., equation (1), would be expected.

(5) Constraint

It has been pointed out (1) that a small crack is essentially contained in the surface plane stress field until it penetrates some distance into the body of the material, where part of the crack front begins to experience plane strain conditions. Unfortunately it is by no means clear whether a crack in a plane stress field will grow faster or slower than a plane strain crack; experimental results are conflicting (e.g., compare (26) and (27)).

Following Knott (28), it can be assumed that the plane stress field extends into the material from the surface for a distance approximately equal to the plastic zone size. This is shown schematically in Fig. 6 for a semi-circular crack.

Applying the condition of equation (2), i.e., putting $a = 10r_p$, simple geometry shows that approximately 94 per cent of the crack front will be out of the plane stress region; 6 per cent experiencing plane stress conditions. This

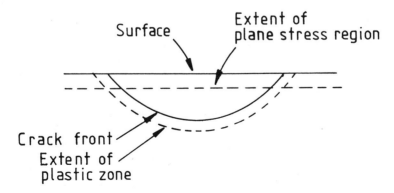

Fig 6 Schematic illustrating the amount of the crack front length which is contained within the surface plane stress region

would seem enough to ensure that the plane stress region was unimportant; hence, equation (2) is again shown to be a sufficient condition on the value of l_2.

Summary of the above discussion

It has been possible to show that the initial hypothesis used in this paper, that l_2 can be estimated as the larger of $10d$ and $10r_p$, constitutes a sufficient condition for each of the five effects considered. Since we are seeking a conservative prediction, a sufficient condition is all that is required. The experimental results shown in Fig. 2, however, suggest that the condition used here is generally either correct or slightly conservative only, so it seems that no severe overestimate of l_2 is likely.

Methods of production of short cracks

When examining experimental data on this subject, it is important to consider the method used by the experimenters to produce the short cracks, as this may affect their behaviour.

In most experimental studies of short cracks, the cracks are produced by initiation from plain specimens, the threshold values being achieved either by load shedding or by load increase after stress relief heat treatment. The process of stress relief presents some problems (**29**) but the two methods seem to give roughly comparable results.

However, some workers (e.g., (**18**)(**30**)–(**32**)) produce short cracks by machining material from specimens containing long through-cracks. James and Knott (**30**) point out the possible nature of stress history effects and show that, in an alloy steel, even if stress relief is attempted, a much higher value of l_2 would be deduced from these short through-cracks than from surface cracks of more common form. Incomplete stress relief, leaving residual closure stresses, may be involved. Similar results were obtained by Breat *et al.* (**18**) and in an aluminium alloy by Zeghloul and Petit (**32**).

A conservative design approach

The emphasis throughout this paper has been to formulate and justify a simple method of predicting an appoximate and conservative value of l_2. It is believed that this addresses a growing need for designers and failure analysts attempting to use fracture mechanics in fatigue situations.

It is to be hoped that in the near future, analytical methods will be developed to predict the growth behaviour of cracks shorter than l_2; some progress has been made in that direction using a statistical approach (**37**), thus extending our predictive capacity into this difficult area. Until this has been achieved, a conservative approach must be used, such as setting the defect size to l_2 in the case where the real defect size is smaller than l_2.

Finally, it should be recognized that the number of situations in practice where small crack behaviour dominates is small. The most widely quoted example of the short crack problem is the jet engine turbine blade, a situation in which the extreme demands placed on the material require a very small inherent defect size. Indeed this is probably the only commonly-occurring case of a crack of length less than l_2 being subjected to stresses well in excess of its propagation threshold. However a number of other cases arise which involve short cracks stressed close to their thresholds; one example would be precision machine parts such as engine pistons, which require very good surface finish and may be made from materials with quite large l_2 values. It is unlikely that NDT crack inspection techniques can be used to help monitor short cracks in service, as the necessary detection accuracy cannot be achieved by any NDT methods presently in use.

Conclusions

(1) Cracks less than some critical length, denoted l_2, show anomalous growth behaviour which cannot be described by linear elastic fracture mechanics.
(2) Difficulties in the experimental measurement of l_2 call for some method of predicting this parameter for any required material. The predicted value of l_2 should be conservative, i.e., an overestimation, in order to be useful for design purposes.
(3) It is proposed that l_2 can be estimated to be the larger of $10d$ and $10r_p$, d being the effective grain size and r_p the cyclic plastic zone size.
(4) It is shown by comparison with experimental data that this hypothesis gives a reasonable prediction of l_2 for many different materials.
(5) The use of this simple predictive model is justified in detail by examining the various current theories on short crack behaviour.

References

(1) SURESH, S. and RITCHIE, R. O. (1983) The propagation of short fatigue cracks, *Rep. No. UCB/RP/83/1014*, University of Berkeley, California.
(2) LEIS, B. N., KANNINEN, M.F., HOPPER, A. T., AHMAD, J., and BROEK, D. (1983) A critical review of the short crack problem in fatigue, *Rep. No. AFWAL-TR-83-4019*, Air Force Wright Aeronautical Laboratories.
(3) JAMES, M. N. (1983) PhD thesis, University of Cambridge.
(4) TAYLOR, D. and KNOTT, J. F. (1981) Fatigue crack propagation behaviour of short cracks; the effect of microstructure, *Fatigue Engng Mater. Structures*, **4**, 147.
(5) TAYLOR, D. and KNOTT, J. F. (1982) Growth of fatigue cracks from casting defects in nickel–aluminium bronze, *Met. Tech.*, **9**, 221.
(6) TAYLOR, D. (1982) Euromech colloquium on short fatigue cracks, *Fatigue Engng Mater. Structures*, **5**, 305.
(7) YODER, G. R., COOLEY, L. A., and CROOKER, T. W. (1981) A critical analysis of grain size and yield-strength dependence of near-threshold fatigue-crack growth in steels, *NRL Memo. Rep. 4576*, Washington D.C.
(8) ALLEN, R. J. and SINCLAIR, J. C. (1982) The behaviour of short cracks, *Fatigue Engng Mater. Structures*, **5**, 343.
(9) SURESH, S. (1983) Crack deflection: implications for the growth of long and short fatigue cracks, *Met. Trans*, **14A**, 2375.

(10) TAYLOR, D. (1982) *Fatigue crack propagation in nickel–aluminium bronze castings*, PhD thesis, University of Cambridge.

(11) FROST, N. E. (1958) A relation between the critical alternating propagation stress and crack length for mild steel, and its significance in the interpretation of plain and notched fatigue results, *MERL Rep. No. PM246 Div. No. 8/58*.

(12) LANKFORD, J. (1980) On the small crack fracture mechanics problem, *Int. J. Fracture*, **16**, R7.

(13) USAMI, S. and SHIDA, S. (1979) Elastic–plastic analysis of the fatigue limit for a material with small flaws, *Fatigue Engng Mater. Structures*, **1**, 471.

(14) OHUCHIDA, H., NISHIOKA, A., and USAMI, S. (1973) Elastic–plastic approach to fatigue crack propagation and fatigue limit of material with crack, *Proc. 3rd Int. Conf. on Fracture*, Munich, Vol. 6, paper V-422/A.

(15) JAMES, M. R. and MORRIS, W. L. (1983) Effect of fracture surface roughness on growth of short fatigue cracks, *Met. Trans*, **14A**, 153.

(16) ROMANIV, O. H., SIMINKOVITCH, V. N., and TKACH, A. N. (1981) Near-threshold short fatigue crack growth, *Fatigue Thresholds* (Edited by Backlund, J., Blom, A. F., and Beevers, C. J.) (EMAS, Warley, UK).

(17) LANKFORD, J. (1982) The growth of small fatigue cracks in 7075-T6 aluminium, *Fatigue Engng Mater. Structures*, **5**, 233.

(18) BREAT, S. L., MUNDRY, F., and PINEAU, A. (1983) Short crack propagation and closure effects in A508 steel, *Fatigue Engng Mater. Structures*, **6**, 349.

(19) ELSENDER, A., GALLIMORE, R., and POYNTON, W. A. (1977) The fatigue behaviour of macroscopic slag inclusions in steam turbo-generator rotor steels, *Proc. ICF4*, Waterloo, Canada.

(20) BEEVERS, C. J. (1980) Micromechanisms of fatigue crack growth at low stress intensities, *Met. Sci.*, **14**, 418.

(21) TAYLOR, D. (1981) A model for the estimation of fatigue crack threshold stress intensities in materials with various different microstructures, *Fatigue Thresholds* (Edited by Backlund, J., Blom, A. F., and Beevers, C. J.) (EMAS, Warley, UK).

(22) TAYLOR, D. (1984) An analysis of data on fatigue crack propagation thresholds, *Proc. Fatigue '84*; Birmingham (EMAS, Warley, UK).

(23) DEHOFF, R. T. and RHINES, F. N. (1968) *Quantitative microscopy,* McGraw-Hill, New York (New York).

(24) ZUREK, A. K., JAMES, M. R., and MORRIS, W. L. (1983) The effect of grain size on fatigue growth of short cracks, *Met. Trans*, **14A**, 169.

(25) KITAGAWA, H., YUUKI, R., and OHIRA, T. (1975) Crack morphological aspects in fracture mechanics, *Engng Fracture Mech.*, **7**, 515.

(26) JACK, A. R. and PRICE, A. T. (1972) *Acta Met.*, **20**, 857.

(27) PICKARD, A. C. and KNOTT, J. F. (1977) Effect of overloads on fatigue crack propagation in aluminium alloys, *Met. Sci.*, **11**, 399.

(28) KNOTT, J. F. (1973) *Fundamentals of fracture mechanics* (Butterworths, London).

(29) TAYLOR, D. (1985) *Compendium of fatigue thresholds and crack growth rates* (EMAS, Warley, UK).

(30) JAMES, M. N. and KNOTT, J. F. (1986) An assessment of crack closure and the extent of the short crack regime in Q1N (HY80) steel, *Fatigue Engng Mater. Structures*, to be published.

(31) CHAUHAN, P. and ROBERTS, B. W. (1979) Fatigue crack growth behaviour of short cracks in a steam turbine rotor steel – an investigation, *Metall. Mater. Technol.*, **11**, 131.

(32) ZEGHLOUL, A. and PETIT, J. (1986) Environmental sensitivity of small crack growth in 7075 aluminium alloy, *Fatigue Engng Mater. Structures*, to be published.

(33) TAYLOR, D. (1984) The effect of crack length on fatigue threshold, *Fatigue Engng Mater. Structures*, **7**, 267.

(34) BROWN, C. W. and HICKS, M. A. (1983) *Fatigue Engng Mater. Structures*, **6**, 67.

K. J. Miller,* H. J. Mohamed,* and E. R. de los Rios*

Fatigue Damage Accumulation Above and Below the Fatigue Limit

REFERENCE Miller, K. J., Mohamed, H. J., and de los Rios, E. R., **Fatigue Damage Accumulation Above and Below the Fatigue Limit**, *The Behaviour of Short Fatigue Cracks*, EGF Pub. 1 (Edited by K. J. Miller and E. R. de los Rios) 1986, Mechanical Engineering Publications, London, pp. 491–511.

ABSTRACT Damage accumulation in fatigue, initially incorporating cycles at stresses below the fatigue limit, was investigated in continuously increasing shear stress amplitude tests on a fully annealed 0.4 per cent carbon steel. Surface damage was observed throughout the test using a plastic replica technique. Fatigue cracks were observed at preferential sites coincident with persistent slip bands at stresses of the order of 85 per cent of the fatigue limit stress. Predictions of fatigue life based on classical models due to Palmgren–Miner, Corten–Dolan, and Marsh are shown to be non-conservative for increasing stress amplitude situations starting from stress levels below the fatigue limit. Satisfactory predictions are obtained, however, by combining crack growth rate expressions derived for short and long cracks. It is shown that a distinction must be made between the accumulation of fatigue damage in the short crack region and that in the long crack region if accurate life predictions are to be obtained. Damage accumulation at stress levels below the fatigue limit can have a significant effect on subsequent damage accumulation above the fatigue limit.

Notation

a	Surface crack length
a_f	Failure crack length
a_o	Initial surface roughness
a_t	Transition length between short and long cracks
d	Microstructure length parameter
A, B, C, m, n	Constants
N	Number of cycles
N_f	Number of cycles to failure in constant amplitude tests
N_{ff}	Number of cycles to failure in increasing stress amplitude tests
N_L	Number of cycles in the long crack growth phase
N_S	Number of cycles in the short crack growth phase
z	Slope of an hypothetical S–N curve
α	Increase in stress amplitude per cycle
$\Delta\gamma_p$	Plastic shear strain range
τ	Shear stress amplitude
ϕ	Increase in plastic shear strain range per cycle

* Mechanical Engineering Department, University of Sheffield, Mappin Street, Sheffield S1 3JD, UK.

Subscripts

CDM	Corten–Dolan–Marsh
f	Failure
i	Intermediate
FL	Fatigue limit
o	Initial
PM	Palmgren–Miner

Introduction

Possibly the earliest realization that fatigue damage could develop below the fatigue limit came about by examining laboratory specimens and machine elements operating at stresses in the vicinity of the fatigue limit. Non-propagating cracks generated at stress levels below the fatigue limit at geometrical discontinuities would be but one example (**1**). These and similar findings generated much laboratory research involving tests which tried to simulate practical conditions, i.e., two-step and multi-step stress levels with sudden or gradually increasing or decreasing changes in the stress level.

The most simple and popular criterion adopted for the analysis of results from such tests was the Palmgren–Miner hypothesis (**2**)(**3**) which is expressed as

$$\sum (N/N_f)_i = 1 \tag{1}$$

Despite its popularity this hypothesis has two major shortcomings. First it assumes that damage accumulates in the same manner at all stress levels which implies that the mechanism of damage is unchanged throughout lifetime. However, by testing at one stress level for a certain number of cycles and then changing to a higher stress level for the remaining portion of the specimen's life, it has been shown that the summation expressed by the left-hand side of equation (1) is greater than unity because the accumulation of damage at low stress levels is much slower than at high levels (**4**)–(**6**) and involves different mechanisms. The second major shortcoming is the failure of equation (1) to consider damage in the form of cracking below the fatigue limit. In a variable stress range situation where some of the cycles are at a stress level below the fatigue limit (and hence N_f is necessarily infinite) the experimental summation term will be less than unity which indicates that a prediction of life based on the Palmgren–Miner hypothesis is dangerously optimistic.

To obviate these and other difficulties and to predict more accurately the life of a specimen or component that is subjected to a loading sequence involving some cycles at stress levels below the fatigue limit, some investigators have used a modified Palmgren–Miner rule based on a S–N curve derived from two-stress-level, single-block fatigue tests (**7**)(**8**). A similar approach had been introduced by Corten and Dolan (**9**) who proposed that such an S–N curve could be determined from certain explicit assumptions on damage accumulation. Such a curve was shown to have a steeper slope than the constant stress amplitude S–N curve, but it apparently facilitated a more precise estimation of the contribution to fatigue damage at low stress ranges just above the fatigue limit.

Marsh (**10**) applied the Palmgren–Miner and the Corten–Dolan hypotheses to the results of fatigue tests in which stress cycles were contained within a symmetrical saw tooth envelope. While the former theory over-estimated the specimen life and erred by a factor of more than five in some cases, the latter theory agreed, at least qualitatively, with some of the experimental results. Further analysis showed that the Corten–Dolan prediction could be improved by considering cycles to be damaging at stress levels as low as 80 per cent of the fatigue limit. This new damaging limit, when used in conjunction with a linear summation rule, produced a more accurate life-prediction estimate.

This latter method does not identify or quantify damage accumulation below the fatigue limit. One aim of the present investigation is to study damage at stresses below the fatigue limit via a surface replication technique; damage being equated to the formation and growth of fatigue cracks. A second aim is to derive a model based on crack growth that can give accurate predictions of fatigue lifetime of plain specimens subjected to a continuously increasing stress amplitude and which incorporates stress levels below and above the fatigue limit.

Previous work

In a previous report (**11**) on the same material, a 0.4 per cent carbon steel, cracks were shown to form during constant amplitude tests at stresses below the fatigue limit; they grew to a certain length which was related to the extent of localized plasticity, usually of the order of a microstructural unit, and they were finally arrested at crystallographic barriers to plastic flow. According to the strength of the barrier the cracks were either temporarily or permanently arrested at the first barrier. Weak barriers were the ferrite grain boundaries while the pearlite regions were the strong impenetrable barriers. Nevertheless, eventual arrest was always attained at stresses below the fatigue limit. Some of these non-propagating or dormant cracks were seen to re-activate and continue to propagate as soon as the stress level was raised to a value higher than the fatigue limit.

Even at stresses above the fatigue limit an effect of the microstructure on crack growth rate was noticeable when the crack was small. In this short crack phase the growth rate showed a deceleration as the tip of the crack approached the microstructural barriers. This phase extended approximately to a crack length equivalent to the linear elastic fracture mechanics (LEFM) threshold. The following phase was the long crack region where crack growth was best characterized by an elastic–plastic fracture mechanics analysis.

Experimental details

The 0.4 per cent carbon steel used in the present investigation had a ferrite grain size which varied between 50 and 100 μm. The structure was banded and had a mixture of about 70/30 pearlite/ferrite. Full details of the heat treatment, microstructure, and specimen dimensions are given in reference (**11**). Fully reversed torsion fatigue tests were first carried out at constant shear stress amplitudes and then a series of tests were performed during which the shear stress amplitude was continuously and uniformly increased throughout a test until failure. The description of the machine and specimen calibration are given elsewhere (**12**). Plastic replicas were taken at specified intervals during tests on selected specimens. The replicas were subsequently observed, in the reverse order, in an optical microscope. The following information was thus obtained: number of cycles to the first sign of damage, number of cycles to the observation of cracks, crack length, crack orientation, and crack growth rate.

Constant amplitude tests

Constant amplitude tests at different shear stress amplitudes, τ, gave the results shown in Fig. 1. The data follow a linear relationship on a log–log plot up to the fatigue limit of $\tau = 122.5$ MPa with a confidence level of ± 2.27 MPa. Linear

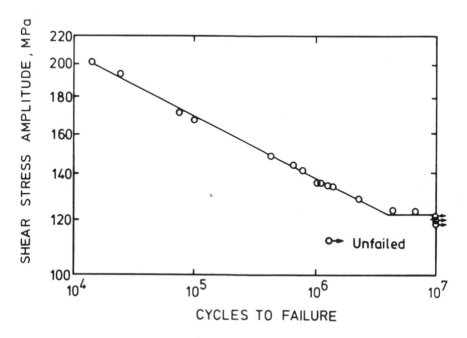

Fig 1   Constant stress amplitude fatigue life test data

regression analysis was used to find the best equation through the experimental points, i.e.

$$\tau = 429.57 N_{\mathrm{f}}^{-0.08219} \tag{2}$$

Replicas taken from some of the specimens tested at different values of plastic strain range $\Delta\gamma_{\mathrm{p}}$ allowed for the determination of the crack growth rate equations of the form presented by Miller (13) and Hobson (14). The equations which best fit the present experimental data are

$$\frac{da}{dN} = A(\Delta\gamma_{\mathrm{p}})^{n}(d - a); \quad \text{for short cracks} \tag{3}$$

with $A = 6$ and $n = 2.24$, and

$$\frac{da}{dN} = B(\Delta\gamma_{\mathrm{p}})^{m} a - C; \quad \text{for long cracks} \tag{4}$$

with $B = 17.4$, $m = 2.68$, and $C = 8.26 \times 10^{-4}$. In these expressions da/dN is given in μm/cycle.

Figure 2 shows the comparison of equations (3) and (4) with a set of experimental data. In the case of constant amplitude tests the value of d in equation (3) which best fits the data is 330 μm, which is of the same order of the crack length threshold, i.e., $a_{\mathrm{th}} = 300$ μm, calculated for this material (11) from published values of ΔK_{th}. This value of 300–330 μm is the same as the average distance between the hard barriers of pearlite which are invariably separated in this banded structure by several grains of ferrite.

Increasing amplitude tests

In these tests a specimen is subjected to fatigue loading commencing at 0.7 of the fatigue limit stress, i.e., at a level sufficiently low where no damage is expected to be observed in a constant amplitude test (11); however, this initial stress amplitude continuously increases at a pre-determined rate, α, until failure.

By taking plastic replicas at intervals throughout the test a clear picture emerges of the initiation, type, and accumulation of damage. Figures 3 and 4 show two examples of a sequence of four replicas at various fractions of fatigue life for different values of α, one showing damage leading to a transverse crack, Fig. 3, the other leading to a longitudinal crack, Fig. 4. In both cases the sequence of events is the same, namely, the first sign of damage detected by this technique is the appearance of persistent slip bands (psbs) at approximately 0.15–0.3 of the fatigue life. Crack growth is only positively identified at about 0.85 of the fatigue limit stress amplitude. It is impossible to say which psbs will contain major cracks, because the plastic replica technique is not able to discern closed stage I, mode II cracks in single crystals, be they transverse or longitudinal cracks. At some stage the cracks break through into the adjacent grain

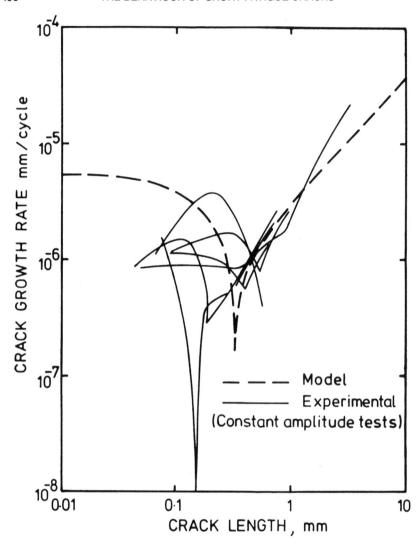

Fig 2 Experimentally determined surface crack growth rates at constant stress amplitude and predicted surface crack growth rates using equations (3) and (4)

and it is at this stage that the existence of a crack is certain and precise crack length measurements are possible. The early uncertainty as to the actual length of a closed stage I crack associated with a particular slip band is reflected by the symbols in Figs 5 and 6, which show psb and crack length measurements related to the fraction of fatigue life. With regard to the predictive model illustrated by the dashed curve, the assumption is that the crack that developed at the slip band in the first grain would propagate at a rate determined by equation (3) but

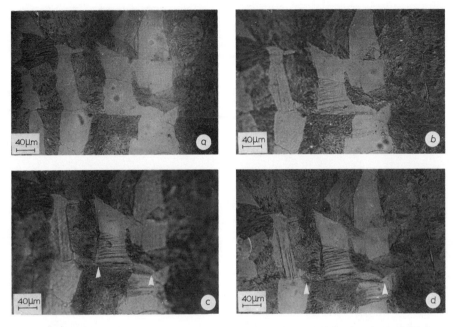

Fig 3 Replicas of fatigue damage and a transverse crack at various fractions of lifetime; $N_{\mathrm{ff}} = 1\,477\,500$ cycles.

(a) $N/N_{\mathrm{ff}} = 0$; $a = 0\ \mu\mathrm{m}$ (b) $N/N_{\mathrm{ff}} = 0.6$; $a = 112\ \mu\mathrm{m}$

(c) $N/N_{\mathrm{ff}} = 0.8$; $a = 112\ \mu\mathrm{m}$ (d) $N/N_{\mathrm{ff}} = 0.9$; $a = 164\ \mu\mathrm{m}$

with the value of d now equal to 100 μm, since it is the largest ferrite grain which controls the development of the failure crack, especially at stresses below the fatigue limit (see later discussion).

In Fig. 5, at 0.28 N_{ff}, only psbs of various length are observed because the cracks are closed. As the stress amplitude increases it eventually attains, at 0.58 N_{ff}, a value of about 0.85 of the fatigue limit stress amplitude when several cracks are clearly discernible. In this test the fatigue limit stress amplitude is not reached until 0.78 N_{ff}, but considerable damage has already accumulated. Figure 6 shows similar characteristics, but because the rate of increase of stress amplitude is higher than in Fig. 5 there is a greater tendency for the crack development phases discussed previously to be accelerated. In subsequent theoretical calculations a failure surface crack length of 1.0 mm was assumed and Figs 5 and 6 show experimental life termination at 1.0 mm to permit comparisons between experiments and the model now to be discussed.

Life prediction methods and the crack growth model

Of the numerous methods published to assess cumulative damage in fatigue, two of the most popular methods will be applied to the result of the present

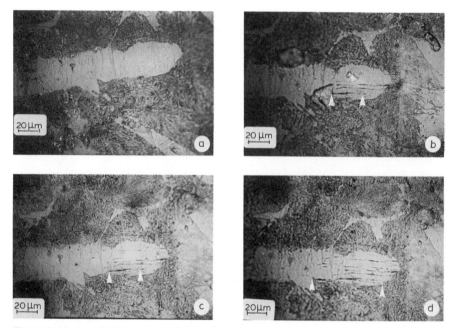

Fig 4 Replicas of fatigue damage and a longitudinal crack at various fractions of lifetime;
$N_{ff} = 8\,964\,414$.

(a) $N/N_{ff} = 0$; $a = 0\ \mu m$ (b) $N/N_{ff} = 0.51$; $a = 30\ \mu m$
(c) $N/N_{ff} = 0.57$; $a = 30\ \mu m$ (d) $N/N_{ff} = 0.62$; $a = 70\ \mu m$

experiments. Finally a new method is derived based on the short and long crack growth rate expressions, equations (**3**) and (**4**). This method is compared with the two classical prediction methods and the experimental fatigue lifetime results.

The Palmgren–Miner hypothesis

Even though this hypothesis has been proved erroneous on several occasions, it is still in common use because of its simplicity. For complicated loading patterns or when a substantial part of the cycles are at stress levels below the fatigue limit, this hypothesis is known to frequently overestimate lifetime.

When applying the hypothesis to the continuously increasing load pattern of the present test, it can be shown (see Appendix A) that

$$\sum \left(\frac{N}{N_f} \right)_i = \frac{1}{\alpha} \int_{\tau_{FL}}^{\tau_f} \frac{1}{N_{f_i}} \, d\tau_i \tag{5}$$

The integration of this equation requires an expression for N_f as a function of τ which can be obtained from constant amplitudes tests; see equation (2).

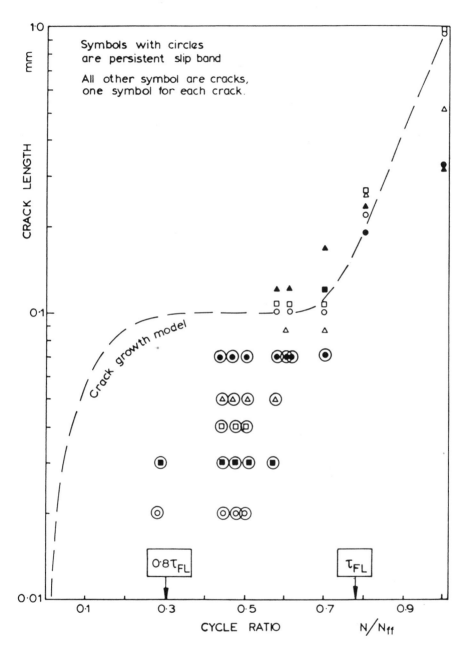

Fig 5 A typical fatigue damage versus fraction of fatigue life graph for one specimen: $N_{ff} = 8\,964\,414$; $\alpha = 5.687 \times 10^{-6}$ Nmm^{-2}/cycle. Different symbols identify different crack systems

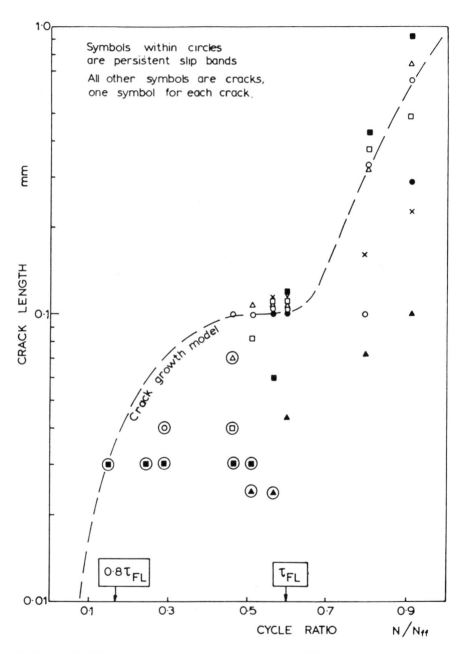

Fig 6 **A typical fatigue damage versus fraction of fatigue life graph for one specimen:** $N_{ff} = 1\,477\,500$; $\alpha = 3.808 \times 10^{-5}$ **Nmm^{-2}/cycle. Different symbols identify different crack systems**

Table 1   Fatigue lifetimes: predictions and experiment

Specimen	N_{ff} Palmgren–Miner	N_{ff} Corten–Dolan– Marsh	N_{ff} from equations (3) and (4)	N_{ff} Expt
1F	1 181 729	1 088 650	524 051	560 330
2D	2 665 370	2 327 891	856 412	981 485
3E	4 034 376	3 337 194	1 269 919	1 373 775
2X	4 958 005	3 978 989	1 474 681	1 477 500
2E	5 230 679	4 060 845	2 431 022	2 781 724
1E	9 848 154	6 720 333	4 873 670	8 964 414

Introducing this relation into equation (5) and making the summation equal unity, integrating and replacing α by the term $(\tau_f - \tau_o)/N_{ff}$, the predicted number of cycles to failure, N_{ff}, can be obtained from the expression

$$N_{ff}(\tau_f^{13.166} - \tau_{FL}^{13.166}) = 1.428 \times 10^{33}(\tau_f - \tau_o) \qquad (6)$$

From Fig. 1, the value of τ_{FL} is 122.5 MPa. Values of N_{ff} for six specimens are given in Table 1 and plotted in Fig. 7, as the Palmgren–Miner curve.

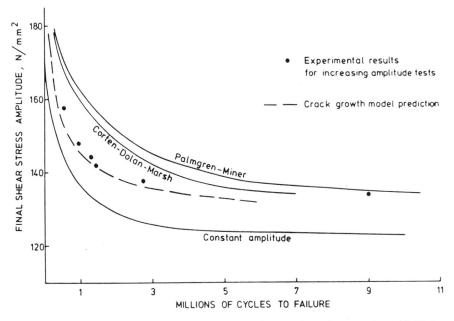

Fig 7   Increasing stress amplitude test results for different values of α, together with lifetime predictions from three models: (i) Palmgren–Miner, (ii) Corten–Dolan–Marsh, (iii) short and long crack growth rate equations

The Corten–Dolan–Marsh hypothesis

Corten–Dolan based their ideas on a theory of damage visualized as the nucleation of submicroscopic voids which developed into cracks, these cracks then propagated at a rate which depends on the stress level. The expression derived for multi-stress conditions was

$$N_g = N_h \Big/ \sum \beta_i \left(\frac{\sigma_i}{\sigma_h}\right)^z \tag{7}$$

where

N_g is the number of cycles to failure under a multi-stress programme
N_h is the number of cycles to failure at a constant stress amplitude, σ_h, equal to the highest stress achieved in the multi-stress test
β_i is the fraction of cycles at a stress level of σ_i
z is the inverse slope of an hypothetical log σ/log N relation which allows for the interaction effects of high and low stress levels

Marsh applied this expression to rotating–bending fatigue tests subjected to a triangular stress block and found that life predictions could be improved if stress cycles down to about 0.8 of the experimental fatigue limit were considered in the calculations. The new fatigue limit was interpreted as being that of the hypothetical S–N curve.

If the Corten–Dolan approach coupled with the developments due to Marsh is applied to the present experimental tests, the following expression is obtained (see Appendix A)

$$\frac{N_{ff}}{N_f} = \frac{(1 - y)(z + 1)}{1 - r^{z+1}} \tag{8}$$

In this expression

$$y = \tau_o/\tau_f \text{ and } r = 0.8\tau_{FL}/\tau_f$$

where

τ_o is the initial stress amplitude
τ_f is the final stress amplitude
$0.8\tau_{FL}$ is 80 per cent of the fatigue stress limit

In order to predict the number of cycles to failure, N_{ff}, a value of the hypothetical S–N curve inverse slope z has to be determined (see Appendix A).

Figure 7 and Table 1 show that the lifetime predictions using the Palmgren–Miner rule are dangerously non-conservative, overestimating the specimen life in some cases by a factor of two. The Corten–Dolan–Marsh method, which considers the interactive effects between high and low stress levels and also the damage from cycles below the fatigue limit stress down to a minimum level of $0.8\tau_{FL}$, gives a slightly better approximation; however the predictions are still non-conservative.

Fatigue life prediction from crack growth considerations

The main cause of error in both of the above mentioned classical methods is that they consider damage to accumulate according to a parabolic law of defect growth at both high and low stresses. It is now known that this presumption is not true; cracks less than a certain threshold length, a_{th}, propagate at an entirely different rate to cracks longer than α_{th}. For example, crack length measurements taken during constant stress amplitude tests lead to the development of the fundamentally different equations (3) and (4) for the short and long crack regions, respectively.

Since the variation of $\Delta\gamma_p$ with N is known for each test having an increasing amplitude of stress, equations (3) and (4) can be integrated to find the number of cycles in both the short and long crack stages; N_S and N_L, respectively. Their simple addition gives a predicted number of cycles to failure N_{ff}. The short crack region exends from $a_o \cong 0$ to a_t where a_t is the crack transition length between the short and long crack growth phases.

From equations (3) and (4) it is also possible to determine values of a_t, for situations in which the strain amplitude is increasing at a cyclic rate of ϕ from an initial plastic strain amplitude of $\Delta\gamma_{p_o}$. The equations for a_t, N_S, and N_L are simple to derive since, at a_t and N_S, the rate of growth da/dN for both the short and long cracks, are identical. However, the required equations are cumbersome and so are presented in Appendix B. Table 1 lists the values of the predicted life and Fig. 7 shows the same predictions in graphical form. The predictions using the crack growth equations are very close to the experimental fatigue lives while the two classical methods considered previously are non-conservative for the reasons already given. Figure 7 shows for comparative purposes the lifetime for constant amplitude tests which are reported in a separate publication (15) but it should be noted that this curve also is similarly and accurately predicted by combining the short and long crack growth equations with a tendency to slight conservatism.

Discussion

The Corten–Dolan–Marsh method shows an improvement when compared to the Palmgren–Miner method when predicting fatigue lives in situations of increasing stress amplitude cycles, but it is still a non-conservative method of prediction. The slight improvement of predictive capability via the Corten–Dolan hypothesis is due to account being taken of the interaction effect of low and high stresses (i.e., the switching to different deformation substructures at different stress levels) because crack growth rate depends on the flow stress which is a function of the substructure developed in the plastic zone. This stress history effect was introduced by Corten–Dolan by invoking a hypothetical *S–N* curve with a slope greater than the experimental one. The other factor which improves the prediction, as pointed out by Marsh, is defining a lower limit (80 per cent of the fatigue limit stress) for the initiation of damage. This is in accordance with recent work showing that fatigue cracks can be initiated by cycling at stresses below the fatigue limit (11). The lower limit of $0.8\tau_{FL}$ is somewhat arbitrary and was obtained by Marsh to fit particular test data; no physical interpretation was attempted at that time.

The microstructural observations in the present work suggest that damage is initiated by the formation of persistent slip bands (psbs) in favourably orientated ferrite grains. The psbs are formed well before the applied stress approaches the level of the fatigue limit, in fact well developed psbs are observed even below the $0.8\tau_{FL}$ stress level; see Figs 5 and 6. How many of these slip bands contain significant cracks is impossible to say with the present experimental technique, but recent advances in acoustic microscopy (16) show great potential in the detection of very small closed cracks and it is expected that this new technique will provide valuable data. Nevertheless the plastic replica techniques will identify cracks as soon as they propagate into surrounding grains, although it should be noted that not all psbs are able to propagate a crack into the next grain. Limited quantitative information from these tests reveal that only between 5 and 10 per cent of psbs are able to extend cracks into neighbouring regions. As the stress approaches the level of the fatigue limit in an increasing amplitude test, coalescence of microcracks may take place and by the time cracks approach the critical size for long crack propagation only a very few cracks have attained a length that can propagate as long cracks to failure.

The dashed curves of Figs 5 and 6 show that the crack growth model tends to be the upper bound on observed damage and that, although cracks are only seen when they are about 50–100 μm in length, it is reasonable to assume that the psbs are disguising the presence of closed cracks in single grains. This is in sympathy with the hypothesis of Miller (13) that cracks grow immediately from the first cycle, but because they are initially smaller than psbs and are closed Stage I cracks they are exceedingly difficult to detect. It is also noticeable from Figs 5 and 6 that the higher the value of α (the cyclic rate of increase in stress amplitude) the closer are the two phases of crack growth; expressed another way there is a reduced tendency for cracks to decelerate.

Figure 7 shows that a much better prediction of lifetime is obtained if both the crack growth rate functions are available for the two main phases of crack growth – the short and the long crack regions. The method of prediction proposed here employs the two crack growth equation derived from experimental data obtained from constant amplitude tests. The short crack equation, equation (3), follows the model given by Miller (13) and Hobson (14) except that the plastic strain range is used to complement the long crack law, equation (4), as proposed by Miller and Ibrahim (6).

There is one major difference between the constant amplitude tests on the same material at stresses around the fatigue limit and the present series involving increasing stress amplitudes. In the former tests discussed in reference (11), the microstructural parameter, d, in equation (3) was of the order of the pearlite–pearlite distance (330 μm) and the short cracks only decelerated to be arrested at those strong microstructural barriers. Alternatively one may say that the fatigue limit stress amplitude was always sufficiently high to quickly overcome the weaker barriers of the ferrite grain boundaries observed at lower stress levels. In the increasing amplitude tests, a considerable proportion of the lifetime was spent at relatively low stress levels at which even the ferrite grain boundaries provided sufficiently strong barriers to crack growth. Figures 5 and 6 indicate that at stress levels well below the fatigue limit the major barrier occurs at approximately 100 μm (i.e., the largest ferrite grain size) at about 60 per cent of lifetime; thereafter the crack accelerates and enters the long-crack growth phase given by equation (4). By the time the crack reaches a length of 300–330 μm (the average pearlite band separation distance) the applied stress level is above the fatigue limit and the pearlite barrier in this situation is not as effective as in a constant stress amplitude test equal to the fatigue limit stress level. This interesting phenomenon of stress level dependent barriers to crack growth will be discussed at some length in a future publication (17) and it is sufficient to report here that by equating d to the ferrite grain size in equation (3) a predictive model is derived that gives good correlation with the experiment data of Figs 5 and 6 and also a model that can be employed to give an excellent, slightly conservative estimate of fatigue lifetime as witnessed in Fig. 7.

In the present tests, therefore, the short crack region extends from a value of $a = a_o$ (the initial roughness) to a value of $a = a_t$ corresponding to a value at which the short and long crack growth rates are identical; values of a_t are given in Table B1 in Appendix B. The long crack region extends from $a = a_t$ to $a = a_f = 1.00$ mm, where a_f is the surface crack length at failure.

Finally, it appears from Fig. 7 that the entire lifetime of specimens is spent in propagating a crack from the first cycle, and that if any crack birth period exists it is negligible and can be omitted from lifetime calculations; a similar result to that given in reference (13). However, it appears that cracks can be arrested for a considerable period at strong barriers. In a recent paper (18) the temporary arrest of a short crack at a pearlite zone was shown to extend the fatigue lifetime by a factor varying linearly from one to two over fatigue lifetimes of one million to six million cycles, i.e., three million cycles were spent out of a total of six million cycles for the short crack to circumnavigate and/or cut through the pearlite.

Conclusions

(1) Classical fatigue life prediction methods, e.g., the Palmgren–Miner method or the Marsh modification to the Corten-Dolan method, are not satisfactory for predicting life in a continuously increasing stress amplitude situation due to their inability to take account of the complex crack growth behaviour of materials close to the fatigue limit where cracks decelerate before accelerating to failure.

(2) Fatigue damage is initiated well below the fatigue limit stress and can be
     observed as persistent slip bands (psbs) and microcracks. Few of these
     psbs contain cracks that can extend beyond the first grain.
(3) Two distinct equations for fatigue crack growth rate, one for the short
     crack phase and the other for the long crack phase, can be suitably
     combined to accurately predict the fatigue life of specimens tested at
     increasing stress amplitudes.

Appendix A: Calculation of damage summation terms

(1) *The Palmgren–Miner hypothesis applied to increasing stress amplitude tests*

For a linear increasing stress situation, the summation of damage requires to
consider each cycle, hence

$$\Sigma \left(\frac{N}{N_f}\right)_i = \frac{N_1}{N_{f_1}} + \frac{N_2}{N_{f_2}} + \cdots \tag{A1}$$

where $i = 1, 2, 3, \ldots$ and N is one cycle.

Equation (A1) can be written in the form, $\Sigma\, (1/N_f)_i$ or, more generally

$$\int_{\tau_{FL}}^{\tau_f} \frac{1}{N_{f_i}}\, dx_i \tag{A2}$$

where

    τ_{FL} is the fatigue limit stress obtained from constant amplitude tests
    τ_f is the stress at failure during an increasing amplitude test
    N_{f_i} is the fatigue life corresponding to τ_i

and, from a consideration of slopes in Fig. A1,

$$x_i = \frac{N_{ff}(\tau_i - \tau_o)}{(\tau_f - \tau_o)} \tag{A3}$$

where

    N_{ff} is the number of cycles at failure for an increasing amplitude test, and
    τ_o is the initial stress amplitude.

Differentiating equation (A3) and subsituting for dx_i in equation (A2) gives

$$\Sigma \left(\frac{1}{N_f}\right)_i = \int_{\tau_{FL}}^{\tau_f} \left(\frac{1}{N_f}\right)_i \frac{N_{ff}\, d\tau_i}{(\tau_f - \tau_o)} \tag{A4}$$

However, the rate of increase of stress, α, is a controlled test-variable and so
equation (A4) can be summarized as

$$\Sigma \left(\frac{1}{N_f}\right)_i = \frac{1}{\alpha} \int_{\tau_{FL}}^{\tau_f} \frac{1}{N_{f_i}}\, d\tau_i \tag{A5}$$

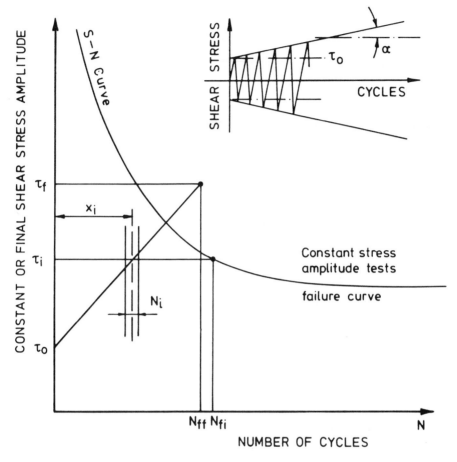

Fig A1 Schematic to illustrate experiment and model parameters

If the relationship between τ and N_f is known, i.e., the conventional S–N curve, the above equation can be integrated to give the required summation of damage according to Palmgren–Miner.

(2) *The Corten–Dolan–Marsh method applied to increasing amplitude tests*

Marsh (**10**) formulated an expression for the summation of fatigue damage, based on the work of Corten and Dolan (**9**), the latter leading to the prediction of the number of cycles to failure as

$$N_{ff} = \frac{N_f}{\sum \beta_i (\tau_i / \tau_f)^z} \tag{A6}$$

where

N_{ff} is the predicted life at failure for an increasing stress amplitude
N_f is the number of cycles to failure at constant stress amplitude, τ_f
β_i is the fraction of cycles at τ_i
z is an experimentally determined parameter

One of the conclusions that can be derived from the work of Marsh (10) is that the parameter z can be determined by making equation (A5) equal to unity and integrating it between the limits $0.8\tau_{FL}$ and τ_f in order to account for damage below the fatigue limit, hence

$$\sum \left(\frac{1}{N_f}\right)_i = \frac{1}{\alpha} \int_{0.8\tau_{FL}}^{\tau_f} \left(\frac{1}{N_f}\right)_i \, \mathrm{d}\tau_i = 1 \tag{A7}$$

This embodies the proposal of the hypothetical S–N curve, discussed previously in the main text, which can be described for the present tests as

$$\frac{1}{N_f} = \left(\frac{\tau}{429.57}\right)^z \tag{A8}$$

Equation (A8) is similar to the conventional S–N curve equation but with a steeper slope.

Substituting (A8) into (A7) and integrating

$$1 = \left(\frac{1}{429.57}\right)^z \frac{1}{\alpha(z+1)} \left(\tau_f^{(z+1)} - 0.8\tau_{FL}^{(z+1)}\right) \tag{A9}$$

If the stress rate, α, is known, the value of z can be obtained by iteration. In the present case the Newton–Raphson formula was used. Values of z for different α values are given in Table A1.

Now in equation (A6) β_i is equal to $\mathrm{d}\tau_i/)\tau_f - \tau_o)$ and so the term

$$\sum \beta_i \left(\frac{\tau_i}{\tau_f}\right)^z = \int_{0.8\tau_{FL}}^{\tau_f} \frac{\mathrm{d}\tau_i}{(\tau_f - \tau_o)} \left(\frac{\tau_i}{\tau_f}\right)^z$$

$$= \frac{\tau_f^{-z}}{(\tau_f - \tau_o)(z+1)} \left(\tau_f^{(z+1)} - 0.8\tau_{FL}^{(z+1)}\right) \tag{A10}$$

From equation (A10) it follows that equation (A6) can be rewritten as

$$\frac{N_{ff}}{N_f} = \frac{(z+1)(1-y)}{1 - r^{z+1}} \tag{A11}$$

where

$$r = \frac{0.8\tau_{FL}}{\tau_f}; \quad y = \frac{\tau_o}{\tau_f}; \quad z = 11.60, \text{ the average value in Table A1.}$$

It follows that if the conventional S–N curve and the stress at failure in an increasing amplitude test are known, then N_{ff} can be calculated from the above

Table A1

Specimen	α $(Nmm^{-2}/cycle)$	z
1F	1.232×10^{-4}	11.52
2D	6.348×10^{-5}	11.335
3E	$3.85 \ \times 10^{-5}$	11.395
2X	3.808×10^{-5}	11.263
2E	1.825×10^{-5}	11.748
1E	5.687×10^{-6}	12.359

equation; this life prediction being based on the assumptions of the Corten–Dolan–Marsh theory of damage accumulation. The fraction N_{ff}/N_f is the summation of damage term which is always greater than unity.

Appendix B: Derivation of equations to determine the lifetimes of the short and long crack growth phases

Experimental crack growth data obtained from constant amplitude tests are of the form

$$da/dN = A(\Delta\gamma_p)^n(d - a); \quad \text{for short cracks} \tag{B1}$$

and

$$da/dN = B(\Delta\gamma_p)^m a - C; \quad \text{for long cracks} \tag{B2}$$

Here

A, n, B, m, and C   are material constants
a is the crack length
d is a microstructural parameter related to the distance between obstacles to crack growth
$\Delta\gamma_p$ is the plastic shear strain range

Equation (B1) is valid from a_o (surface roughness) to a_t, the crack transition length between the short and long crack growth phases.

In an increasing amplitude test the plastic strain range may be considered to vary approximately linearly with number of cycles, and so

$$\Delta\gamma_p = \Delta\gamma_{p_o} + \phi N \tag{B3}$$

where

$\Delta\gamma_{po}$ is the initial plastic shear strain range
ϕ is the rate of increase of plastic shear strain range

Integrating equation (B1) and assuming a_o is approximately zero, since $a \approx 0 - 2 \ \mu m$ has little effect on lifetime prediction (**13**)

$$\int_{a_o}^{a_t} \frac{da}{d - a} = A \int_0^{N_s} (\Delta\gamma_{po} + \phi N)^n \ dN$$

or

$$-\ln (1 - a_t/d) = A\left\{(\Delta\gamma p_o + \phi N)^{n+1}\right\}_0^{N_s}/(n + 1)\phi \tag{B4}$$

but since

$$\Delta\gamma_{p_s} = \Delta\gamma_{p_o} + \phi N_s \tag{B5}$$

where N_s is the number of cycles to complete the short crack propagation phase, then from equations (B4) and (B5)

$$\Delta\gamma_{p_s} = \{\Delta\gamma_{p_o}^{n+1} - (n + 1)\phi A^{-1} \ln (1 - a_t/d)\}^{1/(n+1)} \tag{B6}$$

Also when the crack length is equal to a_t then equation (B1) is equal to equation (B2), hence

$$a_t = \frac{Ad\,(\Delta\gamma_{p_s})^n + C}{A(\Delta\gamma_{p_s})^n + B(\Delta\gamma_{p_s})^m} \tag{B7}$$

When substituting (B6) into (B7), a value of a_t can be determined for a given rate of increase of plastic shear strain range ϕ, determined from experiment. The number of cycles for the short crack propagation phase, N_s, can then be obtained from equations (B5) and (B6) as

$$N_s = [\{\Delta\gamma_{p_o}^{n+1} - (n + 1)\phi A^{-1} \ln (1 - a_t/d)\}^{1/(n+1)} - \Delta\gamma_{p_o}]/\phi \tag{B8}$$

To predict the number of cycles corresponding to the long crack propagation phase, N_L equation (B2) can be integrated from a_t to a_f (here $a_f = 1000\ \mu m$). Assuming $C = 0$ for simplicity

$$\int_{a_t}^{a_f} \frac{da}{a} = \int_{N_s}^{N_s+N_L} B(\Delta\gamma_p)^m \, dN \tag{B9}$$

But

$$\Delta\gamma_p = \Delta\gamma_{p_s} + \phi(N - N_s) \tag{B10}$$

Substituting equation (B10) into equation (B9) and integrating

$$N_L = [\{\Delta\gamma_{p_s})^{m+1} + \phi(m + 1)B^{-1} \ln a_f/a_t\}^{1/(m+1)} - \Delta\gamma_{p_s}]/\phi \tag{B11}$$

Here $\Delta\gamma_{p_s}$ is determined from equation (B6).

Table B1

Specimen	$\phi \times 10^9$	a_t (μm)	N_S	N_L	$N_S + N_L$	N_{ff} Expt
1F	9.89	81.09	266 386	257 665	524 051	560 330
2D	4.76	90.36	463 366	393 046	856 412	981 485
3E	2.948	95.75	767 077	502 842	1 269 919	1 373 775
2X	2.188	98.039	900 952	573 729	1 474 681	1 477 500
2E	1.369	99.767	1 749 987	681 035	2 431 022	2 781 724
1E	0.3557	99.99	3 329 599	1 544 071	4 873 670	8 964 414

Table B1 gives values of N_S and N_L for various test conditions. The same table also gives values of a_t calculated by combining equations (B7) and (B6) and solving iteratively via a simple computer program.

The above data have not been reduced to three or four significant digits. They are either computed values or, in the case of experiments, data taken from the fatigue machine cycle counter.

Acknowledgements

Grateful thanks are extended to the Science and Engineering Research Council of the UK, Rio Tinto Zinc, and the Iraqi government for funding this programme of research.

References

(1)   FROST, N. E., POOK, L. P., and MARSH, K. J. (1974) *Metal fatigue* (Clarendon Press, Oxford).
(2)   PALMGREN, A. (1924) The fatigue life of ball bearings (in German) *ZVDI*, **68**, 339–341.
(3)   MINER, M. A. (1945) Cumulative damage in fatigue, *J. Appl. Mech.*, **12**, A159–A164.
(4)   MILLER, K. J. and ZACHARIAH, K. P. (1977) Cumulative damage laws for fatigue crack initiation and stage I propagation, *J. Strain Analysis*, **12**, 262–270.
(5)   IBRAHIM, M. F. E. and MILLER, K. J. (1980) Determination of fatigue crack initiation life, *Fatigue Engng Mater. Structures*, **2**, 351–360.
(6)   MILLER, K. J. and IBRAHIM, M. F. E. (1981) Damage accumulation during initiation and short crack growth regimes, *Fatigue Engng Mater. Structures*, **4**, 263–277.
(7)   SEKI, M., TANAKA, T., and DENOH, S. (1971) Estimation of the fatigue life under programme load including the stresses lower than endurance limit, *Bull. Jap. Soc. mech. Engrs*, **14**, 183–190.
(8)   MISAWA, H. and KODAMA, S. (1981) Fatigue crack propagation behaviour by cyclic overstressing and understressing, *Mem. Fac. Tech.*, *Tokyo Metropolitan University*, No. 31, 2967–2980.
(9)   CORTEN, H. T. and DOLAN, T. J. (1956) Cumulative fatigue damage, *Proc. Int. Conf. Fatigue of Metals* (Institution of Mechanical Engineers, London), pp. 235–246.
(10) MARSH, K. J. (1965) Cumulative fatigue damage under a symmetrical sawtooth loading programme, *Mech. Engng Sci.*, **7**, 138–151.
(11) DE LOS RIOS, E. R., MOHAMED, H. J., and MILLER, K. J. (1985) A micromechanics analysis for short fatigue crack-growth, *Fatigue Fracture Engng. Mater. Structures*, **8**, 49–63.
(12) MILLER, K. J., MOHAMED, H. J., and BROWN, M. W. (1985) A new fatigue facility for studying short fatigue cracks, *J. Fatigue Fracture Engng Mater. Structures*, in press.
(13) MILLER, K. J. (1985) Initiation and growth rates of short fatigue cracks, *Fundamentals of deformation and fracture* (Eshelby Memorial Symposium) (Cambridge University Press, Cambridge), pp. 477–500.
(14) HOBSON, P. D. (1986) *The growth of short fatigue cracks in a medium carbon steel*, PhD thesis, University of Sheffield.
(15) MOHAMED, H. J. (1986) *Cumulative fatigue damage under varying stress range conditions*, PhD thesis, University of Sheffield.
(16) ILETT, C., SOMEKH, M. G., and BRIGGS, G. A. D. (1984) Acoustic microscopy of elastic discontinuities, *Proc. R. Soc. London*, **A393**, 171–183.
(17) MILLER, K. J., MOHAMED, H. J., BROWN, M. W., and DE LOS RIOS, E. R. (1986) Barriers to short fatigue crack propagation at low stress amplitudes in a banded ferrite–pearlite structure, TMS-AIME, Santa Barbara meeting, to be published.
(18) MILLER, K. J. (1986) Introductory lecture, *Fatigue of engineering materials and structures*, Sheffield (Institution of Mechanical Engineers, London).

L. Bouksim* and C. Bathias*

Initiation and Propagation of Short Cracks in an Aluminium Alloy Subjected to Programmed Block Loading

REFERENCE Bouksim, L. and Bathias, C., **Initiation and Propagation of Short Cracks in an Aluminium Alloy Subjected to Programmed Block Loading,** *The Behaviour of Short Fatigue Cracks,* EGF Pub. 1 (Edited by K. J. Miller and E. R. de los Rios) 1986, Mechanical Engineering Publications, London, pp. 513–526.

ABSTRACT The initiation and propagation of microcracks have been studied in 7175 T7351 aluminium alloy using $100 \times 20 \times 10$ mm smooth test pieces subjected to three point bending. It has been shown that the initiation, determined by an acoustic emission method and corresponding to a 200 μm crack, involves a not negligible amount of crack propagation. The loading pattern chosen is a ground–air–ground spectrum consisting of two loadings each with a constant amplitude of stress ratio $R = 0$ and 0.7. The number of small cycles (n) of low amplitude and high R ratio per block, is variable. The experimental crack growth rate data can explain the influence of the small cycles in a block and also their influence on the propagation of short cracks of the scale of the grain size.

Introduction

In 1975, Pearson (**1**) examined short cracks in aluminium alloys. Whilst in 1979 Topper (**2**) and Kim (**3**) studied steel. This was followed by Lankford (**4**)(**5**) and other authors (**6**)–(**8**) who showed that, for a single material, if the propagation rate of long cracks conforms to the Paris law then this was not the behaviour of small fatigue cracks whose growth can take place below the threshold stress intensity factor of long cracks and whose propagation rates could be slower, and even stopped contrary to expectations based on linear elastic fracture mechanics (LEFM). The growth of these kinds of short cracks cannot, therefore, be forecast on the basis of LEFM.

The above mentioned authors hinted that the local microplasticity caused quick propagation, but that the presence of grain boundaries in plastically deformed areas created a transitory deceleration and some arrests in crack propagation. In order to understand the physical conditions that occur when the material is being damaged, we translated the problem of the initiation of a crack into the nucleation phase of a microcrack followed by the propagation of a short crack limited to 200 μm in length on the surface; this value has been chosen as an initiation criterion which can be detected by acoustic emission and has been considered acceptable technologically.

Several papers (**9**)–(**13**) on the initiation of fatigue cracks under a given load

* Laboratoire de Mécanique de l'UA 849 du CNRS – Université de technologie de Compiègne, France.

spectrum were used as a basis for this investigation, whose object is to answer the following questions

(1) How is a fatigue crack initiated under programmed block loading?
(2) What are the parameters to consider in order to model the initiation process under programmed block loading?
(3) How can the accumulation of damage under programmed block loading be expressed?

Mode of operation

The material studied was a 7175 T7351 aluminium alloy which is often used for thick pieces in aeronautical structures because of its good mechanical properties, particularly, resistance to fatigue, tensile strength, etc. The chemical constitution (% wt) was 0.08 Si, 0.17 Fe, 1.48 Cu, 2.64 Mg, 0.07 Mn, 0.19 Cr, 5.94 Zn and 0.04 Ti.

The mechanical properties of the material are, in the (L) longitudinal direction 456 MPa yield stress, 532 MPa tensile strength, and 10 per cent elongation to fracture. The corresponding data in the (T) Transverse direction were, respectively, 428 MPa, 505 MPa, and 9.4 per cent.

The microstructure as revealed by a Keller etching agent is shown in Fig. 1. The main directions are marked with the letters 'LD' (longitudinal direction), 'TD' (transverse direction), and 'ND' (normal direction); the size of the grains according to these three directions is estimated at 500 μm, 200 μm, and 20 μm, respectively. The test pieces were machined in the longitudinal direction so that the crack propagated in the normal direction. The specimen were three point bend smooth test bars measuring $100 \times 20 \times 10$ mm.

The bending area was polished mechanically with abrasive paper, followed by aluminium oxide of 0.3 μm size and finally the specimen was given an electrolytic polishing with a buffer. The type of loading chosen was a repeated single spectrum typical of those found for aeroplanes (see Fig. 4). It consists of a ground–air–ground cycle having a zero loading for the stress at 'ground' level and a high stress at the 'air' level, e.g., $R = 0$. At the 'air' level, lower stress amplitude cycles, but with a high R ratio of 0.7, are added to simulate the perturbations during the flight of the aeroplane. During 'air flight' the maximum stress level attained is the same as that in the ground to air cycle. This spectrum is named n, with n being the number of small cycles with a high R ratio; n has a value 1, 5, 20, or 80 in the current tests. All the fatigue tests were made at room temperature on electrohydraulically pilot-operator machines of the Mayes brand. The machine is operated and controlled with the help of an external controller, either a PDP 11/10 digital microcomputer or an Apple II microcomputer. All the tests were at 10 Hz frequency.

The detection of the initiation of fatigue cracks is made with an acoustic

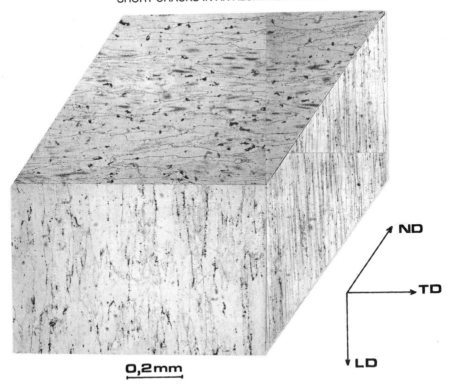

Fig 1   7175 T 7351 alloy; a three dimensional appreciation of the microstructure

emission instrument which consists of a pick up on the surface of the specimen. A preamplifier set below the pick up enables the matching of the impedence; the whole is connected with the aquisition chain of the signal (**14**). The initiation detected by acoustic emission corresponds to a semi-elliptical crack about 200 μm long. It should also be noted that the technique of plastic replication was used to observe the initiation and the propagation of short cracks until the cracks was 200 μm long. Such cellulose acetate replicas have also been used by Dowling (**15**) on A 533 B steel smooth test pieces.

The three point bending tests have been carried out with a maximum stress of 365 MPa and R ratio of 0.7 for the small cycles and R equal to 0.01 for the basic cycles. The tests are frequently interrupted while the maximum stress value is maintained in order to take prints from the surface of the damaged specimen. The replicas so obtained are first observed under the light microscope in order to perceive the damaged areas and possibly the growing of a crack. Cathodic metal spraying with gold facilitates further studies under the scanning electron microscope.

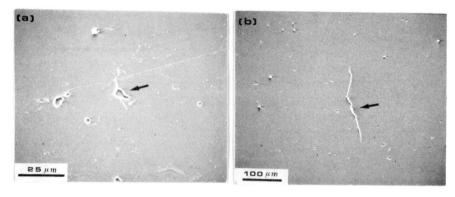

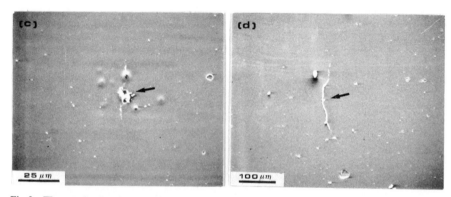

Fig 2   Three point bending results: (a) $n = 1$ type test – 25 μm long crack for 13000 blocks; (b) $n = 1$ type test – 200 μm long crack for 39000 blocks; (c) $n = 5$ type test – 37 μm long crack for 11000 blocks; (d) $n = 5$ type test – 197 μm long crack for 27000 blocks

Results and discussion

Fractographic observations

The microscopic examination of the surfaces shows that the initiation of microcracks occurs in the centre of specimen surface at Mg_2Si inclusions whose greatest dimension on the plane of observation is between 10 μm and 30 μm. Table 1 shows the results for such a crack a few microns long and gives the number of blocks required to get a semi-elliptical crack 200 μm long. This value is chosen as a criterion of initiation.

In general, the initiation crack grows out of the Mg_2Si deposits and it propagates in the matrix along a direction more or less perpendicular to the longitudinal axis of the test piece. Figures 2 and 3 illustrate the initial and final

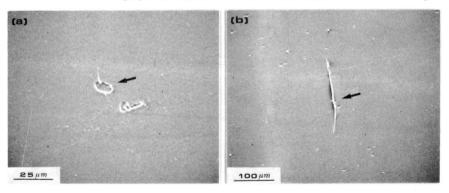

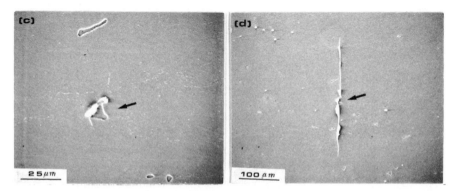

Fig 3   Three point bending results: (a) $n = 20$ type test – 10 μm long crack for 9500 blocks; (b) $n = 20$ type test – 195 μm long crack for 16 880 blocks; (c) $n = 80$ type test – 10 μm long crack for 7200 blocks; (d) $n = 80$ type test – 280 μm long crack for 10 500 blocks

stages of the propagation of a crack for the $n = 1, n = 5, n = 20$, and $n = 80$ types of tests, respectively.

The number of blocks corresponding to these various tests is given in Table 1.

Table 1   Results of the initiation tests

σ_{max} (MPa)	a (μm)	Thousands of cycles for initiation	Thousands of blocks for crack initiation			
		Loading with a constant amplitude $R = 0.01$	$n = 1$	$n = 5$ $R = 0.7$	$n = 20$	$n = 80$
365	200	10	10.5	10	9.5	7.125
		40	39	27	17	9.940

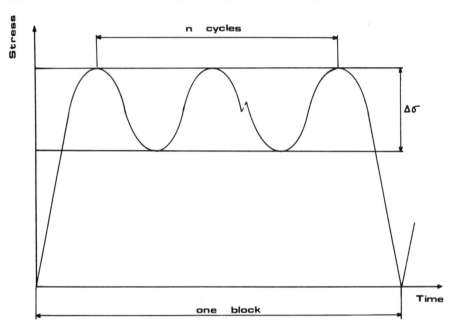

Fig 4   Schematic: when $n = 0$, $R = 0$, and when $n = 20$ or 80, $R = 0.7$

Figure 5 illustrates the results of the various tests in terms of n, the number of cycles for each block. One can note that the number of blocks, N, necessary to nucleate a very short crack, about 10 μm long, is almost constant whatever the number of small cycles per block may be. On the contrary, the further propagation of the short crack depends both on the main cycles and on the small cycles of the block. In the latter case, the experiment shows that there is a linear relation between the number of blocks to initiation, Na, and the number of small cycles per block, n for a given maximum stress.

The two straight lines of Fig. 5 divide the (Na, n) domain in two areas: first, an area in which only the main cycles of the block are damaging, and, secondly, an area in which the small cycles contribute to the damage as soon as a microcrack appears. It should be remembered that the small cycles of the block are not damaging when they are applied alone because their amplitude is only 110 MPa. As has been demonstrated in other works concerning initiation at a notch tip (12)(13), under programmed block loading the accumulation of damage done by the small cycles and the main ones within a repeated block can be linearly expressed by equations (1), (2), and (3) below. Of course, one can write

$$\text{(Initiation)} = \text{(Nucleation)} + \text{(short crack growth)} \tag{1}$$

The equation of the first straight line giving the *nucleation* for our experimental results is

$$N = -1.7 \times 10^3 \log n + 11 \times 10^3 \tag{2}$$

Similarly, the number of blocks at the point of initiation is written

$$Na = -1.67 \times 10^3 \log n + 4 \times 10^4 \tag{3}$$

Thus we infer the equation of the propagation of the short cracks resulting from the simultaneous damage from the small cycles and the main cycles of the block

$$Np = -1.5 \times 10^4 \log n + 2.9 \times 10^4 \tag{4}$$

Crack growth rates

It is well known that the growth behaviour of a short crack is different from that of a long crack. The propagation curves that we obtained are now compared in Figs 6 and 8 for tests with a constant amplitude and with a variable amplitude of the *C/n* type, *n* being equivalent to 1, 5, 20, and 80. These results of quick propagation of short cracks below the threshold confirm the work of Pearson (1) and Lankford (4) on the aluminium alloy. In Fig. 6 the empirical curve

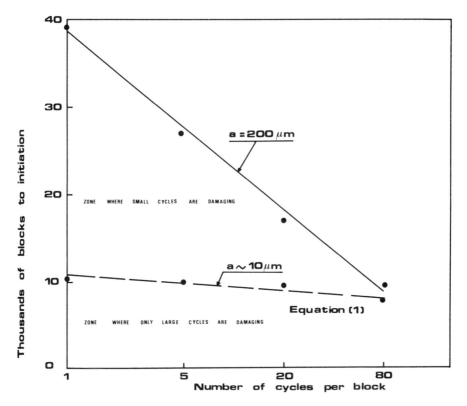

Fig 5   Number of blocks to nucleation and initiation as a function of the number of cycles per block during three point bending tests; $\sigma_{\text{max}} = 365$ MPa

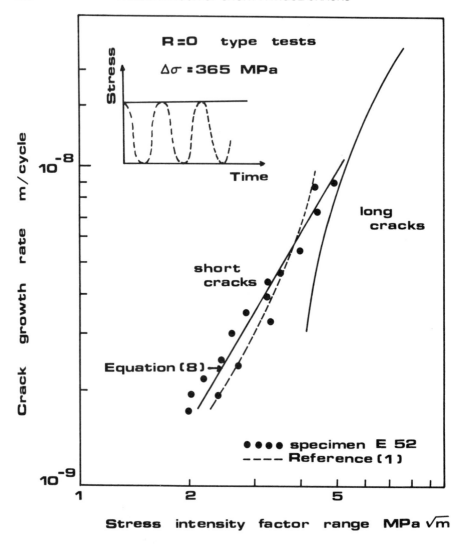

Fig 6   Three point bend test results for repeated load cycles, $R = 0$. Stress range 365 MPa

derived from Pearson's work is in good agreement with the empirical points of our tests with a $\Delta\sigma$ of 365 MPa.

The ΔK stress intensity factor in these figures was calculated with Newman's formula (16) for a surface elliptical crack in bending.

It may be written

$$\Delta K = HSb\sqrt{\pi \frac{a}{Q}} F\left(\frac{a}{t}, \frac{a}{c}, \frac{a}{b}, \theta\right) \tag{5}$$

with the following conditions

$$0 < \frac{a}{c} \leqslant 1$$

$$0 < \frac{a}{t} \leqslant 1$$

$$c/b < 0.5$$

and

$$0 \leqslant \theta < \pi$$

Here Sb = bending stress, a = depth of the crack, c = 1/2 length of the crack on the surface and H, F, and Q = functions of a/t and a/c, t = 20 and $\theta = \pi/2$.

It must be remembered that the $2c$ length of the surface cracks is measured on replicas with a light microscope. When $2c$ is between 10μ and 20μ, as a first approximation the outline of the crack is considered to be semi-circular. Moreover, the crack speed during each block is determined by applying a linear accumulation between the basic cycle and the superimposed small cycles. The principle of the calculation is that two basic relationships can be written as

$$da/dN = 0.41 \times 10^{-9}(\Delta K)^{1.88} \text{ for large stress range cycles, } R = 0 \qquad (6)$$

$$da/dN = 0.41 \times 10^{-9}(\Delta K)^{2.24} \text{ for the small stress range cycles, when } n = 1 \qquad (7)$$

By superposition da/dN can be determined in terms of the above two laws of propagation for any value of n (1, 5, 20, 80, etc.), see Fig. 7. The relation for a block where n = any number becomes

$$da/dN_n = n \ da/dN_{n=1} - (n-1)da/dN_{R=0} \qquad (8)$$

The results that we obtain are added to Fig. 8 for n = 5, 20, and 80; the straight lines determined from equation (8) coincide rather well the experimental points for the various spectra.

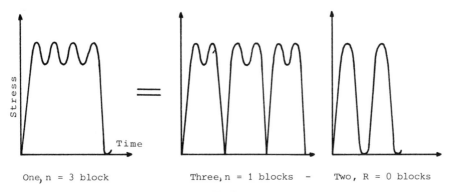

One, n = 3 block   Three, n = 1 blocks   -   Two, R = 0 blocks

Fig 7

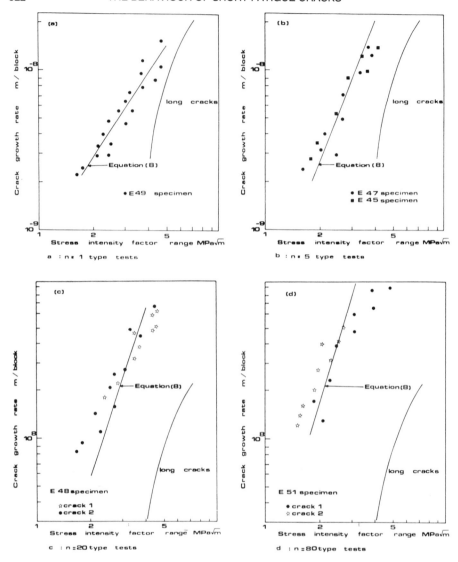

Fig 8   Three point bend test results on 7175 T 7351 alloy

It can be noticed that da/dN increases simultaneously with n for a given K and that the threshold for non-propagation is very much affected by the small n type cycles.

Mechanism of propagation for the short cracks

Observations under a light microscope on prints of the damaged area show that the microcrack grows irregularly and comes to a standstill periodically during a

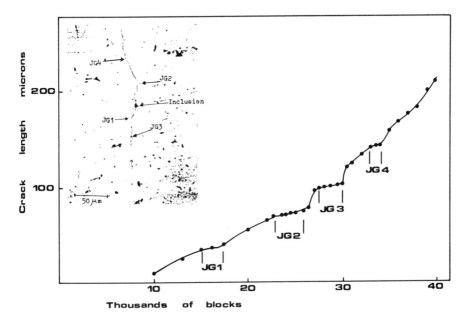

Fig 9 Fatigue crack length as a function of the number of blocks for 7175 T 7351 alloy subjected to three point bending of the $n = 1$ type

few cycles before starting its progress again. In order to explain this phenomenon we have written down the length of the crack on the surface, see Figs 9, 10, and 11, according to the number of blocks recorded for the n-type tests for n equal to 1, 20, and 80. One can note that a microcrack which initiated from an inclusion grows until it reaches the first grain boundary.

It is then slowed down by the presence of the latter; one can also note a definite time necessary for the crack to get through the grain boundary and grow into the next grain. In so far as the plastic area at the tip of the crack is limited by the nearest grain boundary, crossing the boundary implies exceeding a critical value of cumulative plastic deformation and of shearing. When the crack reaches an adequate size of between 100 μm and 130 μm it then grows continuously without being impeded by the grain boundaries through which it propagates; this aspect has also been observed by Fathualla *et al.* (**17**) on aluminium alloys.

As for the $n = 80$ test, we observed the propagation of several microcracks; two of the microcracks become predominant, while the others stopped. Figure 11 shows the nucleation of these two cracks (for a number of blocks equivalent to 7200) and their subsequent propagation. The crack that became the main one grew much more quickly and reached a length of 280 μm. The secondary crack seems to have been stopped when only 94 μm long after 10 500 blocks.

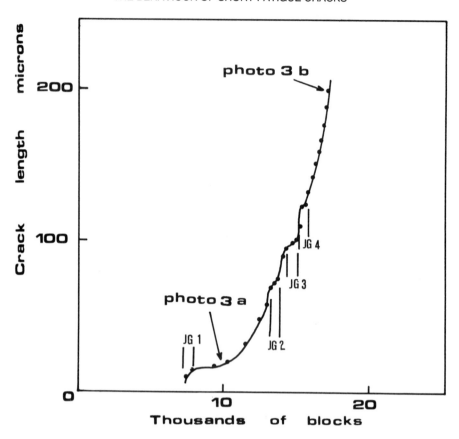

Fig 10   Fatigue crack length as a function of the number of blocks for 7175 T 7351 alloy subjected to three point bending of the $n = 20$ type

Conclusions

This work concerns the initiation of fatigue cracks in a 7175 T7351 aluminium alloy under complex loading.

(1)   The process of initiation is one of continuous damage development whose physical nature depends on the number of cycles and the nature of the complex loading cycle.

       The acoustic emission technique used in this study has made it possible to divide the initiation phase into a nucleation of a microcrack phase and a phase for the propagation of a short crack up to 200 μm in length. This criterion of initiation provides both a scientific and a technological significance.

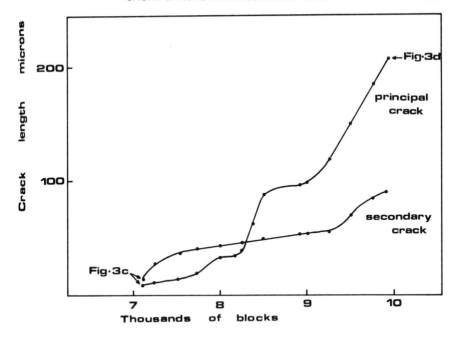

Fig 11   **Fatigue crack length as a function of the number of blocks for 7175 T 7351 alloy subjected to three point bending of the $n = 80$ type**

(2)   The crack initiation phase under a ground-air-ground type of programmed loading shows that the small cycles which are limited in their
damaging potential when applied alone become very damaging when they
are associated with the major cycles of the block.

(3)   The significance of the small cycles is related to the propagation of the
short cracks on the scale of the grain size.

(4)   Under programmed block loading the growth rate of short cracks, up to
the point of macro-initiation, accrues in a way very different from the one
anticipated by Paris's law for long cracks.

  The influence of the n-type small cycles in each block is clearly revealed
by a law of propagation under programmed block loading which can be
described for a given loading ratio, by

$$da/dN_n = n \; da/dN_{n=1} - (n - 1) \; da/dN_{R=0}$$

Knowing the growth rate of the short crack for a n-type block and the
number of cycles necessary to nucleate a crack, we can calculate the
number of cycles to initiation according to: number of cycles to initiation
= number of cycles to nucleation + number of cycles for propagation of
the short crack.

Acknowledgements

This study was financed by the Centre National de la Recherche Scientifique (CNRS), and by the Aerospace industry to whom we are most grateful.

References

(1) PEARSON, S. (1975) Initiation of fatigue cracks in commercial aluminium alloys and the subsequent propagation of very short cracks, *Engng Fracture Mech.*, **7**, 235–247.
(2) TOPPER, T. H. and EL HADDAD, M. H. (1979) Fracture mechanics analysis for short fatigue cracks, *Canad. Metall. Q.*, **18**, 207–217.
(3) KIM, Y. H., MURA, T., and FINE, M. E. (1978) Fatigue crack initiation and microcrack growth in 4140 steel, *Metall. Trans, A*, **9A**, 1679–1684.
(4) LANKFORD, J. (1982) The growth of small fatigue cracks in 7075 T 6 aluminium, *Fatigue Engng Mater. Structures*, **5**, 233–248.
(5) KUNG, C. Y. and FINE, M. E. (1979) Fatigue crack initiation and microcrack growth in 2024 T 4 and 2124 T 4 aluminium alloys, *Metall. Trans, A*, **10A**, 603–610.
(6) SHELDON, E. P., COOK, T. S., JONES, J. W., and LANKFORD, J. (1981) Some observations on small fatigue cracks in a superalloy, *Fatigue Engng Mater. Structures*, **3**, 219–228.
(7) BROWN, C. W. and HICKS, M. A. (1983) A study of short fatigue crack growth behavior in titanium alloy IMI 685, *Fatigue Engng Mater. Structures*, **6**, 67–76.
(8) DOWLING, N. E. (1982) Growth of short fatigue cracks in an alloy steel. Westinghouse R and D Center, paper 82-107–STINE–P1.
(9) GABRA, M. (1982) *Contribution à la modélisation de l'endommagement des alliages d'aluminium en fatigue par blocs programmés*, Thèse, Université de Technologie de Compiègne.
(10) GABRA, M. and BATHIAS, C. (1984) Fatigue crack initiation in aluminium alloys under programmed block loading, *Fatigue Engng Mater. Structures*, **7**, 13–27.
(11) BATHIAS, C., GABRA, M., and ALIAGA, D. (1982) Low cycle damage accumulation of aluminium alloys, *ASTM STP 770*, pp. 23–44.
(12) BOUKSIM, L. and BATHIAS, C. (1984) Etude de l'amorçage à fond d'entaille sous spectre de charge dans un alliage léger à haute résistance, *Journées Internationales de Printemps* (Société Française de Métallurgie, Paris), p. 289.
(13) BOUKSIM, L. (1985) *Amorçage des fissures de fatigue sous spectre de charge dans un alliage d'aluminium à fond d'entaille et sur des barreaux lisses.* Thèse, Université de Technologie de Compiègne.
(14) HOUSSNY EMAM, M. (1981) *Etude de l'émission acoustique associée à la déformation plastique des métaux sous sollicitation cyclique et sous l'action de l'environnement.* Thèse, Université de technologie de Compiègne.
(15) DOWLING, N. E. (1977) Cyclic stress–strain and plastic deformation aspects of fatigue crack growth, *ASTM STP 637*, p. 93.
(16) NEWMAN, J. W. Jr and RAJU, I. S. (1981) An empirical stress intensity factor equation for the surface crack, *Engng Fracture Mech.*, **15**, 185.
(17) FATHULLA, A., WEISS, B., and STICKLER, R. (1984) Initiation and propagation of short crack under cyclic loading near threshold in technical alloys, *Journées Internationales de Printemps* (Société Française de Métallurgie, Paris), p. 182.

ACOUSTIC MICROSCOPY

G. A. D. Briggs,* E. R. de los Rios,† and K. J. Miller†

How to Observe Short Surface Cracks by Acoustic Microscopy

REFERENCE   Briggs, G. A. D., de los Rios, E. R., and Miller, K. J. **How to Observe Short Surface Cracks by Acoustic Microscopy**, *The Behaviour of Short Fatigue Cracks*, EGF Pub. 1 (Edited by K. J. Miller and E. R. de los Rios) 1986, Mechanical Engineering Publications, London, pp. 529–536.

ABSTRACT   In the search for a technique for the observation of short fatigue cracks, scanning acoustic microscopy offers exceptional potential. The scanning acoustic microscope has now been developed to the stage where it may routinely be operated with a resolution better than a micron. It is generally used in reflection imaging of surfaces, and its unique advantage lies in the ability to show how acoustic waves interact with the elastic properties of the specimen. More specifically, factors that affect the propagation of Rayleigh waves may be observed with great sensitivity. This enables surface cracks to be detected and imaged even when they are only a few microns long and deep because, although they may be much less than a wavelength wide and therefore undetected by waves that are geometrically reflected from the surface, the Rayleigh waves excited within the surface can strike the cracks broadside and therefore be strongly scattered. This leads to greatly enhanced contrast from such cracks in acoustic micrographs, and gives exceptional sensitivity to short cracks at the earliest stages of their formation and growth. The theory of the contrast from cracks has been developed and experimentally verified, and this makes quantitative analysis and measurement possible. The acoustic microscope also reveals grain structure and second phases without any need for etching, so that the relationship between these and the development of cracks may be observed directly.

Introduction

Any study of short crack behaviour depends on sensitive detection and observation of the crack. At present the usual method of measuring short cracks is by a replication technique, which involves taking several plastic replicas at intervals during a test. The replicas are then examined in reverse order, starting with the last, in which the crack should be largest and followed by working backwards through the series. Despite the well-known difficulties of the method, and its relative labour intensiveness, some fundamental results have been obtained using the replica technique (1)(2).

The detection of short cracks in a carbon steel demonstrates clearly the limitation of the replication technique. In this case the first sign of fatigue damage capable of imprinting the replica is the formation of multiple persistent slip bands (psbs) which form and extend across a ferrite grain (3). A small proportion of these psbs contain cracks which initially grow along the psb plane. Within this stage of growth the cracks are indistinguishable from the slip bands until, at a much later stage, the crack begins to propagate into the

* Department of Metallurgy and Science of Materials, University of Oxford, Oxford OX1 3PH.
† Department of Mechanical Engineering, University of Sheffield, Sheffield S1 3JD.

529

grain. This concealed period of crack growth represents an important part of the lifetime, e.g., between 0.5–0.7 of the total lifetime of a specimen. Much information and understanding of short crack growth could be gained if these small cracks were to be observed and measured with some degree of certainty.

The requirements for examining cracks at the earliest stages of fatigue life are that it must be possible to see cracks that may be only a few microns in length and less than a micron wide. It must be possible to observe them non-destructively, so that afterwards the test may continue and the progress of the same crack followed. It is also desirable to be able to relate the position and development of the crack to the surrounding microstructure.

Scanning acoustic microscopy offers the unique possibility of meeting these requirements directly, without any need for replicas. The principle of scanning acoustic microscopy is becoming increasingly well known (4)(5). It has not proved possible to make an acoustic imaging system that will give an image of all points on an extended object simultaneously, but it is possible to make an acoustic lens that has excellent focussing properties on its axis. This is achieved by grinding a small spherical cavity in the centre of one face of a disk of sapphire, and growing a zinc oxide transducer on the opposite face. A drop of water is placed between the spherical surface and the specimen, so that acoustic waves generated by the transducer propagate through the sapphire, are refracted at the lens surface, and come to a focus in the water. Because of the very high refractive index involved, geometrical aberrations are negligible, and the size of the focussed spot is limited solely by diffraction. Acoustic microscopes can be operated readily at frequencies up to 2 GHz. The velocity of sound in water is $1.5~\mu\mathrm{m~ns}^{-1}$, so that for a lens of large numerical aperture the resolution is comparable with that available using light microscopy. Scanning acoustic microscopes are usually operated in reflection for high resolution work, and for this purpose the transducer is excited with a short pulse at the required frequency. When the resulting acoustic pulse is reflected by the specimen it returns again through the same lens and in turn excites an electrical signal on the transducer. The strength of this signal is measured, and this gives a value for the acoustic reflection from that point on the specimen surface. In order to build up an image, the lens is moved in a raster over the surface, and the signal from each point is fed to a corresponding address in a digital framestore that enables the complete image to be displayed on a television monitor. If the specimen has significant variations in surface height, then the image may simply represent the topography of the specimen. Therefore for most metallographic purposes the surface is polished (but not etched), to a smoothness and flatness better than a wavelength, and much more interesting contrast can then be obtained that is unique to acoustic microscopy.

Contrast

The advantages of using acoustic waves for microscopy come from the unique ways in which they can propagate in solids. One such advantage lies in their ability to penetrate materials that are opaque; it is this ability that is exploited in conventional ultrasonic non-destructive testing. In high resolution scanning acoustic microscopy another advantage is exploited, namely the distinctive origin of the contrast in the interaction of the acoustic waves with the elastic properties of the specimen. In the majority of materials this is dominated by the excitation of surface (or Rayleigh) waves in the specimen. These are waves that contain components of both longitudinal and transverse elastic waves that decay exponentially away from the surface. They can be strongly excited by the acoustic waves arriving from the lens in an acoustic microscope and, by varying the defocus (i.e., the distance, z, between the focal plane of the lens and the surface of the specimen), the contrast in the scanning acoustic microscope can be made very sensitive to factors that affect their propagation (5). Such factors may change due to material composition (for example, if the lens is scanning over a second phase) or perturbation of the surface wave velocity by a surface layer, or attenuation. Grain structure, too, can affect the contrast, because the propagation of surface waves on elastically anisotropic materials depends on their crystallographic orientation, and this varies from one grain to another. Grain structure is therefore revealed without any need for etching. Particularly strong contrast can be obtained from surface boundaries and cracks, because the Rayleigh waves can be strongly scattered by these. It is important to realise that this can happen even when the cracks are very thin. This means that even when such cracks are much too fine to be imaged by waves that are simply geometrically reflected from the surface, because they are much narrower than the resolution spot size (whether it be acoustic or optical), they can nevertheless give strong contrast because of the way that they scatter the Rayleigh waves that are generated in the surface and then strike them from the side.

Images

A comparison of optical and acoustic images of fatigue cracks is presented in Fig. 1. The specimen was a plain bearing with an Al–20% Si alloy that was being developed for increased fatigue resistance. The specimen had been subjected to a fatigue test and was being examined for incipient failure. The images shown are of a cross section through the bearing, with a steel substrate on the left and the alloy occupying the main area in the centre of the picture. The magnifications and the areas imaged are identical in the two pictures. It is quite difficult to find all the fatigue cracks in the optical picture, though perhaps with etching they could be revealed more clearly. But in the acoustic image, without the need for any special preparation beyond the initial polishing, it is rather easier to find the cracks. In this example an area of the specimen was chosen where the cracks could be found in the optical image, to faciliate comparison, but the scanning acoustic microscope also gives good contrast from the cracks in cases where they cannot be found at all optically (6).

Experience in imaging such cracks with the scanning acoustic microscope suggests that there are two important features of such images that must be

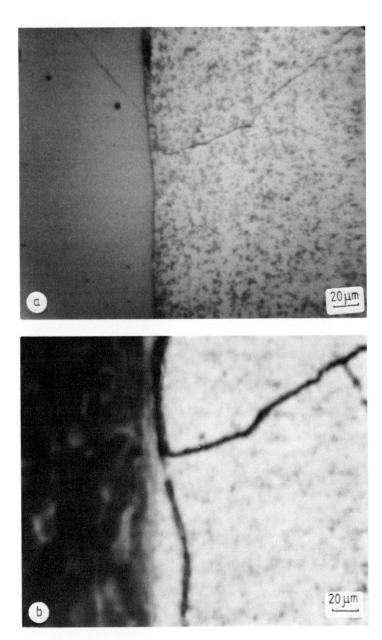

Fig 1   An Al–20%Si plain bearing which has been subjected to fatigue (The material on the left of
the picture is a steel substrate): (a) optical micrograph; (b) acoustic micrograph (0.73 GHz)

understood. The first is that the width of the acoustic image of the crack may be much wider than the true width of the mouth of the crack in the specimen. This is apparent from the image of the fatigue crack in Fig. 1, and has been confirmed by SEM studies of cracks in TiN coatings, where cracks that gave strong contrast in the acoustic microscope at 0.73 GHz ($\lambda = 2 \mu$m) appeared in the SEM to be about 0.1 μm wide. The second is that the contrast that is seen depends sensitively on the amount of defocus employed; indeed, even relatively small changes in the defocus can lead to complete reversal of contrast. These effects have now been accounted for in terms of a theoretical model that has been developed for the images of cracks in the scanning acoustic microscope (7), and this has now been extensively verified experimentally (8). Although the theory is at present restricted to two-dimensions, it seems to give a good account of the phenomena that are found using a full spherical lens.

One of the tests of a good theory is that it should indicate how to obtain better results. Cracks are not the only features that give contrast in acoustic microscopy: contrast is also obtained from grain structure. If the elastic anisotropy is small, as for example with an aluminium alloy, then such contrast may be relatively weak. But for alloys such as stainless steel the grain contrast can be very strong. Two aspects of this must be distinguished. First there is the contrast between grain-and-grain, due to the differing surface wave propagation within a grain itself (9). This contrast is quite different from that due to a crack, because it is extended over an area, and will give different contrast either side of a boundary. Second there is the contrast due to a grain boundary (6). This has contributions both from the change in surface-wave impedance across the boundary, which leads to scattering, and from the change in velocity across the boundary, which leads to different Rayleigh angles. The contrast from a grain boundary can look very similar to contrast from a crack, and this can make it difficult to distinguish them. However, at a crack the scattering is in general much stronger than it is at a grain boundary, and the theory indicates that at grain boundaies the effect of scattering is in general much less than the effect of the change in Rayleigh angle. This in turn suggests that the contrast due to grain structure will be greatly reduced by imaging a specimen at positive defocus (i.e., with the specimen further from the lens than the focal plane) rather than in the more usual mode of negative defocus (i.e., closer than the focal plane) (10). When the microscope is used in this way the grain boundaries should disappear, and the cracks should be revealed by a pattern of interference fringes either side of them (11).

This problem and its solution are illustrated in Fig. 2. The specimen was a section through 316 stainless steel that had been subjected to fatigue loading. Figure 2(a) was taken at 0.37 GHz with $z = -20 \mu$m, i.e., the surface of the specimen moved five wavelengths towards the lens relative to focus. In this image the grain structure dominates the contrast, and it is not at all easy to find any cracks. Figure 2(b) was taken with $z = +32 \mu$m, and the picture looks very different. The grain contrast has almost completely disappeared, and there are

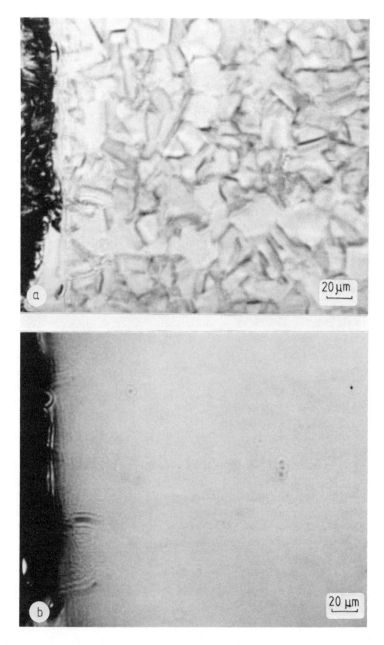

Fig 2   Fatigued 316 stainless steel, acoustic micrographs (0.37 GHz): (a) $z = -27$ μm;
(b) $z = +32$ μm

three sets of fringes on the left, indicating three fatigue cracks that grew in from the original surface of the specimen before it was sectioned. Once they have been identified, these cracks can then be seen by looking back at Fig. 2(a), where they are present but camouflaged by the grain structure.

Discussion

Both Figs 1 and 2 present images of sections through specimens that contain fatigue cracks. However, it is equally possible to image the outside surface of a fatigue specimen. The preparation that is necessary is to polish, to ordinary metallographic standards, the area to be studied; this can be done before any fatigue cycling begins. There is no need for any etching because, as shown in Fig. 2a, grain contrast shows up well without it. By using a suitable scanning and fatigue cycling system it should be possible to study the formation and growth of short fatigue cracks in situ during the progress of a fatigue test. From the evidence presented here it will be possible to measure the length of cracks along the surface, and also their relationship to grain structure and second phases. It would also be desirable to be able to measure the depth of a crack into a specimen, without sectioning. That is a harder problem, and probably requires the development of swept frequency techniques to find the wavelength dependence of scattering by the crack.

Acknowledgement

The development of acoustic microscopy for materials research at Oxford is in collaboration with AERE Harwell.

References

(1) MILLER, K. J. (1984) Initiation and growth rates of short fatigue cracks, *IUTAM Eshelby Memorial Conference*, Sheffield, April 1984 (Cambridge University Press).

(2) DE LOS RIOS, E. R., MOHAMED, H. J., and MILLER, K. J. (1985) A micromechanics analysis for short fatigue crack growth, *Fatigue Fracture Engng Mater. Structures*, **8**, 49–64.

(3) MILLER, K. J., MOHAMED, H. J., and DE LOS RIOS, E. R. (1986) Fatigue damage accumulation above and below the fatigue limit, *The Behaviour of Short Cracks*, EGF Pub. 1 (Edited by K. J. Miller and E. R. de los Rios) (Mechanical Engineering Publications, London), pp. 491–511. (This publication.)

(4) LEMONS, R. A. and QUATE, C. F. (1979) Acoustic microscopy, *Physical Acoustics* (Edited by W. P. Mason and R. N. Thurston) (Academic Press, London), Vol. 14, pp. 1–92.

(5) BRIGGS, G. A. D. (1985) *An introduction to scanning acoustic microscopy* (Oxford University Press and the Royal Microscopical Society, UK).

(6) ILETT, C., SOMEKH, M. G., and BRIGGS, G. A. D. (1984) Acoustic microscopy of elastic discontinuities, *Proc. R. Soc. Lond.*, **A393**, 171–183.

(7) SOMEKH, M. G., BERTONI, H. L., BRIGGS, G. A. D., and BURTON, N. J. (1985) A two-dimensional imaging theory of surface discontinuities with the scanning acoustic microscope, *Proc. R. Soc. Lond.*, **A401**, 29–51.

(8) ROWE, J. M., KUSHIBIKI, J., SOMEKH, M. G., and BRIGGS, G. A. D. (1986) Acoustic microscopy of surface cracks, *Phil. Trans. R. Soc. Lond.*, **A** (in press).

(9) SOMEKH, M.G., BRIGGS, G. A. D., and ILETT, C. (1984) The effect of anisotropy on contrast in the scanning acoustic microscope, *Phil. Mag.*, **49**, 179–204.

(10) SOMEKH, M. G., BERTONI, H. L., and BRIGGS, G. A. D. (1984) Applications of two dimensional Green's function model to image interpretation in SAM, *IEEE Ultrasonics Symposium* (Institute of Electrical and Electronic Engineers).
(11) YAMANAKA, K. and ENOMOTO, Y. (1982) Observation of surface cracks with scanning acoustic microscope, *J. appl. Phys.*, **53**, 846–850.

B. London,* J. C. Shyne,* and D. V. Nelson†

Small Fatigue Crack Behaviour Monitored Using Surface Acoustic Waves in Quenched and Tempered 4140 Steel

REFERENCE London, B., Shyne, J. C., and Nelson, D. V., **Small Fatigue Crack Behaviour Monitoried Using Surface Acoustic Waves in Quenched and Tempered 4140 Steel**, *The Behaviour of Short Fatigue Cracks*, EGF Pub. 1 (Edited by K. J. Miller and E. R. de los Rios) 1986, Mechanical Engineering Publications, London, pp. 537–552.

ABSTRACT A surface acoustic wave ultrasonic technique employed to monitor the growth of small surface fatigue cracks is briefly described. The technique allows accurate measurement of crack depth and rapid determination of the crack closure stress. The closure stress determined acoustically is shown to correlate well with that measured by monitoring the crack tip opening displacement versus applied stress in an SEM. Semi-circular surface cracks from 70 to 250 μm in depth and long through cracks in compact tension samples were studied in high purity 4140 steel. Tempering temperature was used to vary the yield strength and microstructure of the specimens. Samples were tempered at 200,400,550, or 700°C producing a range of yield strengths from 875 to 1600 MPa, respectively. Small surface cracks were initiated from a 25 to 40 μm deep surface pit in specially designed cantilevered bending fatigue specimens. The effects of tempering temperature on both long and small fatigue crack growth are presented. Crack closure was monitored for the small surface cracks and found to decrease with crack growth in these tests.

Introduction

The study of the growth behaviour of small surface fatigue cracks has taken a major role in the fatigue literature within the past decade. The so-called 'small crack effect', in which crack growth occurs below the long crack threshold, has been documented in many engineering materials including 7075-T6 aluminium (1)(2), nickel aluminium bronze (3), iron 3% silicon (4) and IMI 685 titanium (5). However, there has been very little study of the characteristics of small crack growth in steels, especially in the heat treated, or quenched and tempered, condition (6)–(8). Since steels represent the most widely used structural material, it is essential to examine the similarities and differences between long and small fatigue crack growth in these materials. This paper presents the results of work in progress on small and long cracks in quenched and tempered 4140 steel.

The most common methods used to monitor the growth of small fatigue cracks are optical or scanning electron microscopy of the crack itself or of a surface replica. Unfortunately, only information about the intersection of the crack with the surface is obtained. The surface acoustic wave (SAW) technique employed in the present study allows crack behaviour to be monitored in depth.

* Department of Materials Science and Engineering.
† Department of Mechanical Engineering, Stanford University, Stanford, CA 94305

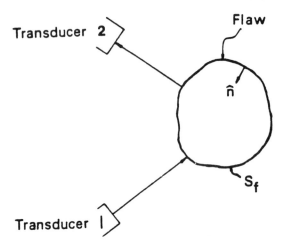

Fig 1   Schematic of generalized scattering of acoustic waves from a flaw. Transducer 1 emits the
waves which are received at transducer 2 (14)

The technique is based on the acoustic scattering theory of Kino (**9**) and Auld
(**10**) which treats the reflection of acoustic waves from a flaw within a material
(Fig. 1). The ratio of the amplitude of the scattered signal received at
transducer 2 to that input to the sending transducer 1 is known as the reflection
coefficient S_{21}, and

$$S_{21} \equiv \frac{A_2}{A_1} = \frac{j\omega}{4P} \int_{S_f} u_j \sigma_{ij} \hat{n}_i \, dS \tag{1}$$

For an explanation of the terms in the acoustic equations see Appendix 1.

Equation (1) is greatly simplified when considering a surface acoustic wave
travelling in the z direction normal to a semi-elliptical surface crack in the xy
plane whose depth, a, is much smaller than the acoustic wavelength. In this case
the crack resides in a SAW stress field with the component of stress σ_{zz}. The
other components of stress can be neglected relative to σ_{zz}. Equation (1) then
reduces to

$$S_{21} = \frac{j\omega}{4P} \int_{S_C} \Delta u_z \sigma_{zz} \, dS \tag{2}$$

So, the reflection coefficient for a surface crack can be measured from
information regarding the stress field of the surface acoustic wave along with a
few experimental parameters.

In order to apply the acoustic theory to an actual surface fatigue crack more
development is necessary. Resch *et al.* (**11**)–(**14**) have shown that the crack
depth and crack closure stress can be measured using the acoustic technique.
The crack depth can be determined from a measurement of the surface length,

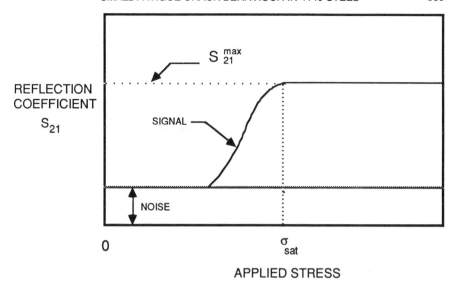

Fig 2   Acoustic measurement of crack closure stress. The stress at which the acoustic signal saturates (σ_{sat}) corresponds to the closure stress

$2c$, and the reflection coefficient, S_{21}. The crack closure stress can be measured as described below. The accuracy of crack depth prediction has been shown to be within 20 per cent of the actual crack depth for cracks 50–200 μm in depth in a range of materials (**14**). Details concerning implementation of the acoustic technique can be found in reference (**14**).

The above discussion shows that, theoretically, there is a direct relationship between the reflection coefficient and the crack depth. An empirical correlation between the reflection voltage, a quantity proportional to the reflection coefficient and measured directly on the laboratory oscilloscope, and crack depth has also been developed for the material, microstructure, and sample geometry of this study. This aids prediction of crack depth in the laboratory.

The measurement of crack closure stress may be an even more valuable aspect of the technique than measurement of depth. Figure 2 shows the behaviour of the acoustic signal as a tensile stress is applied to a specimen containing a surface crack. If a closure stress is present, the crack will be tightly closed in the absence of an applied stress. The reflection from such a closed crack is below the 'noise' level of the measurement system. This 'noise' is caused by electronic amplification of the signal and scattering of the surface waves by the microstructure (e.g., grain boundaries) of the material. As stress is applied, the crack begins to open. This appears as an increase in the acoustic reflection signal. The signal continues to increase with stress until it saturates. This saturation stress (σ_{sat}) corresponds to the stress at which the crack is completely open in depth, i.e., the crack closure stress. This is illustrated in

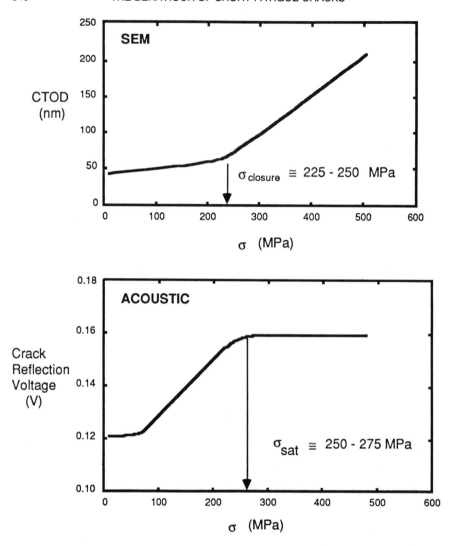

Fig 3　Comparison of the acoustic measurement of the crack closure stress with that determined by crack tip compliance measured in a scanning electron microscope (SEM)

Fig. 3 where the dependency of CTOD measured in the SEM and of acoustic response to the applied stress are compared. The stress at which the crack tip compliance changes corresponds to the crack closure stress (**15**). This value correlates well with σ_{sat}.

To summarize the use of surface acoustic waves to monitor small crack behaviour some advantages and disadvantages can be cited. Crack depth and closure stress can be accurately measured and monitored continuously during crack growth. The technique is extremely sensitive with a measurement resolution of 3 μm crack growth. It is viable for many materials including quenched and tempered high strength steel, age hardened aluminium, and ceramics. There are some disadvantages to the technique. Only one isolated fatigue crack can be monitored in a given sample. An array of cracks would cause multiple reflections which the present theory cannot treat. Also, only a limited range of crack sizes from 70 to 200 μm can be studied at present. The reflections from smaller cracks are below the 'noise' of the measurement system and larger cracks are beyond the limits of the theory for the SAW frequency used. Working within these limitations, however, the acoustic technique is a valuable method for the study of small surface cracks.

Experimental procedures

A schematic diagram of the major components of the surface acoustic wave system is shown in Fig. 4. A function generator produces a sine wave at 3 MHz which is then gated to 3 wavelengths duation. The signal is amplified and serves as the input to the sending transducer. Each transducer consists of a small slice of piezoelectric PZT crystal held at a specific angle by an RTV silicone wedge. The wedge is cast into a hard polymer block and the entire ensemble measures only 13 × 12 × 11 mm. A low viscosity diagnostic medical ultrasonic fluid is used as a coupling medium. The incoming signal causes the PZT crystal to produce a longitudinal wave which travels through the RTV silicone. The angle of the crystal is such that when the longitudinal wave encounters the specimen surface it is converted to a surface, or Rayleigh, wave. The wave packet then travels away from the transducer in a narrow beam approximately 4 mm wide. The beam reflects from the microcrack, if the crack is sufficiently open as described above, and travels back to the receiving transducer. This signal is then amplified and monitored on an oscilloscope.

The material used in this study was AISI 4140 steel of chemical composition (per cent weight) C 0.38, Mn 0.86, P 0.004, S 0.004, Si 0.23, Cr 0.90, and Mo 0.18. It was chosen because of the ease with which it could be heat treated to various microstructures and due to its wide industrial use. This particular grade of 4140 has very low phosphorous and sulfur contents producing a very clean microstructure.

All heat treatments of the fatigue specimens were performed in vacuum to prevent surface decarburization which could affect small crack growth. Samples were austenitized at 850°C for 1 hour and oil quenched. The specimens were then tempered for 1 hour at either 200, 400, 550, or 700°C to produce a range of yield strengths and microstructures. The microstructures varied from very lightly tempered martensite (200°C temper, $R_c = 54$, $\sigma_y = 1600$ MPa) to a heavily tempered, slightly spheroidized structure (700°C temper, $R_c = 26$, $\sigma_y = 875$ MPa). The intermediate Rockwell hardness levels were 45 and 38 R_c.

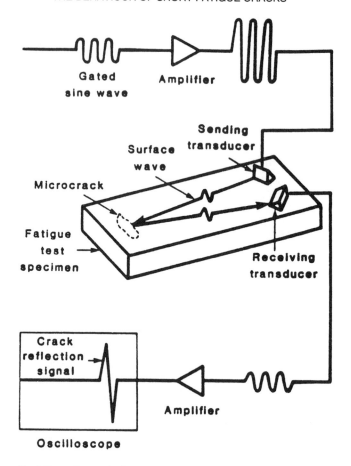

Fig 4   Schematic of the surface acoustic wave system used to monitor the growth of small surface fatigue cracks (16)

Each structure possessed the same prior austenite grain size of approximately 10–15 μm.

In order to utilize the acoustic technique to study small cracks in the heat treated 4140 steel, a special sample design and crack initiation technique had to be developed (**16**). Several major characteristics of the cantilevered bending fatigue sample (Fig. 5) are as follows. The design is compatible with the acoustic technique, meaning no unwanted reflections from specimen boundaries are produced that could interfere with the crack reflection signal. The tapered cross-section produces a small region of high stress on the specimen surface and minimizes corner cracking that could occur in a rectangular cross-section. The overall sample size is small enough to fit into a typical SEM chamber to allow high resolution studies of the cracks. A single, isolated fatigue

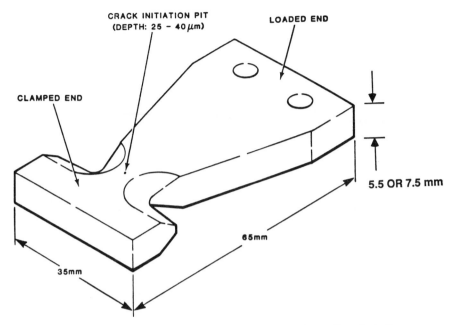

Fig 5   Cantilevered bending fatigue specimen used for the small crack studies

crack is initiated from a small surface pit 25–40 μm in depth. The pit is produced by an electro-spark cutting machine using an extremely sharp electrode. Samples were tempered following pitting to minimize any residual stresses introduced by the spark machining. The transducers are placed at the large end of the sample pointing toward the pit and initiated crack. A simple 'C-clamp' device holds the transducers on the specimen for the entire testing period, allowing crack behaviour to be monitored continuously. A strain gauge on the back face of the sample measures the strain from which the stress was calculated. The compliance of the specimens was not affected by the small cracks on the top surface.

All small and long crack tests were performed on a servo-hydraulic testing machine. The small crack samples were cycled under $R = 0$ conditions at 3–8 Hz, depending on the loading required, in stroke control. Cracks could only be initiated when a maximum cyclic stress was 70–85 per cent of the yield strength of the material. Cracks allowed to grow at this stress level would do so at well above threshold rates. Since the interesting region of small crack growth occurs near-threshold, the load was shed in 20 or 10 per cent increments, depending on the stress level, until near-threshold growth rates were achieved. The cracks were then allowed to grow under constant maximum stress conditions. Data on small crack growth was taken following the load shedding procedure. During load shedding, care was exercised to insure that the crack had grown beyond the plastic zone of the previous loading step. The crack depth and closure stress (σ_{sat}) were measured acoustically. The value of ΔK for the small cracks was calculated using $\Delta K = 1.32\Delta\sigma\sqrt{a}$ as in reference (2).

Long crack tests were done with compact tension samples ($63.5 \times 61 \times 6.35$ mm) under $R = 0.05$ conditions. Near-threshold growth data and the threshold values (ΔK_{th}) were obtained using K-decreasing test procedures (17). Crack length was measured optically using a travelling microscope.

Results

Figure 6 shows the da/dN–ΔK behaviour for long through-cracks in compact tension samples at near threshold growth rates for the four tempering temperatures. The threshold values are given in Table 1. Increasing the tempering temperature is associated with an increase in ΔK_{th}. This effect has also been documented in 300M steel by Ritchie (18).

Table 1   Effect of tempering on long crack thresholds

Tempering temperature (°C)	ΔK_{th} (MPa$\sqrt{m}$)
200	2.80
400	3.50
550	7.20
700	8.50

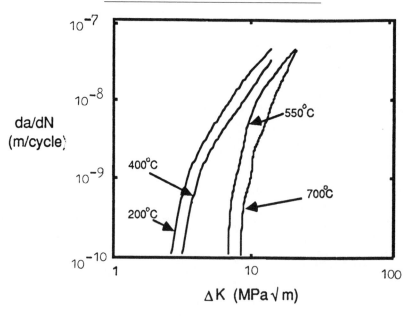

Fig 6   Long crack growth near the threshold region for the four tempering temperatures indicated

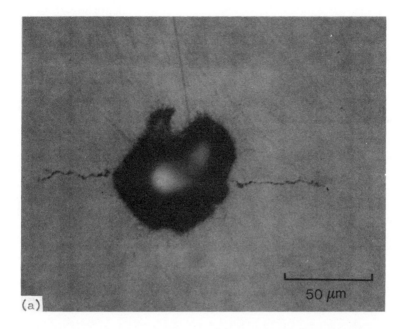

Fig 7   The appearance of small surface cracks (a) initiated from the starter pit on the surface and
         (b) as an oxidized profile showing a nearly semi-circular crack front

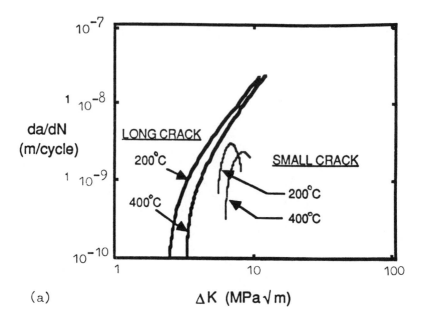

(a)

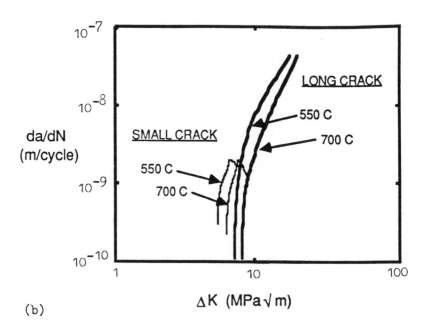

(b)

Fig 8   Comparison of long crack growth with small crack growth (a) for the lower tempering temperatures, and (b) for the higher tempering temperatures

Before presenting the small crack growth results, it is worthwhile to examine the appearance of these cracks on the surface and in depth (Fig. 7). On the surface the cracks are shown to grow from both sides of the small pit. By oxidizing a cracked sample at 300°C for 20 minutes in air and then fatiguing the sample in two, the profile of the crack is marked by a thin oxide layer. The cracks grow in a semi-circular fashion. This crack profile, also seen in other materials (1), persists up to a depth of 250 μm in this particular sample design. Once the crack is initiated, the pit does not seem to affect crack growth.

The small crack growth results are best presented in comparison to the long crack data. At the lower tempering temperatures of 200 and 400°C (Fig. 8(a)) the small cracks grow at ΔK levels significantly above the long crack threshold. For these samples, the small cracks are actually growing slower than their long crack counterparts. The higher tempering temperatures, 550 and 700°C, show the opposite behaviour (Fig. 8(b)). Small cracks grow at ΔK levels below their respective long crack threshold values; small crack growth for these tempering treatments shows the so-called 'small crack effect'. The small and long crack curves did not superpose for any of the tempering treatments at near-threshold growth rates. An interesting finding is that all the small crack da/dN–ΔK curves show a growth rate maximum. This is presently under investigation; it occurs for all the tempering treatments.

The crack closure results for the small surface cracks show another interesting trend. Figure 9 shows the crack closure history for a single crack in a sample

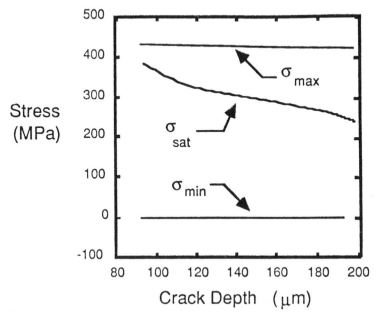

Fig 9   Crack closure results for a single growing small surface crack. σ_{max} is the maximum cyclic stress, $\sigma_{min} = 0$ and σ_{sat} is the acoustically measured crack closure stress

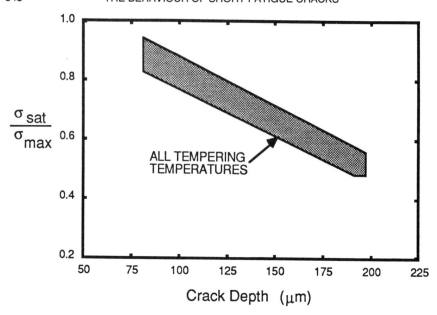

$$\frac{\sigma_{sat}}{\sigma_{max}}$$

Fig 10 Normalized closure stress versus crack depth for all tempering temperatures studied

tempered at 550°C. The bending stress gradient is taken into account in the σ_{max} and σ_{sat} curves. The closure stress decreases with crack growth for the crack size regime (75–200 μm depth) studied. Similar behaviour was documented for all tempering treatments (Fig. 10). The closure stress begins as a very high percentage of σ_{max} decreasing to about 50 per cent at a depth of 200 μm. This high level of crack closure causes a low value of the effective stress range and therefore ΔK_{eff}. Experiments are currently underway in order to compare the da/dN–ΔK_{eff} behaviour of long cracks to small cracks.

Discussion

The two major aspects of the surface acoustic wave technique are the prediction of crack depth and measurement of crack closure stress. Since the small cracks in these tests grew as half circles, one could have used optical methods to determine the crack depth. However, many times cracks are very difficult to see optically whereas they are readily detected acoustically. Thus, even when the depth of the small cracks is simply half the surface length, acoustic detection still has advantages over optical methods. The rapid and accurate determination of crack closure is perhaps the major contribution of the acoustic technique to fatigue research. In a matter of minutes the acoustic method can measure what it takes other methods (e.g., compliance measurement in the SEM) hours to do.

The results for long crack growth are not unexpected, however, there is not a simple explanation for the increase in ΔK_{th} with increasing tempering temperature (decreasing yield strength). The increase in the ductility of the structures tempered at higher temperatures may be a major factor. This could cause the mode of failure to change drastically from, for example, cleavage to ductile rupture. Work is currently underway to document the fracture morphology for each tempering treatment. Variation in the amount of crack closure could also contribute to the different thresholds measured. Experiments to measure the level of crack closure for the long cracks are in progress. The curves for all four tempering treatments begin to merge at higher growth rates. This follows the well-documented trend of a decreased influence of microstructure on long crack growth well above the threshold (19).

The small crack growth results show several interesting features. The trend observed above in the long cracks is preserved, that is, increasing resistance to crack growth with increasing tempering temperature. However, the difference between 200°C and 700°C tempers is much less than would be expected based on the long crack studies. This may indicate that diffent factors control the growth of long cracks compared to small cracks in this material. Further study is underway to investigate this.

There are two results from the small crack investigations which are particularly puzzling; the maximum in the da/dN–ΔK curves and the decreasing level of crack closure. The reasons for the growth rate maximum observed in small crack growth are not presently known. It is not caused by an increase in the crack closure stress since this decreases as the crack grows. A changing crack profile is not responsible since post fracture studies indicate a semi-circular crack front up to the maximum depth studied. It may be merely a transient which develops when the small crack curve begins to merge with the long crack curve. This seems to occur for the 500° and 700°C tempers. The continuously decreasing level of crack closure with crack growth was a consistent finding for all the tempering temperatures studied. Many small crack studies in the literature have documented the opposite effect – increasing closure stress with crack growth (4)(6)(20)–(22). The fact that the crack sizes studied here were somewhat larger that the studies referenced above may be involved with the explanation for the decreasing closure stress. It is not yet clear why closure should decrease as crack growth occurs. Detailed studies are presently underway to determine the mechanical or microstructural variables responsible for these effects.

The importance of rapid small crack growth below the long crack threshold has been well-documented in the literature (1)–(8). Using long crack growth behaviour in design may yield non-conservative crack growth estimates. In quenched and tempered 4140 steel this effect occurs as the yield strength is decreased (samples tempered at higher tempering temperatures). The magnitude of this small crack effect is not as large as reported in other alloys (1)(2)(5). It is, however, important to realize that this effect can be produced in the same alloy system by varying the microstructure and, in this case, the yield strength. The underlying reasons may be based on the variation in microstructure from very lightly tempered martensite to a nearly spheroidized structure. Identification and characterization of the relevant microstructural features controlling crack growth are currently underway. It is hoped that these studies will shed more light on the reasons for the small crack effect in this alloy system.

Conclusions

(1) A surface acoustic wave ultrasonic technique was shown to be capable of measuring both the crack depth and closure stress for small surface cracks in quenched and tempered 4140 steel.

(2) ΔK_{th} for long through-cracks in compact tension samples increased from 2.8 to 8.5 MPa$\sqrt{m}$ as the tempering temperature increased from 200 to 700°C. The increase in tempering temperature caused a decrease in yield strength from 1600 to 875 MPa.

(3) Small surface crack growth in specially designed cantilevered bending fatigue samples showed slower growth rates at the same ΔK as the tempering temperature was increased (decreasing yield strength). The differences in small crack growth with tempering temperature were not as great as would be expected based on the long crack studies.

(4) Comparison between long cracks and small cracks showed the following trends. The small crack and long crack growth curves did not superpose for any of the tempering treatments.
  (a) At the lower tempering temperatures (200, 400°C), small cracks grew at ΔK levels above their corresponding long crack thresholds.
  (b) At the higher tempering temperatures (550, 700°C), small cracks grew at ΔK levels below their corresponding long crack thresholds.

(5) The level of crack closure for small surface cracks decreased from 85–90 per cent of the maximum applied stress to 50–60 per cent when crack growth occurred from 75 to 200 μm in depth under constant maximum cyclic stress conditions.

Acknowledgements

The support of the following people and organizations is greatly appreciated: Dr A. D. Wilson of the Lukens Steel Company, Coatesville, Pennsylvania for donating the high quality 4140 steel used for this investigation, the IBM corporation, San Jose, California for the use of a high resolution SEM with which to monitor the crack tip compliance, the Office of Naval Research for providing a fellowship to Mr London, and the US Department of Energy for a research contract, with Dr Stanley Wolf as technical monitor, under which this study was done.

Appendix: Explanation of terms in acoustic equations

S_{21}	reflection coefficient
A_2, A_1	amplitude of acoustic waves scattered from the flaw and input to the sending transducer measured at the transducer terminals
j	$\sqrt{-1}$
ω	angular frequency of the acoustic wave (rad/sec)
P	power input to the sending transducer
S_f	flaw surface
u	acoustic displacement field at the flaw surface
σ	acoustic stress field
$\hat{n}$	inward directed normal to the flaw surface
S_c	crack surface

References

(1)   LANKFORD, J. (1982) The growth of small fatigue cracks in 7075-T6 aluminum, *Fatigue Engng mater. Structures*, **5**, 233–248.

(2)   LANKFORD, J. (1983) The effect of environment on the growth of small fatigue cracks, *Fatigue Engng mater. Structures*, **6**, 15–31.

(3)   TAYLOR, D. and KNOTT, J. F. (1981) Fatigue crack propagation behaviour of short cracks; the effect of microstructure, *Fatigue Engng Mater. Structures*, **4**, 147–155.

(4)   TANAKA, K., HOJO, M., and NAKAI, Y. (1983) Fatigue crack initiation and early propagation in 3% silicon iron, *ASTM STP 811*, p. 207.

(5)   BROWN, C. W. and HICKS, M. A. (1983) A study of short fatigue crack growth behavior in titanium alloy IMI 685, *Fatigue Engng Mater. Structures*, **6**, 67–76.

(6)   BREAT, J. L., MUDRY, F., and PINEAU, A. (1983) Short crack propagation and closure effects in A508 steel, *Fatigue Engng Mater. Structures*, **6**, 349–358.

(7)   DOWLING, N. E. (1983) Growth of short fatigue cracks in an alloy steel, American Society of Mechanical Engineers 4th National Congress on Pressure Vessel and Piping Technology, Portland, Oregon.

(8)   JAMES, M. N. and SMITH, G. C. (1983) Crack closure and surface microcrack thresholds – some experimental observations, *Int. J. Fractures*, **5**, 75.

(9)   KINO, G. S. (1978) The application of reciprocity theory to scattering of acoustic waves by flaws, *J. Appl. Phys.*, **49**, 3190.

(10) AULD, B. A. (1979) General electromechanical reciprocity relations applied to the calculation of elastic wave scattering coefficients, *Wave Motion*, **1**, 3.

(11) RESCH, M. T., SHYNE, J. C., KINO, G. S., and NELSON, D. V. (1982) Long wavelength Rayleigh wave scattering from microscopic surface fatigue cracks, *Review of Progress in Quantitative Nondestructive Evaluation* (Edited by Thompson, D. O. and Chimenti, D. E.) (Plenum Press, New York), vol. 1, p. 573.

(12) RESCH, M. T., NELSON, D. V., SHYNE, J. C. and KINO, G. S. (1983) Acoustic monitoring of small surface fatigue crack growth, *Advances in Crack Length Measurement* (Edited by Beevers, C. J.) (EMAS, West Midlands).

(13) RESCH, M. T. (1982) *Nondestructive evaluation of small surface cracks using surface acoustic waves*, Doctoral dissertation, Stanford University, Stanford, CA.

(14) RESCH, M. T., LONDON, B., RAMUSAT, G. F., YUCE, H. H., NELSON, D. V., and SHYNE, J. C. (1985) A surface acoustic wave technique for monitoring the growth behavior of small surface fatigue cracks, *J. Nondestructive Eval.* (in press).

(15) MORRIS, W. L. (1977) The early stages of fatigue crack propagation in aluminum 2048. *Metall. Trans A*, **8A**, 589.

(16) LONDON, B. (1985) New specimen design for studying the growth of small fatigue cracks with surface acoustic waves, *Rev. Sci. Instrum.*, **56**, 1632.

(17)   American Society for Testing and Materials (1978) Tentative test method for constant-load-amplitude fatigue crack growth rates above 10^{-8} m/cycle, ASTM Specification E647–78T. Also: Proposed ASTM test method for measurement of fatigue crack growth rates (specification under development).

(18)   RITCHIE, R. O. (1977) Influence of microstructure on near-threshold fatigue-crack propagation in ultra-high strength steel, *Metal Science*, Aug./Sept., 368.

(19)   IMHOF, E. J. and BARSOM, J. M. (1973) Fatigue and corrosion-fatigue crack growth of 4340 steel at various yield strengths, *ASTM STP 536*, p. 182.

(20)   MORRIS, W. L. (1977) Crack closure load development for surface microcracks in aluminum 2219-T851, *Metall. Trans A*, **8A**, 1079.

(21)   TANAKA, K. and NAKAI, Y. (1983) Propagation and non-propagation of short fatigue cracks at a sharp notch, *Fatigue Engng Mater. Structures*, **6**, 315.

(22)   RITCHIE, R. O. and SURESH, S. (1982) Mechanics and physics of the growth of small cracks, Proceedings of the 55th Meeting of the AGARD Structural and Materials Panel on Behavior of Short Cracks in Airframe Components, Toronto, Canada.

Conclusions

This volume has presented the results and opinions of many researchers and a few designers of engineering plant. The first conclusion therefore is to stress the need for a greater degree of collaboration between researchers in industry and academia, designers who are responsible for initially assessing the safety and durability of engineering components and structures, and finally, the scientists, engineers, and insurance agents who have to analyse the serious but frequently avoidable mistakes made during the design stage, the manufacturing phase, and the operating lifetime of engineering plant.

The behaviour of short fatigue cracks is truly an interdisciplinary topic requiring the skills and experience of many persons, but unless the information, so painstakingly gathered, sometimes at great expense, is not rapidly and efficiently disseminated throughout the breadth of the engineering profession, more fatigue failures will result that perhaps could have been avoided.

Researchers themselves also need to pay greater attention to detail. For example, short crack growth is obviously affected by microstructural features and by high stress levels that can induce macro-plasticity. It follows that scientific papers should present information on all aspects of the microstructure of the material being studied, including phase distributions, grain sizes, inclusion/precipitate sizes with their orientation and distribution, and so on. Heat treatment, surface profile, and surface preparation details should also be presented. Finally, surface micro-hardness values should be quoted. Wherever possible, cyclic as well as static mechanical properties should be given, including the cyclic yield stress and the cyclic stress–plastic strain behaviour equation. It is only with this information that designers can use experimental data with confidence and evaluate the worthiness of the data they wish to use.

The papers presented in this volume leave many questions for future analysis. For example, can the linear elastic stress intensity factor approach be adapted to describe short crack behaviour? Is the short crack growth period synonymous with 'initiation'? – a question which probably leads to the more important problem; is the short crack growth period important in fatigue life studies? After reviewing all of the papers written for this volume, the editors believe it is their responsibility to give answers to these questions, if only to offer tentative guidelines for future research. Should the answers given below be proved wrong, it is hoped that the editors themselves will be the first to know; such is the frailty of the human kind.

The first problem to be resolved concerns the modelling of crack growth in the regime between (a) the microstructure dominated processes where linear elastic fracture mechanics (LEFM) is not applicable, a conclusion reached by most if not all authors involved in these proceedings, and (b) the phase of low-stress, long-crack growth in which small scale yielding applies, a phase that all contributors acknowledge as being characterized by LEFM. There is increasing evidence both qualitatively and quantitatively, that LEFM is not applicable in the intermediate region, despite the fact that in some instances modifications of the ΔK parameter appear to rationalize some short crack growth studies. Obviously experimental studies close to the boundaries of the two above mentioned widely-separated regions can be accommodated within the commonly agreed mathematical formulations pertinent to the physical processes involved in those regions, but what obscures the problem in many cases is the nature of the log–log Kitagawa–Takahashi diagram which does not display the scatter of data as clearly as a linear plot of stress range against crack length. Similarly the log–log plot of the traditional $S–N$ curve does not present a clear picture of the statistical nature of the near fatigue limit behaviour of plain specimens. What is required is a clearer understanding of the physical processes occurring in the region described as the physically small crack growth region in the Introduction to this book.

Many authors have modified LEFM to account for a part of this region, using the concept of an effective stress intensity factor range, a parameter that takes account of a possibly ineffective period when a crack is closed during a loading cycle. Although crack closure undoubtedly occurs and researchers will continue to study this phenomena to advantage, it will, in our opinion, be of little use to the designer attempting to understand the whole range of physically small crack growth. A major difficulty with the crack closure approach is the dependence of closure on stress range, mean stress, crack length, stress state (plane stress/plane strain), and environment, all of which change their individual order of priority in terms of effectiveness throughout the fatigue lifetime, which means that no engineering designer will be able to make quantitative use of the crack closure concept with confidence.

An alternative approach which will be more attractive to designers is to use the more easily evaluated parameters of applied strain range and crack length. The first parameter can be transformed into the plastic strain range and so be related to the microstructural short crack growth phase, or it can be transformed into the stress range and, hence, be related to the low-stress, long-crack growth phase via the use of the conventional linear elastic stress intensity factor range. No doubt this issue will be discussed in conference halls for some considerable time.

Finally a clear picture is now beginning to emerge concerning the meaning of crack initiation. Experimental data indicate that fatigue cracks propagate immediately upon loading even at stress levels below the fatigue limit. In that sense crack initiation is, in reality, the propagation of a microstructurally short crack. However, when the crack reaches a barrier such as a twin or triple point boundary, a precipitate or inclusion, a stronger phase or an unfavourably oriented grain, etc., then the crack either stops growing or it grows in a discontinuous manner until the barrier is overcome. Clearly it is the zone between non-propagation and exceedingly slow discontinuous extension of the crack that will provide new grounds for research by metal physicists. In this respect it is expected that acoustic microscopy will play an important role. Additionally, conventional short crack growth studies will be extended to cover wider aspects of all the problems discussed in this book, but new studies into fretting, wear, and thermal shock will be given impetus by the improved understanding of the behaviour of short cracks provided by all the contributions presented in this collection of papers.

K. J. Miller
E. R. de los Rios

Index

Acoustics:
  emission, 515
  image contrast, 531, 533
  lens, 530
  microscopy, 527, 529, 537
  scattering theory, 538
  surface wave technique, 537, 541
  wave propagation, 530
  wave scattering, 533
Adsorbed layer, 173
Aerospace structures:
  durability, 27, 32
  safety, 27, 30
Aluminium alloys:
  Al-4.5%Cu, 203, 204
  Al-20%Si, 531
  L99, 216
  2017, 285
  2024, 40, 324, 354
  6061, 194
  7010, 102
  7175, 514
  7475, 40, 164
Anisotropy, 83
Anodizing, 194

Bailon, J. P., 387–404
Basquin Law, 217
Bathias, C., 295–307, 513–526
Baxter, W. J., 193–202
Beevers, C. J., 203–213
Blom, A. F., 37–65
Bolingbroke, R. K., 101–114
Bouksim, L., 513–526
Brass, 285
Briggs, G. A. D., 529–536
Brown, M. W., 309–321, 423–439, 441–459
Byrne, J., 337–351

Carbon steels, *see* ferritic steels
Chemisorption, 173
Compliance, 165
Copper (OFHC), 147
Corten–Dolan hypothesis, 491–511
Crack(s): (*See also* Initiation, Propagation, Short cracks, Long cracks, Stage I and II cracks, Mode I, II, and III cracks)
  arrest, 166, 505
  branching, 349, 485
  closure, 47, 110, 140, 166, 237, 251–252, 316, 329, 341, 370, 371, 389, 484, 547, 549
  contact zone, 385
  extension force, 416–417
  length measuring system, 391, 538
  obstacles, 436, 445, 493
  opening displacement, 173, 267
  opening load, 136, 391, 468
  path, 104
  shape, 103, 167, 175, 206, 220, 277, 289, 295, 297, 299, 515, 521, 547
  stress fields, 377
  sub-critical, 415
  tip blunting, 349
Crack growth rate:
  environment, 158, 166
  law, 225, 463, 464, 468, 469, 495, 509, 521
  notch geometry, 90
  retardation, 206, 208, 445
  tempering temperature, 547, 549
Critical crack length, 55, 480, 481
Crystallographic cracking, 104, 122, 172, 232, 428
Cyclic stress–strain curve, 432

Damage accumulation, 129, 492
Dimensional analysis, 445
Directional solidification, 229
Defocusing distance, 531, 533
Design:
  considerations, 13, 15, 27, 37, 243
  damage tolerant, 22, 27
  Fail-safe, 22, 27
  guidelines, 23
  philosophies, 29, 488
Dislocations:
  cell structure, 194, 199
  density, 415
  ladder structure, 194
  matrix structure, 199, 200
  planar slip, 230